Sustainable Civil Infrastructures

Sustainable Infrastructure impacts our well-being and day-to-day lives. The infrastructures we are building today will shape our lives tomorrow. The complex and diverse nature of the impacts due to weather extremes on transportation and civil infrastructures can be seen in our roadways, bridges, and buildings. Extreme summer temperatures, droughts, flash floods, and rising numbers of freeze-thaw cycles pose challenges for civil infrastructure and can endanger public safety. We constantly hear how civil infrastructures need constant attention, preservation, and upgrading. Such improvements and developments would obviously benefit from our desired book series that provide sustainable engineering materials and designs. The economic impact is huge and much research has been conducted worldwide. The future holds many opportunities, not only for researchers in a given country, but also for the worldwide field engineers who apply and implement these technologies. We believe that no approach can succeed if it does not unite the efforts of various engineering disciplines from all over the world under one umbrella to offer a beacon of modern solutions to the global infrastructure. Experts from the various engineering disciplines around the globe will participate in this series, including: Geotechnical, Geological, Geoscience, Petroleum, Structural, Transportation, Bridge, Infrastructure, Energy, Architectural, Chemical and Materials, and other related Engineering disciplines.

More information about this series at http://www.springer.com/series/15140

Hesham El-Naggar · Khalid Abdel-Rahman
Bengt Fellenius · Hany Shehata
Editors

Sustainability Issues for the Deep Foundations

Proceedings of the 2nd GeoMEast International Congress and Exhibition on Sustainable Civil Infrastructures, Egypt 2018 – The Official International Congress of the Soil-Structure Interaction Group in Egypt (SSIGE)

Springer

Editors
Hesham El-Naggar
University of Western Ontario
Western Ontario, ON, Canada

Khalid Abdel-Rahman
Leibniz University of Hannover
Hannover, Germany

Bengt Fellenius
Professional Global Expert
of Geotechnical Engineering
Ottawa, ON, Canada

Hany Shehata
Soil-Structure Interaction Group
in Egypt (SSIGE)
Cairo, Egypt

ISSN 2366-3405 ISSN 2366-3413 (electronic)
Sustainable Civil Infrastructures
ISBN 978-3-030-01901-3 ISBN 978-3-030-01902-0 (eBook)
https://doi.org/10.1007/978-3-030-01902-0

Library of Congress Control Number: 2018957408

This Springer imprint is published by the registered company Springer Nature Switzerland AG
The registered company address is: Gewerbestrasse 11, 6330 Cham, Switzerland

Contents

Earth Pressure Distribution on Rigid Pipes Overlain by TDA Inclusion 1
Mohamed A. Meguid

Dynamic Nonlinear Horizontal Resistance of Pile Group Considering Soil Types and Pile–Soil Contact Conditions 14
Takaharu Nakano and Yuji Miyamoto

Response of Four Pile Group Subjected to Eccentric Loading 26
M. Padmavathi, L. Koteswara Rao, V. Padmavathi, M. R. Madhav, and P. Sanjeeva

Negative Skin Friction Distribution on a Single Pile - Numerical Analysis 36
Talal Awwad, Salma Al Kodsi, and Alexey Shashkin

The Proposed Use of Generalized Regression Neural Network to Predict the Entire Static Load Test 49
A. K. Alzo'ubi and Farid Ibrahim

Case Study: Micropiles as Integral Part of an Alternative Slope Stabilization System 61
Thomas Hangartner and Christophe Balg

Experimental Study on the Behavior of Pile and Soil at the Frost Condition 69
Talal Awwad, Zh. A. Shakhmov, R. E. Lukpanov, and S. B. Yenkebayev

Analysis of Driven Pile Bearing Capacity Results by Static and Dynamic Load Tests 77
Talal Awwad, S. B. Yenkebayev, D. V. Tsigulyov, and R. E. Lukpanov

Bearing Capacity of an Incompletely End-Supported Pile Based on Rigid Plastic Finite Element Method (RPFEM) . . . 85
Hamadoum Hamidou and Koichi Isobe

Influence of Cap Rigidity on the Performance of Piled-Cap Foundation . . . 104
Lan Lin, Anup Sinha, and Adel Hanna

Geotechnical Issues of Testing Piles on Construction Site of Astana . . . 116
A. S. Tulebekova, A. Zh. Zhussupbekov, T. Mussabayev, and S. Mussina

An Appraisal of Pile Capacity Interpretation Methods Using Model Piles . . . 125
Abideen Adekunle Ganiyu, Ahmad Safuan A. Rashid, Mohd Hanim Osman, and Atef Badr

Bearing Behaviors and Reasonable Rock-Socketed Depth of Bridge Pile Foundation in Karst Areas . . . 141
Yunxiu Dong, Zhongju Feng, Yumeng Hao, Hongsheng Yao, and Qi Wan

Analytical Approaches to Predict Pile Settlement in London Clay . . . 162
Elia Voyagaki, Jamie Crispin, Charlotte Gilder, Paul Nowak, Nick O'Riordan, Dinesh Patel, and Paul J. Vardanega

Energy Formulation for Axial Pile Head Stiffness in Inhomogeneous Soil . . . 181
Jamie Crispin, Huize Su, Katherine Round, and George Mylonakis

Indirect In-situ Tests During the Execution of Deep Foundations: Relating the Excavation Energies to the Young's Moduli of the Surrounding Soils . . . 191
Luan C. S. M. Ozelim, Darym J. F. Campos, José Camapum de Carvalho, and Andre L. B. Cavalcante

Vibration Wave Displacement Introduced by Driving Steel Open Ended Pipe and H-Piles . . . 206
Mahmood R. Mahmood and Shaimaa M. Abdulrahman

Development of Negative Skin Friction and Dragloads on Piles and Pile Groups in Soft Clay . . . 215
Manal Salem, A'laa Adel, and Mohamed Amer

Shaft Resistance of Displacement Piles in Overconsolidated Cohesionless Soils . . . 232
Yasir M. Alharthi and Adel M. Hanna

Influence of Exposure Temperature on Shaft Capacity of Steel Piles in Ice-Poor and Ice-Rich Frozen Soils . . . 247
Abdulghader A. Aldaeef and Mohammad T. Rayhani

Effect of Scour Depth on Ultimate Capacity and Settlement of Piled Foundation 258
Wael Nashaat Abd El Samee and Mohamed Kassem Elsamny

Author Index 285

About the Editors

M. Hesham El Naggar

Professor

PhD - Civil Engineering, University of Western Ontario (1994)

MSc - Civil Engineering, Alexandria University (1989)

BSc - Civil Engineering, Alexandria University (1982)

My Research Group

Honours and Awards

- Geosynthetics Award, Canadian Geotechnical Society (2016)
- Engineering Medal: Research and Development, OPSE and PEO (2015)
- Fellow of American Society Of Civil Engineers (2014)
- Fellow of Engineering Institute of Canada (2012)
- Engineering Prize for Achievement in Research, Faculty of Engineering, Western Engineering (2010)
- A. G. Stermac Award, Canadian Geotechnical Society (2007)
- Certificate of Appreciation, Canadian Geotechnical Society (2007)
- Faculty Scholar Award, University of Western Ontario (2006–2008)
- G. G. Meyerhof Award, Canadian Geotechnical Society (2002)

- Colloquium Speaker, Canadian Geotechnical Conference (2002)
- Faculty of Engineering Science Prize for Outstanding Teaching, Western (2000–2001)
- Outstanding Teaching Assistant Award recipient, Civil Engineering Department, Western (1992–1993)
- Special University Scholarship, Western (1990–1993)
- Ontario Fees Bursary, Western (1990–1993)
- Alexandria University Scholarship for Outstanding Undergraduates (1977–1982)
- Alexandria University MSc Award (1989)

Professional Registration

- Licensed Professional Engineer, Province of Ontario (P.Eng.)
- Engineers Nova Scotia
- Association of Professional Engineers and Geoscientists of British Columbia
- Member of Engineers Syndicate, Egypt

Scientific and Professional Societies Membership

- Canadian Geotechnical Society (CGS)
- Canadian Society for Civil Engineers (CSCE)
- American Society Of Civil Engineers (ASCE)

Dr. Abdel-Rahman got his bachelor degree from Faculty of Civil Engineering, "Ain Shams University", Cairo, Egypt, in 1986. Afterwards, he went to Germany at the "Leibniz University of Hannover", where he got a master degree with "distinction" in "Geotechnique and Infrastructure Engineering". He completed his PhD study under the excellent supervision of "Prof. Achim Hettler" at the Technical University of Dortmund, Germany, in 1999 with innovative results related to the "Numerical Investigation of Scale Effect by Earth Pressure Problems". After getting his PhD, he joined the team at the Institute for Geotechnical Engineering at "Leibniz University of Hannover" as Senior Lecturer; then, he was nominated by the head of the department

"Prof. Martin Achmus" to be his "Oberingenieur", "Deputy Head".

During his academic career, he published with the team in his department "Institute for Geotechnical Engineering (IGtH) more than 60 publications mainly in the Numerical Modeling of Foundations Systems for Wind Energy Plants (Onshore & Offshore) and supervised more than 80 bachelor and master theses.

Different conference publications have been awarded "Best Conference Paper Award" e.g. "K. Abdel-Rahman & M. Achmus" Behaviour of Monopile and Suction Bucket Foundation Systems for Offshore Wind Energy Plants" in 2006. Last but not least, winning in 2016 "Outstanding Paper Award" from the international journal "Computers and Geotechnics" for a paper with "Prof. Martin Achmus" et al. titled "Behaviour of Monopile Foundations under Cyclic Lateral Loading".

Dr. Bengt H. Fellenius Professional Global Expert of Geotechnical Engineering is a professional engineer specializing in foundation design and studies by participation in project teams, special investigations, instrumented field tests, etc. Services are also provided in regard to construction problems, claims and litigation in collaboration with consultants and contractors, as well as owners. He, Professor of Civil Engineering at the University of Ottawa from 1979 through 1998, is an internationally recognized authority in the field of soil mechanics and foundation engineering, and, in particular, in deep foundations. He has gained a wealth of practical experience during more than 50 years of work at home and overseas through a variety of assignments that encompass foundation, embankment and soil improvement design for water and sewage treatment plants, industrial plants, as well as bridges, highway and airport projects, and marine structures and urban area development projects, some of which he has written up in 300+ technical journal and conference papers, articles, books and chapters. Copies of many of the papers are available for downloading from his website: [www.Fellenius.net] He moved from his native Sweden to Canada in 1972 where he worked on foundation investigations and design and construction projects in North America and overseas. In 1973, he was one of the first to apply geotextile soil separation sheets to stabilize roadbeds and construction surfaces, investigating conventional carpet underlay (Celanese) for this purpose. He was active in promoting to the US market the splicing of prestressed concrete piles by means of mechanical full-strength splices, and he introduced to Canada and the USA ground improvement applications of lime column method for reducing soil compressibility and wick drains (the Geodrain and Alidrain) for accelerating consolidation and stabilizing landslides. He was one of the earliest (1977) to research and use dynamic testing and the Pile Driving Analyzer in actual project design and construction. In 1984, he introduced the Janbu method of determining soil compressibility and analysis of settlement. He has also had a fundamental part of the development of commercial software for analysis of settlement from loads on natural soils and soils subjected to soil improvement methods, design of piled foundations and other software. In 1984, he published the design and analysis method for foundation design known as the "Unified Method of Design for Capacity, Drag Force, Settlement, and Downdrag". He is and has been an active participant in many national and international professional societies and research associations and in Canadian and US Codes and Standards Development, for example, member of the subcommittee for the American Society for Testing and Materials D-4945 Standard for High-Strain Dynamic Testing of Piles; Chairman of the Canadian Geotechnical Society (CGS), Technical Committee on Foundations writing the 1985 Canadian Foundation Engineering Manual; Member of the Ministry of Transportation Committee for the Development of the 1983 and 1992 Ontario Bridge Design Code; Author of three Public Works Canada publications: Marine Division Master Specifications for Piling, Pile Design Guidelines, and Hammer Selection Guide; Past Overseas Correspondent Member to the Geotechnical Engineering Advisory Panel of the Institution of Civil Engineers, ICE (London); and Past Member of Editorial Board for the ASCE Geotechnical Engineering Journal. He has given lectures and courses to several universities and been invited lecturer at international conferences throughout Europe, the Americas, and Southeast Asia.

Hany Farouk Shehata is the Founder and CEO of the Soil-Structure Interaction Group in Egypt "SSIGE". He is a partner and Vice-President of EHE Consulting Group in the Middle East and Managing Editor of the "Innovative Infrastructure Solutions" journal, published by Springer. He worked in the field of civil engineering early, while studying, with Bechtel Egypt Contracting & PM Company, LLC. His professional experience includes working in culverts, small tunnels, pipe installation, earth reinforcement, soil stabilization and small bridges. He also has been involved in teaching, research and consulting. His areas of specialization include static and dynamic soil–structure interactions involving buildings, roads, water structures, retaining walls, earth reinforcement, and bridges, as well as different disciplines of project management and contract administration. He is the author of an Arabic practical book titled "Practical Solutions for Different Geotechnical Works: The Practical Engineers' Guidelines". He is currently working on a new book titled "Soil-Foundation-Superstructure Interaction: Structural Integration". He is the contributor of more than 50 publications in national and international conferences and journals. He served as a Co-Chair of the GeoChina 2016 International Conference in Shandong, China. He serves also as a Co-Chair and Secretary General of the GeoMEast 2017 International Conference in Sharm El Sheikh, Egypt. He was the 2016 Outstanding Reviewer of the ASCE selected by the Editorial Board of International Journal of Geomechanics.

Earth Pressure Distribution on Rigid Pipes Overlain by TDA Inclusion

Mohamed A. Meguid(✉)

Civil Engineering and Applied Mechanics, McGill University, Montreal, Canada
mohamed.meguid@mcgill.ca

Abstract. Rigid pipes installed under high embankments are often installed using the induced trench technique by introducing a compressible zone above the pipe. A new application of tire-derived aggregate (TDA) for induced trench rigid pipes is evaluated in this study. An experimental investigation that has been conducted to measure the earth pressure distribution on a rigid pipe buried in granular material and backfilled with TDA is presented in this study. A setup has been designed and built to allow for the installation of an instrumented pipe in granular material and measuring the contact pressure acting on the pipe wall. Results show that the use of TDA as a compressible material can provide similar beneficial effects to rigid pipes compared to expanded polystyrene (EPS) geo-foam and other commonly used soft inclusions. The earth pressure is found to decrease significantly all around the pipe compared to the conventional installation with aggregate backfill. The average measured earth pressure above the crown of the pipe was found to be as low as 30% of the overburden pressure for installations with granular backfill material. The pressure at the invert also decreased by about 75% with the introduction of the soft TDA zone above the pipe.

Keywords: Soil-structure interaction · Tire-derived aggregate
Buried pipes · Induced trench installation · Soil arching
Contact pressure measurement

1 Introduction

The construction of buried pipes and culverts under high embankment fills can impose significant earth loads on these structures depending on the installation condition. The vertical earth pressure on a rigid conduit installed under earth embankment is generally greater than the weight of the soil above the structure because of negative arching. To reduce the vertical earth pressure on rigid pipes and culverts, the induced trench installation (ITI) method introduced by Marston (1922, 1930) can be used. In this method, the loads are redistributed around the buried structure by introducing a compressible material above the top of the pipe to promote positive arching as depicted in Fig. 1.

H. El-Naggar et al. (Eds.): GeoMEast 2018, SUCI, pp. 1–13, 2019.
https://doi.org/10.1007/978-3-030-01902-0_1

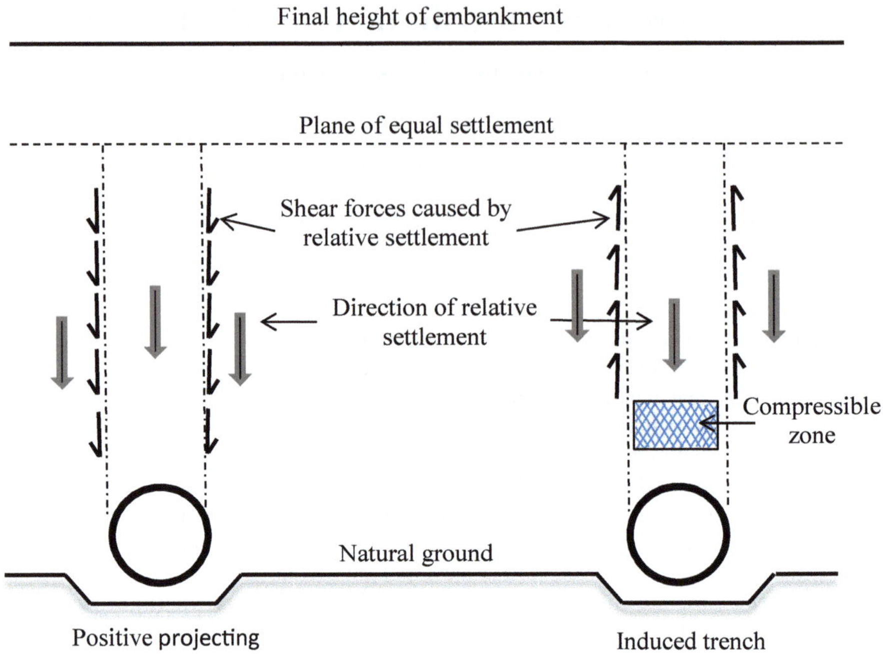

Fig. 1. Schematic of the induced trench method

The ITI installation method for rigid conduits buried under high embankments dates back to the early 1900s. Researchers studied the relevant soil-structure interaction using experimental testing and field instrumentation (e.g. Sladen and Oswell 1988; Vaslestad et al. 1993; Liedberg 1997; Sun et al. 2011; Oshati et al. 2012; Meguid et al. 2017a, b; Meguid and Youssef 2018) as well as numerical modelling (Kim and Yoo 2002; Kang et al. 2008; McGuigan and Valsangkar 2010; Meguid and Hussein 2017) to help understand the method and to address uncertainties associated with this design approach. The material used for the compressible zone mostly consisted of sawdust, wood chips, or EPS geofoam.

Tire-derived aggregate (TDA) is an engineered product made by cutting waste tires into small pieces (25–305 mm) that can be used to replace natural aggregates in civil engineering applications (Humphrey 2004). Extensive research has been done over the past two decades (e.g. Bosscher et al. 1992; Drescher and Newcomb 1994; Tweedie et al. 1998; Strenk et al. 2007; Mills and McGinn 2010) to examine the performance of TDA under different loading conditions. These studies demonstrated that TDA is a suitable fill material with engineering properties that are comparable to conventional aggregates. Applications include the construction of large embankments, bridge abutment backfill, retaining walls and pavement subgrade.

TDA is generally classified, based on size and gradation, into two types: Type-A, with a maximum particle size of 75 mm, and Type-B with a maximum particle size of 305 mm (ASTM D 6270). Direct shear tests performed on Type-A TDA samples

(Humphrey 2008; Xiao et al. 2015) suggested a friction angle that ranges from 36° to about 39° for TDA-sand interface.

In this study, an investigation into the contact pressure distribution on buried pipes is conducted using laboratory experiments. The experimental work involves a thick-walled PVC pipe that is instrumented with tactile pressure sensors and buried in granular material while a distributed load is applied at the surface. The tactile sensors used in this study are adapted from the robotics industry and have been successfully used in geotechnical engineering applications to measure the distribution of normal stresses in granular soils (e.g. Paikowsky et al. 2000; Springman et al. 2002; Ahmed et al. 2015). Due to their limited thickness, tactile sensors possess favourable characteristics with respect to aspect ratio and stiffness over the conventional load cells. In addition, being flexible enables shaping the sensing pads to cover curved surfaces, hence suitable for cylindrical shape structures. Meguid et al. (2017a, b) and Meguid and Youssef (2018) used tactile (TactArray) sensors to measure contact pressure distributions on both circular as well as square shaped structures.

The measured pressures are initially presented for a benchmark condition where only granular backfill material is used and the results are compared with Hoeg's analytical solutions (Hoeg 1968). A series of experiments are then performed by incrementally increasing the surface pressure for two backfill conditions: (1) the pipe is backfilled with only granular material; and (2) a layer of TDA is introduced above the pipe. The experimental setup and test procedure are presented in the next section, followed by the pressure results of the two-backfill conditions.

2 Experimental Setup

The experimental setup consists of a thick-walled pipe embedded in granular backfill material contained in a test chamber. The pipe is instrumented using tactile sensing pads - wrapped around its outer perimeter - covering the area near the middle third of the pipe length. A universal MTS testing machine with a capacity of 2650 kN is used to apply distributed load, utilizing a rigid steel platform. A detailed description of the experimental setup is given below.

2.1 Test Chamber

The test chamber used in the experiments is shown schematically in Fig. 2. The dimensions (1.4 m × 1.0 m × 0.45 m) are selected in order to represent a two-dimensional loading condition. The rigid walls are placed far from the pipe to minimize boundary effects; wherein the distance from the outer perimeter of the pipe to the sidewalls of the tank is more than 4 times the pipe diameter (Leonards and Roy 1976). All steel wall surfaces are epoxy coated and covered with two plastic sheets separated using a thin layer of grease to minimize friction between the backfill material and the rigid walls.

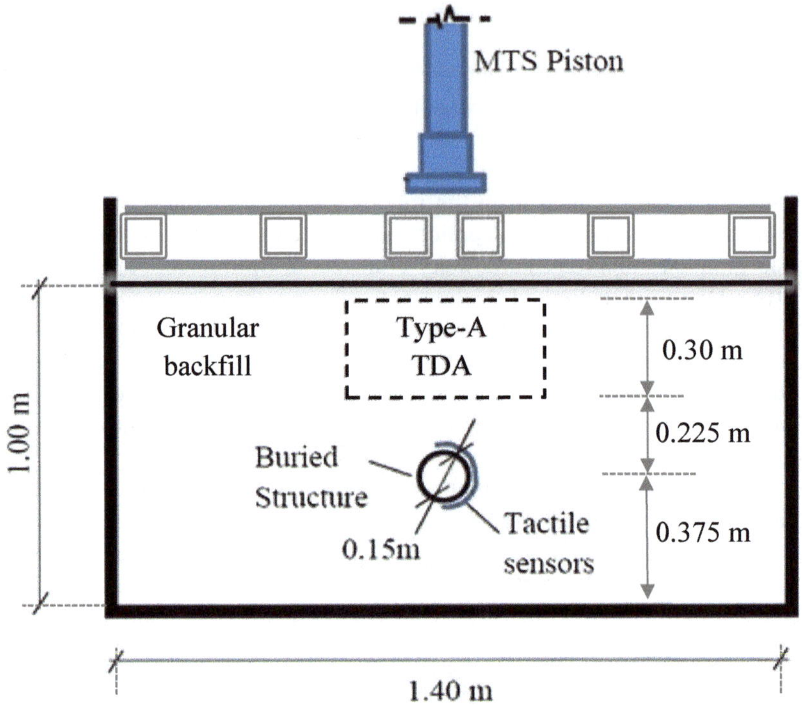

Fig. 2. Schematic of the test chamber

The pipe used for this study (15 cm in outer diameter and 1 cm in wall thickness) is instrumented using two custom-made sensing pads installed directly on the pipe wall. Each sensing pad contains 255 square-shaped sensors with pressure ranging from 0 to 140 kPa. The sensors are protected from backfill abrasion by covering the instrumented pipe with a thin layer of stiff rubber sheet.

Two LVDTs are installed orthogonally inside the pipe to confirm the rigid pipe assumption. The maximum diameter change, at surface pressure of 100 kPa, was found to be less than 0.04 mm, which is considered insignificant.

2.2 Material Properties and Testing Procedure

The dry sandy gravel backfill used in this study has an average unit weight of 16.3 kN/m^3. Sieve analysis, conducted on selected samples, indicated a coarse-grained material with 77% gravel and 23% sand. The friction angle of the backfill - determined using direct shear tests - is found to be 47°. The soil was placed and tamped in layers to form a dense base bedding material below the pipe. This was achieved by raining the sand from a constant height into the chamber in layers. From the base up to the pipe invert, the soil was placed in three layers 100 mm in height. Each layer was first graded to level the surface then tamped using a steel plate with a long handle to reach the base of the test chamber. The sand placement continued up to the pipe invert. Above the invert, the rained sand was gently pushed around the pipe up to the crown to ensure full contact between the sand and the pipe. Then, another layer of sand was added to cover the pipe. The remaining sand required to reach the height above the crown was placed and graded to minimize damage to the pressure sensors. The instrumented pipe was placed over a thin film of sand to improve the contact between the soil and the pipe. Density of the backfill was measured in the rigid box using density cups placed at different locations and collected after the completion of the tests.

The TDA material used in the experiments was obtained from a local tire-recycling centre in Montreal, Canada (Meguid and Youssef 2018). The material is categorized as Type-A with pieces that range in size from 30 to 120 mm. The properties of both the granular backfill and TDA material are summarized in Table 1.

Table 1. Properties of the backfill material

(a) Granular material*	
Specific gravity	2.65
Coefficient of uniformity (C_u)	2.3
Coefficient of curvature (C_c)	1.6
Minimum dry unit weight (γ_{min})	15.1 kN/m^3
Maximum dry unit weight (γ_{max})	17.3 kN/m^3
Experimental unit weight (γ_d)	16.3 kN/m^3
Internal friction angle (ϕ)	47°
Cohesion (c)	0
Elastic Modulus (E)	150 MPa
Poisson's ration (υ)	0.3
(b) Tire-derived aggregate	
Maximum size	75 mm
Unit weight (γ)	6.2 kN/m^3
Uniformity coefficient (C_u)	1.9

*Meguid et al. (2017a, b)

A total of six experiments were conducted including three benchmark tests where the instrumented pipe is surrounded by dry sandy gravel backfill only. For the remaining tests, a layer of TDA is placed above the pipe. For the benchmark tests, the placement of the backfill continued up to a distance 0.45 m above the crown. For experiments involving TDA, the granular backfill was used to make a 15 cm cover above the pipe, followed by the addition of TDA within a zone that measures 45 cm in width and 30 cm in height. The thickness of the aggregate cover placed above the pipe has been chosen to ensure that the protruding steel wires in the TDA (up to 10 cm in length) do not damage the pressure sensors. These dimensions were chosen to represent a compressible layer that has a thickness of about two times the pipe diameter. Two wood planks were used to assist in shaping the TDA zone while the rest of backfill material was added. The surface was finished by adding 10 cm of granular material over the entire backfill and levelled in the test chamber.

Surface load was applied using a loading platform placed under the actuator of the MTS machine. The load was gradually applied under displacement control with a displacement rate of 1.0 mm/min. The test would be stopped when either a surface displacement of 15 mm is reached or the pressures on the tactile sensors exceeded their allowable capacity (140 kPa). These limiting values were chosen in this study to avoid excessive pressure that could damage the tactile sensors. After the completion of each test, the tank was emptied using a vacuum machine connected to a collection barrel. The pipe was then retrieved and the setup was prepared for the next test.

3 Results and Discussions

The measured radial earth pressures are summarized in this section and the results are compared for the two investigated cases: (1) conventional backfill (benchmark tests); and (2) induced trench method using TDA inclusion. To validate the measured pressures in the benchmark tests, the sensor readings recorded following the placement of the granular backfill material are compared to Hoeg's analytical solution. The changes in earth pressure at selected locations on the circumference of the pipe are then presented.

3.1 Initial Condition

Hoeg (1968) developed an analytical solution for the contact pressure acting on buried cylinders. The solution is expressed in terms of two stiffness ratios (the compressibility ratio and the flexibility ratio) and three constants derived for two different boundary conditions, namely, no slippage or free slippage condition. The radial pressure (σ_r) is expressed as follows:

$$\sigma_r = \frac{1}{2}p\left\{(1+K)\left[1-a_1\left(\frac{R}{r}\right)^2\right]-(1-K)\left[1-3a_2\left(\frac{R}{r}\right)^4-4a_3\left(\frac{R}{r}\right)^2\right]\cos 2\theta\right\} \quad (1)$$

where R is the pipe radius, r is the distance from the pipe center to the soil element under analysis, K is the lateral earth pressure coefficient expressed in terms of Poisson's ratio $(\upsilon/1-\upsilon)$, p is the soil vertical stress, θ is the angle of inclination from the springline and a_1, a_2 *and* a_3 are constants. The latter constants for the free slippage condition are given by:

$$a_1 = \frac{(1-2v)(C-1)}{(1-2v)C+1} \quad (2)$$

$$a_2 = \frac{2F + 1 - 2v}{2F + 5 - 6v} \tag{3}$$

$$a_3 = \frac{2F - 1}{2F + 5 - 6v} \tag{4}$$

The above constants are functions of the compressibility and flexibility ratios, which are measures of the extensional and flexural stiffnesses of the soil medium relative to the pipe. These rations are defined as follows:

$$C = \frac{1}{2}\frac{1}{1 - v}\frac{E_s}{\frac{E_p}{1 - v_p^2}}\left(\frac{D}{t}\right) \tag{5}$$

$$F = \frac{1}{4}\frac{1 - 2v}{1 - v}\frac{E_s}{\frac{E_p}{1 - v_p^2}}\left(\frac{D}{t}\right)^3 \tag{6}$$

where, E_s, v are the Elastic modulus and Poisson's ratio of the soil, respectively; E_p, v_p are the Elastic modulus and Poisson's ratio of the pipe, respectively; D is the average pipe diameter and t is the wall thickness. A soil-pipe system with both C and F of zero signifies a perfectly rigid embedded pipe.

The initial earth pressure calculated using Hoeg's analytical solution is demonstrated in Fig. 3. The pressure distribution is characterized by maximum values at the crown and invert which quasi-sinusoidal decrease towards the springline (SL). For the free and no slippage boundary conditions, the pressures at the crown and invert were found to be about 12 kPa and 9 kPa, respectively. These values decreased to 3 kPa and 5 kPa, respectively, at the springline. By superimposing the measured pressure values on the analytical solution, it is evident that the experimental results followed a similar pressure distribution pattern with an increase of about 2 kPa at the invert. This is consistent with additional in-situ pressure at a depth of 1D below the crown. This pressure difference is not usually captured by Hoeg's solution that assumes a deeply buried pipe with vertical pressure that is the same above and below the pipe. It should be noted that the contact pressure measured using the tactile sensors and reported in this study represents the radial earth pressure acting on the pipe.

3.2 Effect of TDA on Contact Pressure

To evaluate the effect of introducing the soft TDA zone on the distribution of earth pressure onto the pipe, a comparison is made between the recorded pressure readings at three selected locations, namely: crown (C), springline (SL), and invert (I).

Figure 4 illustrates the changes in contact pressure at the crown with the increase in applied load up to a maximum surface movement of 15 mm. The benchmark tests

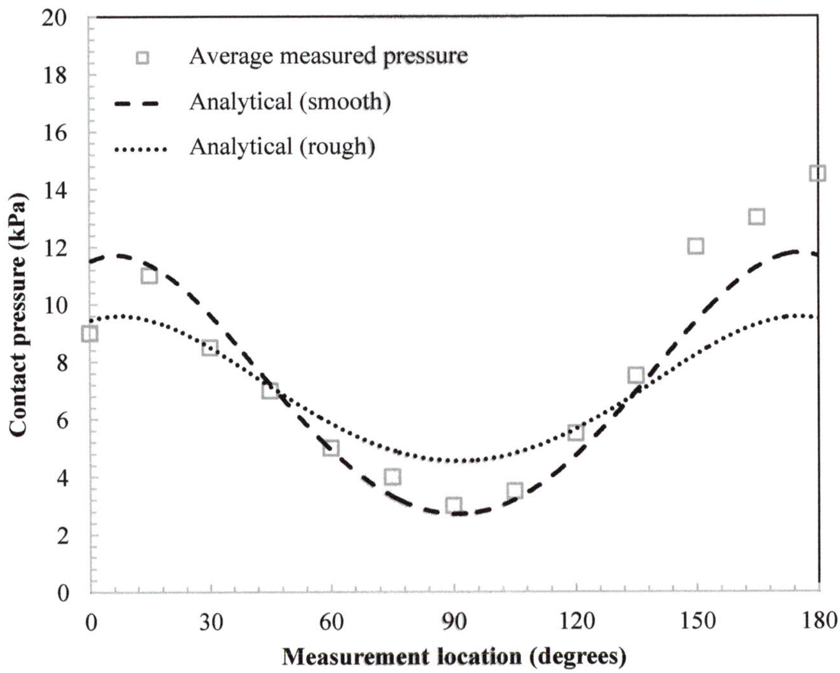

Fig. 3. Measured and calculated pressures on the pipe under initial loading condition

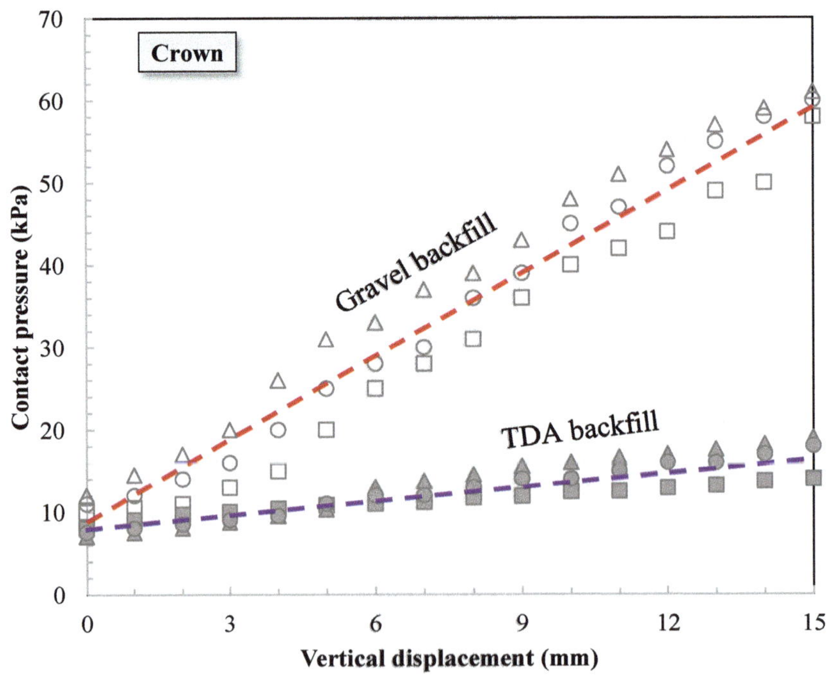

Fig. 4. Measured contact pressures at the crown

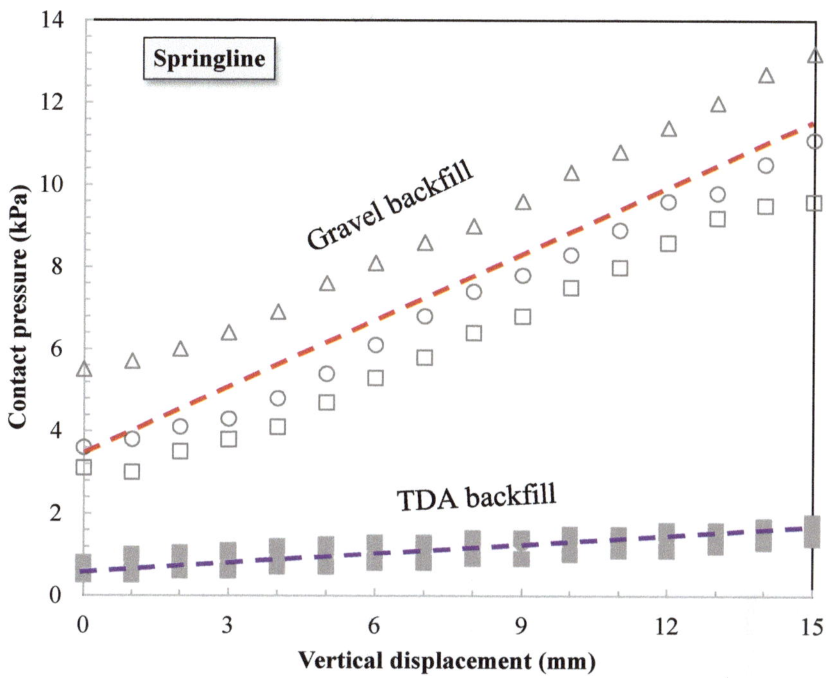

Fig. 5. Measured contact pressures at the springline

showed consistent increase in radial pressure from the initial condition (about 10 kPa) up to a maximum value of about 60 kPa. Tests involving the TDA inclusion above the pipe resulted in a significant decrease in radial pressure at the crown from 60 kPa to as low as18 kPa. This is attributed to the soil arching effect and load re-distribution around the pipe; ultimately a consequence of the soft-zone presence.

The earth pressure at the springline increased from 4 kPa to as high as 11 kPa in the benchmark tests as shown in Fig. 5. Contact pressure decreased from 11 kPa in the benchmark tests, at applied displacement of 15 mm, to less than 2 kPa, using the induced trench method with TDA inclusion.

The highest measured pressure was recorded at the invert of the pipe where pressure reached about 70 kPa when gravel backfill was used (see Fig. 6). This value decreased significantly to less than 20 kPa using TDA inclusion. The pressure distribution on the pipe at all investigated locations is summarized in Fig. 7. The contact pressure is normalized with respect to the benchmarks tests. It is found that contact pressures decreased at all locations as the surface displacement increased from 0 mm (initial condition) to 15 mm. The pressure ratio at the beginning of the test was found to

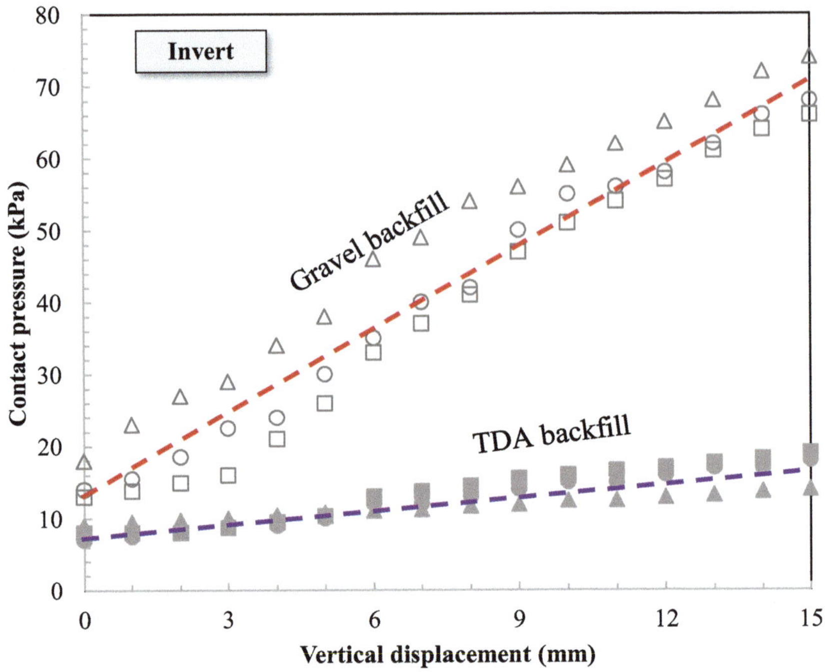

Fig. 6. Measured contact pressures at the invert

be 70%, 52% and 20% at the crown, invert and springline, respectively. As the load is applied and the induced trench took effect, the pressure ratio changed to 30%, 25% and 13% at the crown, invert and springline, respectively when the applied displacement reached 15 mm. This suggests that utilizing TDA above buried pipes can result in a significant reduction in earth load and consequently, an economic design.

Based on the above results, the effect of TDA in reducing radial earth pressure on the pipe is evident. In addition, the effect seems to become more pronounced at high surface loads where sufficient compression develops in the TDA zone and consequently more shear stresses are generated at the boundaries of the induced trench. Given the increasing amount of waste materials that may be disposed in landfills around the world, re-using scrap tires in buried structures seems to be a sound alternative both technically and environmentally.

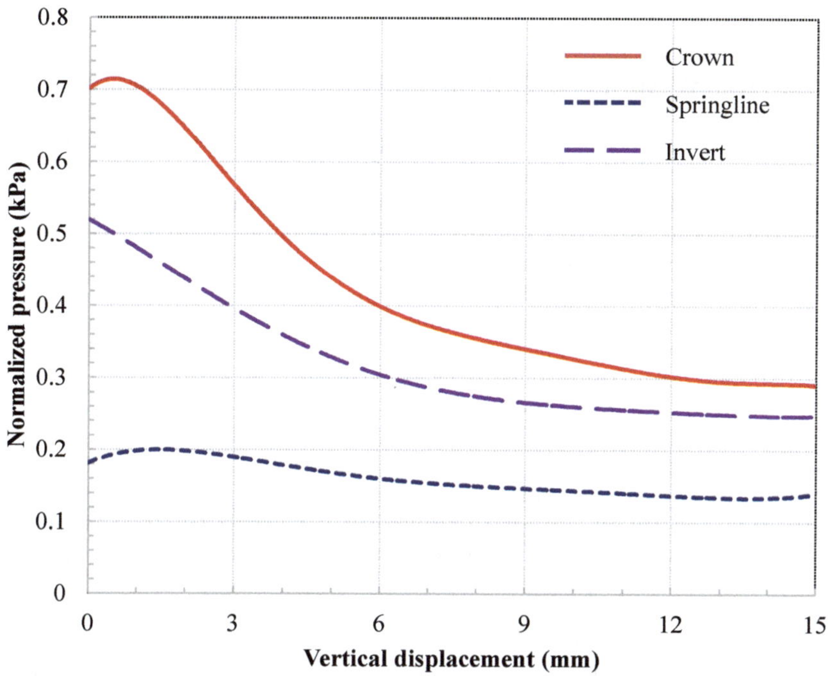

Fig. 7. Average normalized pressures for the case of TDA inclusion with respect to the measured values for gravel backfill

4 Summary and Conclusions

This study investigated the earth pressure distribution on a buried pipe installed using the induced trench method. The contact pressure distribution on the pipe was measured using the tactile sensing technology, which is able to provide a continuous pressure profile on the pipe wall. The effect of installing a soft zone of TDA on the radial pressure distribution on the pipe was examined.

The induced trench installation described in this study was successful in reducing the vertical loads on the buried pipe. The average measured earth pressure above the crown of the pipe was found to be about 30% of the overburden pressure for installations with granular backfill material. Significant reduction in radial pressure was also recorded at the invert with pressure reduction of about 77% with the introduction of the soft TDA inclusion.

This study suggests that using TDA in induced trench construction is both technically and environmentally sound alternative to gravel backfill and EPS geofoam.

Acknowledgements. This research is supported by a research grant from the Natural Sciences and Engineering Research Council of Canada (NSERC). The authors would like to acknowledge the support of Dr. W.D. Cook and Mr. J. Bartczak in conducting these experiments.

References

Ahmed, M., Tran, V., Meguid, M.A.: On the role of geogrid reinforcement in reducing earth pressures on buried pipes. Soils Found. **55**(3), 588–599 (2015)

ASTM Standard D6270: Standard Practice for Use of Scrap Tires in Civil Engineering Applications. ASTM International, West Conshohocken (2008)

Bosscher, P.J., Edil, T.B., Eldin, N.N.: Construction and performance of a shredded waste tire test embankment. Transportation Research Record, 1345, pp. 44–52. Transportation Research Board, Washington, D.C. (1992)

Drescher, A., Newcomb, D.E.: Development of design guidelines for use of shredded tires as a lightweight fill in road subgrade and retaining walls. Report no. MN/RC-94/04, Department of Civil and Mineral Engineering, University of Minnesota, Minneapolis (1994)

Hoeg, K.: Stresses against underground structural cylinders. Am. Soc. Civ. Eng. Proc. J. Soil Mech. Found. Div. **94**(SM4), 833–858 (1968)

Humphrey, D.N.: Effectiveness of design guidelines for use of tire derived aggregate as lightweight embankment fill. Recycled materials in geotechnics (GSP 127). In: Proceedings of ASCE Civil Engineering Conference and Exposition (2004)

Humphrey, D.N.: Tire derived aggregates as lightweight fill for embankments and retaining wall. In: International Workshop on 'Scrap Tire Derived Geomaterials – Opportunities and Challenges', Yokosuka, Japan, pp. 59–81 (2008)

Kang, J., Parker, F., Kang, Y.J., Yoo, C.H.: Effects of frictional forces acting on sidewalls of buried box culverts. Int. J. Numer. Anal. Methods Geomech. **32**(3), 289–306 (2008)

Kim, K., Yoo, C.H.: Design Loading for Deeply Buried Box Culverts. Highway Research Center. Report no. IR-02-03, p. 215. Auburn University, Alabama (2002)

Leonards, G.A., Roy, M.B.: Predicting Performance of Pipe Culverts Buried in Soil. Joint highway research project report JHRP-76-15. Purdue University, West Lafayette (1976)

Liedberg, N.S.D.: Load reduction on a rigid pipe: pilot study of a soft cushion installation. Transp. Res. Rec. **1594**, 217–223 (1997)

Marston, A.: Second Progress Report to the Joint Concrete Culvert Pipe Committee. Iowa Engineering Experimental Station, Ames (1922)

Marston, A.: The theory of external loads on closed conduits in the light of the latest experiments. In: The 9th Annual Meeting of the Highway Research Board, pp. 138–170. Highway Research Board, Washington, D.C. (1930)

McGuigan, B.L., Valsangkar, A.J.: Centrifuge testing and numerical analysis of box culverts installed in induced trenches. Can. Geotech. J. **47**(2), 147–163 (2010)

Meguid, M.A., Hussein, M.G., Ahmed, M.A., Omeman, Z., Whalen, J.: Investigation of soil-geosynthetic-structure interaction associated with induced trench installation. Geotext. Geomembr. (2017a). https://doi.org/10.1016/j.geotexmem.2017.04.004

Meguid, M.A., Ahmed, M.R., Hussein, M.G., Omeman, Z.: Earth pressure distribution on a rigid box covered with U-shaped geofoam wrap. Int. J. Geosynth. Ground Eng. **3**, 11 (2017b). https://doi.org/10.1007/s40891-017-0088-4

Meguid, M.A., Hussein, M.G.: A numerical procedure for the assessment of contact pressures on buried structures overlain by EPS geofoam inclusion. Int. J. Geosynth. Ground Eng. **3**, 2 (2017). https://doi.org/10.1007/s40891-016-0078-y

Meguid, M.A., Youssef, T.: Experimental investigation of the earth pressure distribution on buried pipes backfilled with tire-derived aggregate. Transp. Geotech. **14**(1), 117–125 (2018)

Mills, B., McGinn, J.: Design, construction, and performance of highway embankment failure repair with tire-derived aggregate. Transportation research record 1345, pp. 90–99. Transportation Research Board, Washington, D.C. (2010)

Oshati, O.S., Valsangkar, A.J., Schriver, A.B.: Earth pressures exerted on an induced trench cast-in-place double-cell rectangular box culvert. Can. Geotech. J. **49**(11), 1267–1284 (2012)

Paikowsky, S.G., Palmer, C.J., Dimillio, A.: Visual observation and measurement of aerial stress distribution under a rigid strip footing. Geotech. Spec. Publ. **94**, 148–169 (2000)

Sladen, J.A., Oswell, J.M.: The induced trench method—a critical review and case history. Can. Geotech. J. **25**(3), 541–549 (1988)

Springman, S.M., Nater, P., Chikatamarla, R., Laue, J.: Use of flexible tactile pressure sensors in geotechnical centrifuge. In: Proceeding of the International Conference of Physical Modeling in Geotechnical Engineering, pp. 113–118 (2002)

Strenk, P.M., Wartman, J., Grubb, D.G., Humphrey, D.N., Natale, M.F.: Variability and scale-dependency of tire derived aggregate. J. Mater. Civ. Eng. **19**(3), 233–241 (2007)

Sun, L., Hopkins, T., Beckham, T.: Long-term monitoring of culvert load reduction using an imperfect ditch backfilled with geofoam. Transp. Res. Rec. **2212**, 56–64 (2011)

Tweedie, J.J., Humphrey, D.N., Sandford, T.C.: Full Scale Field Trials of Tire Shreds as Lightweight Retaining Wall Backfill, At-Rest Condition. Transportation Research Board, Washington, D.C. (1998)

Vaslestad, J., Johansen, T.H., Holm, W.: Load reduction rigid culverts beneath high fills: long-term behavior. Transp. Res. Rec. **1415**, 58–68 (1993)

Xiao, M., Ledezma, M., Hartman, C.: Shear resistance of tire-derived aggregate using large-scale direct shear tests. J. Mater. Civ. Eng. **27**(1), 1–8 (2015)

Dynamic Nonlinear Horizontal Resistance of Pile Group Considering Soil Types and Pile–Soil Contact Conditions

Takaharu Nakano(✉) and Yuji Miyamoto

Graduate School of Engineering, Osaka University, Suita, Osaka, Japan
nakano_takaharu@arch.eng.osaka-u.ac.jp

Abstract. Subgrade reactions should be properly evaluated to simulate pile stress under large earthquakes. The subgrade reaction of piles is known to be strongly influenced by soil types (e.g., sand and clay), contact conditions at the pile–soil interface (e.g., separation and friction), and the effect of a pile group (e.g., pile arrangement and configurations). Therefore, this study explains the influence of soil types and contact conditions on the subgrade reaction of single piles and three-pile groups with series arrangement. For this purpose, a dynamic lateral loading analysis of soil–pile systems was conducted based on a nonlinear finite element method. The following conclusions were obtained from the analysis:

(1) The pile arrangements, soil types, and contact conditions influenced the hysteresis curve of the subgrade reactions.
(2) The slip displacement of the subgrade reaction was influenced by the pile arrangement caused by the effect of the pile group. Friction reduced the slip displacement for sand.
(3) The equivalent stiffness and the equivalent damping ratio of the subgrade reaction were influenced by the pile arrangement. Friction and separation influenced these values for sand and clay, respectively.

These findings will contribute to realizing an improved dynamic soil–pile system model to better understand the seismic response of a pile-supported building.

1 Introduction

Piles are important building members that support the superstructure on soft soil deposits. If a building leans because of damage to piles, the piles must be repaired to ensure safety. However, doing so is expensive. Furthermore, design codes for piles under very large earthquakes have not been established because the seismic behavior of piles remains unclear. The seismic response of pile-supported buildings is influenced by the nonlinearity of soil deposits, soil near piles, and structures. Subgrade reactions need to be properly evaluated to simulate pile stress under the lateral loads. The subgrade reaction is known to be strongly influenced by soil types (e.g., sand and clay) and contact conditions at the pile–soil interface (e.g., separation and friction). Moreover, the effect of pile groups (e.g., pile arrangement and configurations) cannot be ignored if two or more piles are located with a narrow space.

H. El-Naggar et al. (Eds.): GeoMEast 2018, SUCI, pp. 14–25, 2019.
https://doi.org/10.1007/978-3-030-01902-0_2

A number of studies have been conducted on the pile group effect through experimental and analytical approaches. For example, Tuzuki et al. (1992) conducted full-scale dynamic loading tests on two-pile and four-pile groups and observed impedance; Lemnitzer et al. (2008) conducted a full-scale static loading test on nine-pile group and obtained pile group coefficient; and Kashiwa et al. (2008) conducted cyclic loading tests on model four-pile groups, and showed that shear force at pile head and bending moment were different between leading and trailing piles. These above-mentioned studies have mainly focused on the characteristics of springs at the pile head.

Meanwhile, some studies have focused on the characteristics of subgrade reaction. For example, Miyamoto et al. (1992) analytically evaluated soil springs along the pile shafts based on the thin layer method to represent the subgrade reaction; Suzuki and Adachi (2003) conducted cyclic loading tests on model pile groups with different pile configurations, and showed that the subgrade reaction–pile displacement relationship of each pile was different depending on the pile location; Hijikata et al. (2007) conducted full-scale vibration tests on a 25-pile group, and showed that the maximum subgrade reaction of the trailing pile was lower than that of the leading pile; and Hirose et al. (2017) conducted shaking table tests on a model 25-pile group, and showed that the subgrade reaction–pile displacement relationship of each pile was different depending on the pile location. Evaluating characteristic of subgrade reaction of each pile in a pile group is necessary in improving the dynamic soil–pile system model.

This study aims to evaluate subgrade reaction of pile groups under lateral load and explain the influence of pile arrangements, soil types, and contact conditions. For this purpose, we conduct a dynamic lateral loading analysis of soil–pile systems based on the finite element method by considering these nonlinearities. In the analysis, we define the slip displacement as well as the equivalent stiffness and equivalent damping ratio to discuss the nonlinear characteristics of the subgrade reaction.

2 Simulation Analysis on the Shaking Table Test

This chapter briefly explains a simulation on the shaking table test based on Hirose et al. (2017). A shaking table test on a model 25-pile group was conducted under normal gravity condition at Osaka University. Figure 1a shows an overview of the model. Acryl cylinders with a diameter of 12 mm and a length of 400 mm represented the piles. The model piles were installed with pile spacing of 2.0*D*, where *D* denotes the pile diameter, in a Toyoura sand deposit in a laminar box. The laminar box had a length of 400 mm, a width of 600 mm, and a height of 400 mm in inside dimension. Pile heads were plugged into the holes of the bottom of the superstructure and fixed using urethane resin. The superstructure was assumed to be a rigid body made of steel with a mass of 54.2 kg. In the test, seismic motions with different magnitudes were applied. Figure 1b shows the vertical distribution of the maximum bending moment of Pile-03, Pile-05, and Pile-13. Pile-03 had a larger pile head moment and a larger maximum moment under the first inflection point than Pile-05 and Pile-13.

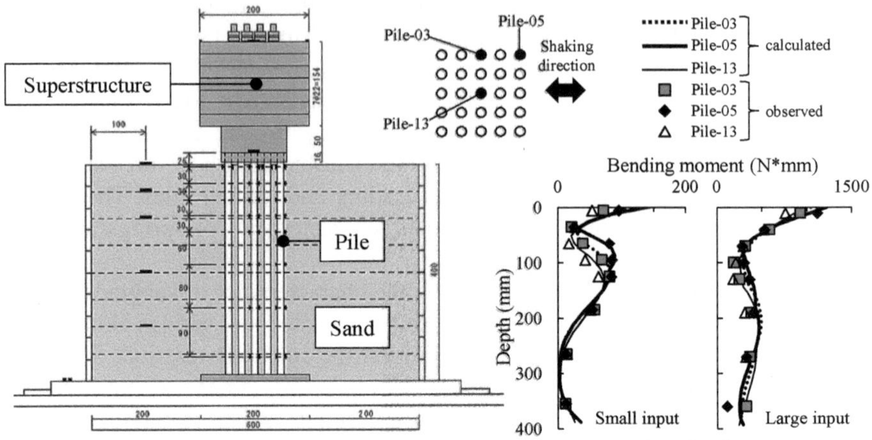

(a) Test model and observation points (b) Maximum bending moment

Fig. 1. Model and result of shaking table test on the 25-pile group (based on Hirose et al. 2017)

A finite element analysis was conducted for the test using LS-DYNA, a software of the three-dimensional finite element method. Eight-node elasto-plastic solid elements considering pressure dependency were used for the soil. Elastic shell elements were used for the piles. The superstructure was represented by a rigid body. Pile heads and tips were rigidly connected to the superstructure and laminar box, respectively. The separation and the slip at the pile–soil interface were considered based on the penalty method. Figure 1b shows the calculated bending moment of Pile-03, Pile-05, and Pile-13 from the analysis. The calculated bending moment was in good agreement with the observed one. Therefore, the finite element method is useful in evaluating the effect of the pile group.

3 Model and Analytical Method for the Parametric Study

The finite element analysis was conducted for single piles and the three-pile groups in sand and clay by considering the nonlinearity of the soil material and the separation at the pile–soil interface. The analysis was performed using LS-DYNA. The piles had a diameter D of 600 mm and a length L of 10 m. The pile spacing of the pile group was assumed to be 2.5D. We described the difference in soil types by the angle of shear resistance ϕ and cohesion c (Table 1).

Table 1. Properties of sand and clay

Soil type	ρ (t/m^3)	Vs (m/s)	Poisson's ratio	ϕ (°)	c (kN/m^2)
Sand	1.8	100	0.33	35	1
Clay	1.5	100	0.45	1	50

Figure 2 shows an example of the model. Full three-dimensional geometric models were developed to represent the soil–pile systems. Eight-node hexahedral elements using the Mohr–Coulomb yield criteria with the non-associated flow rule were used for the soil. The dilation angle was assumed to be zero. The shear modulus was calculated from equation $G = \rho V_S^2$, where ρ is the density, and V_S is the shear wave velocity. The horizontal and vertical displacements of the nodes were restricted for the side and bottom boundaries of the soil, respectively. The size of the soil model was assumed to be large enough to avoid the influence of the boundary. NULL shell elements that do not have stress were used for the piles. In the analysis, the separation and the slip at the pile–soil interface were considered based on the penalty method. Table 2 shows the analytical cases with different contact conditions. First, we assumed the Sand-SP07 and Clay-SP07 cases with a friction coefficient of $\tan\phi_s$ (=0.7) at the interface, where ϕ_s is the angle of shear resistance for sand. We assumed the same friction coefficient for these cases to discuss the influence of the nonlinearity of the soil material. Second, we assumed the Sand-SP00 and Clay-SP00 cases with zero friction coefficient at the interface to discuss the influence of friction. Finally, we conducted the analysis without separation for clay (i.e., Clay-NSP case).

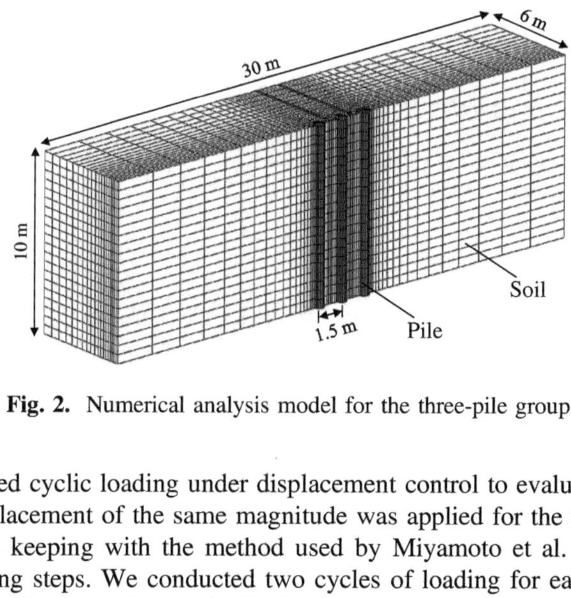

Fig. 2. Numerical analysis model for the three-pile group

We conducted cyclic loading under displacement control to evaluate the subgrade reaction. A displacement of the same magnitude was applied for the full length of all piles (Fig. 3) in keeping with the method used by Miyamoto et al. (1995). Table 3 shows the loading steps. We conducted two cycles of loading for each step with the displacement amplitude u_A that gradually increased.

4 Results and Discussion

4.1 Hysteresis Curves of the Subgrade Reaction

Figure 4 shows the hysteresis curves of the subgrade reaction–displacement relationship for single piles at GL-1.5 m. Figures 4, 5 and 6 show the curves smoothed by a

Table 2. Analytical cases considering soil types and contact conditions

Case	Soil type	Separation	Friction coefficient at interface	Pile arrangement	Diameter of each pile (mm)	Length of each pile (m)
Sand-SP07	Sand	Considered	0.7	Single or three-pile	600	10
Sand-SP00		Considered	0			
Clay-SP07	Clay	Considered	0.7			
Clay-SP00		Considered	0			
Clay-NSP		No				

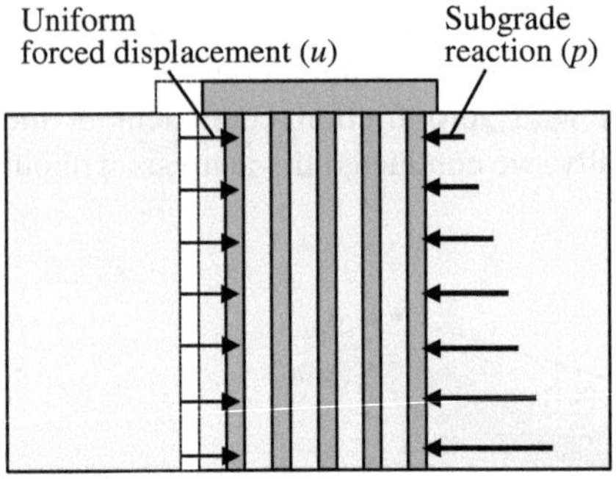

Fig. 3. Loading method to evaluate the subgrade reaction of piles (based on Miyamoto et al. 1995)

Table 3. Loading steps

Step	Amplitude of displacement u_A (mm)	Number of cycles
1	1	2
2	2	2
3	5	2
4	10	2
5	20	2
6	50	2
7	100	2

low-pass filter to remove noise. The hysteresis curve for the Sand-SP07 case had spindle- and quadrangle-like shapes for small and large displacements, respectively, because the sand followed the movement of the pile, and the pile–soil gap reduced because of the low cohesion. A comparison of the Sand-SP07 and Sand-SP00 cases showed that the latter curve had a smaller maximum subgrade reaction and a remarkable slip behavior. Therefore, the friction at the interface influenced the subgrade reaction for sand.

By contrast, the slip behavior for the Clay-SP07 case was remarkable because under the large pile displacement, the clay stayed in almost the same position, in which the separation started. A comparison of the Clay-SP07 and Clay-SP00 cases showed that the latter curve had a smaller maximum subgrade reaction. The hysteresis curve of the Clay-NSP case was spindle-shaped, and had larger maximum subgrade reaction and area under the hysteresis curve compared to the Clay-SP07 because of the tensile resistance of the soil behind the pile. Therefore, both friction and separation at the interface influenced the subgrade reaction for clay.

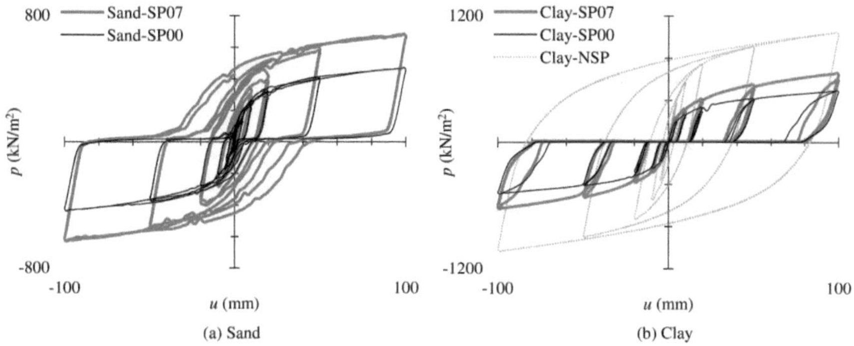

Fig. 4. Comparison of the contact conditions on the subgrade reaction–displacement relationship for single piles at GL-1.5 m

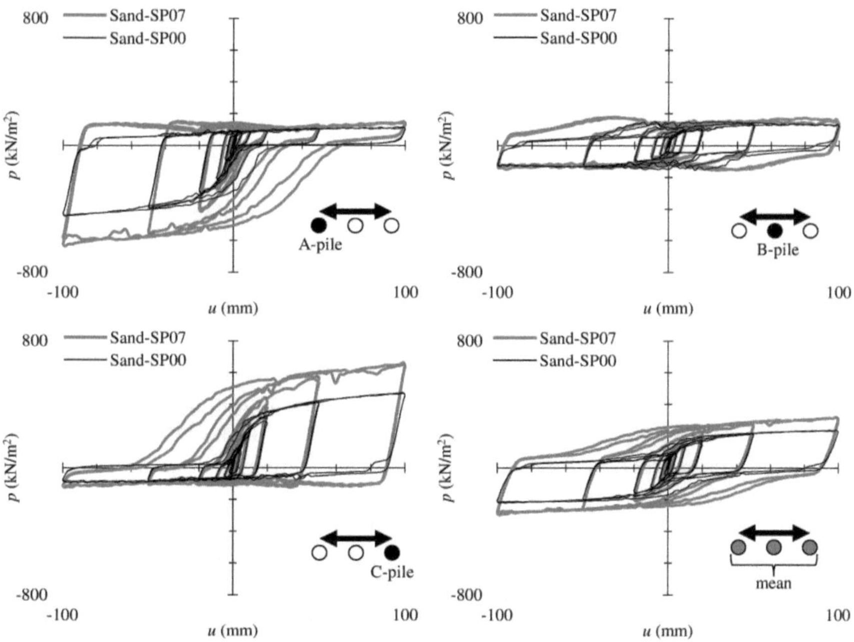

Fig. 5. Comparison of the contact conditions on the subgrade reaction–displacement relationship for the three-pile group in sand at GL-1.5 m

Figure 5 shows the hysteresis curves for the three-pile group in sand. The maximum subgrade reaction of the right-direction loading for the A-pile was smaller than that of the left-direction loading because of the decrease of the compression resistance caused by the existence of the B-pile, and the curve had a triangle shape. The hysteresis curve of the C-pile was a mirror image of that of the A-pile. The maximum subgrade reaction for the B-pile was small because of the existence of the A- and C-piles, and the curve had a quadrangle shape. The hysteresis curve for the mean subgrade reaction of

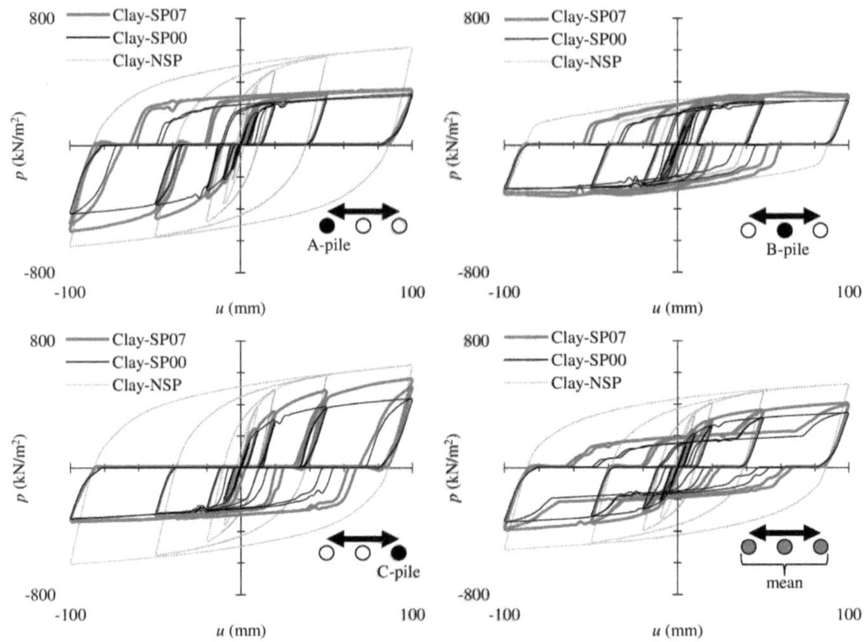

Fig. 6. Comparison of the contact conditions on the subgrade reaction–displacement relationship for the three-pile group in clay at GL-1.5 m

all piles was spindle- and quadrangle-shaped for the small and large displacements, respectively. A comparison of the Sand-SP07 and Sand-SP00 cases showed that for the left-direction loading of the A-pile and the right-direction loading of the C-pile, the latter curve had a remarkable slip behavior and a smaller maximum subgrade reaction.

These trends were similar to those of the single pile. Meanwhile, the difference in maximum subgrade reaction for the B-pile, the right-direction loading of the A-pile, and the left-direction loading of the C-pile was small.

Figure 6 shows the hysteresis curves for the three-pile group in clay. The maximum subgrade reaction of the right-direction loading for the A-pile was smaller than that of the left-direction loading. The slip behavior of the right-direction loading was smaller than that of the left-direction loading because the piles and the soil between the piles moved together, and the gap reduced. The hysteresis curve of the C-pile was a mirror image of that of the A-pile. The maximum subgrade reaction for the B-pile was small. The hysteresis curve for the mean subgrade reaction had a distorted shape because of superposition of the curves with different degrees of the slip behavior. A comparison of the Clay-SP07 and Clay-SP00 cases showed that the maximum subgrade reaction for the left-direction loading of the A-pile and the right-direction loading of the C-pile was smaller in the Clay-SP00 case. This trend was similar to that of the single pile. By contrast, the difference in the maximum subgrade reaction for the B-pile, the right-direction loading of the A-pile, and the left-direction loading of the C-pile was small. The hysteresis curve for the Clay-NSP case had larger maximum subgrade reaction and area under curve compared to those of the Clay-SP07 case.

Therefore, the pile arrangements, soil types, and contact conditions influenced the hysteresis curve of the subgrade reaction. The friction influenced the subgrade reaction for sand because the piles and the soil contacted together owing to the low cohesion.

4.2 Slip Displacement

To discuss the degree of slip behavior, we defined the slip displacement u_{slip} as the length of the range, in which the absolute value of the subgrade reaction is lower than the threshold value in the left-direction loading of the second cycle of each step (Fig. 7). In this analysis, the threshold value was one-tenth of the maximum subgrade reaction of the step.

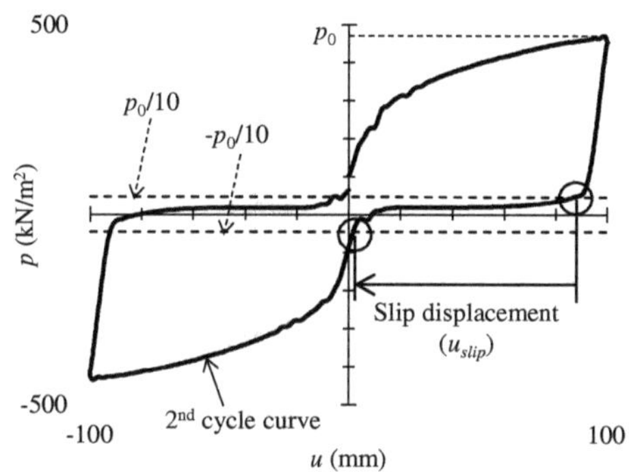

Fig. 7. Definition of the slip displacement u_{slip} for each step

Figure 8 shows the slip displacement–amplitude of the displacement relationship. The slip displacement was normalized by the displacement amplitude. For a single pile, the larger the displacement amplitude, the larger the normalized slip displacement. A comparison of the Sand-SP07 and Clay-SP07 cases showed that the normalized slip displacement was larger in the latter. A comparison of the different friction coefficients showed that the normalized slip displacement of the Sand-SP00 case was smaller than that of the Sand-SP07 case and that the difference between the Clay-SP07 and Clay-SP00 cases was small.

The A-pile of the three-pile group showed these trends. In contrast, the B- and C-piles showed smaller normalized slip displacements than the A-pile because of the interaction of piles.

Therefore, the slip displacement was influenced by the pile arrangement because of the effect of the pile group. The contact conditions also influenced the slip displacement of the subgrade reaction. Friction reduced the slip displacement for sand. Meanwhile, friction had little influence on clay.

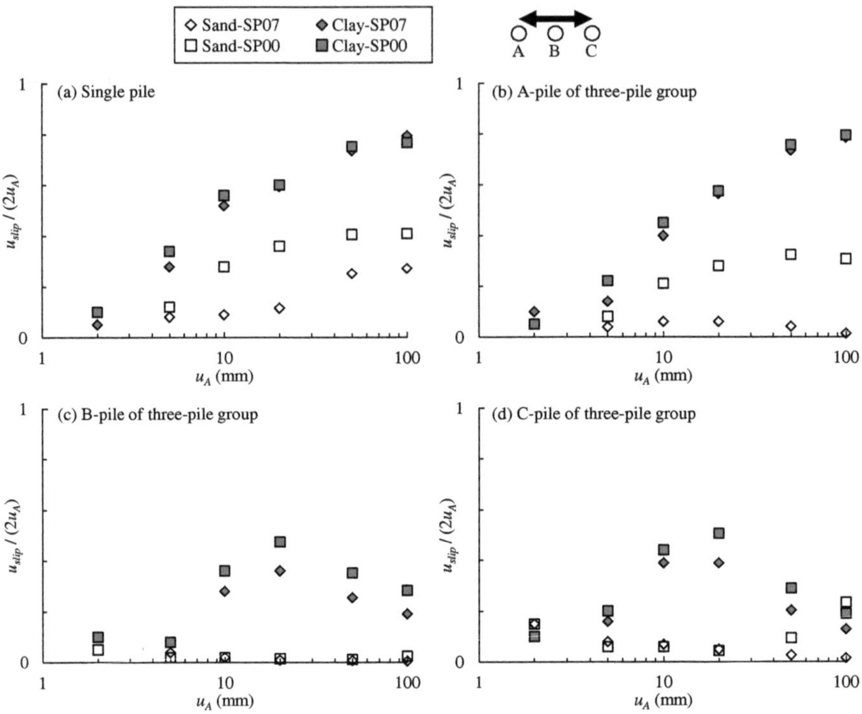

Fig. 8. Comparison of the pile arrangement and the contact conditions on the normalized slip displacement of the subgrade reaction at GL-1.5 m

4.3 Equivalent Stiffness and Equivalent Damping Ratio

The equivalent stiffness k and the equivalent damping ratio h_{eq} of the subgrade reaction were defined in Fig. 9.

Figure 10 shows the equivalent stiffness–displacement and the equivalent damping ratio–displacement relationships. For the equivalent stiffness, the larger the displacement, the smaller the stiffness. A comparison of single piles and three-pile groups showed that the latter had a smaller stiffness. A comparison of different contact conditions showed that for sand, the stiffness of the Sand-SP00 case was smaller than that of the Sand-SP07 case. The difference in the Clay-SP07 and Clay-SP00 cases was small for clay. The stiffness in the Clay-NSP case was larger than that in the Clay-SP07 case.

Generally, the larger the displacement, the larger the equivalent damping ratio. However, the equivalent damping ratio for the Sand-SP00 case decreased for the 10 mm displacement because of the slip behavior. The equivalent damping ratio of the three-pile group was larger than that of the single pile because the subgrade reaction of the piles behind another pile quickly yielded. A comparison of different contact conditions showed that for sand, the damping ratio of the Sand-SP00 case was smaller than that of the Sand-SP07 case. The difference in the Clay-SP07 and Clay-SP00 cases was small for clay. Meanwhile, the damping ratio of the Clay-NSP case was larger than that in the Clay-SP07 case.

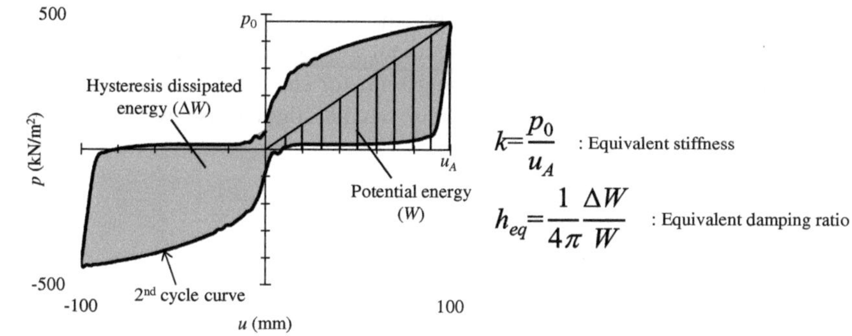

Fig. 9. Definition of equivalent stiffness k and equivalent damping ratio h_{eq} of the subgrade reaction for each step

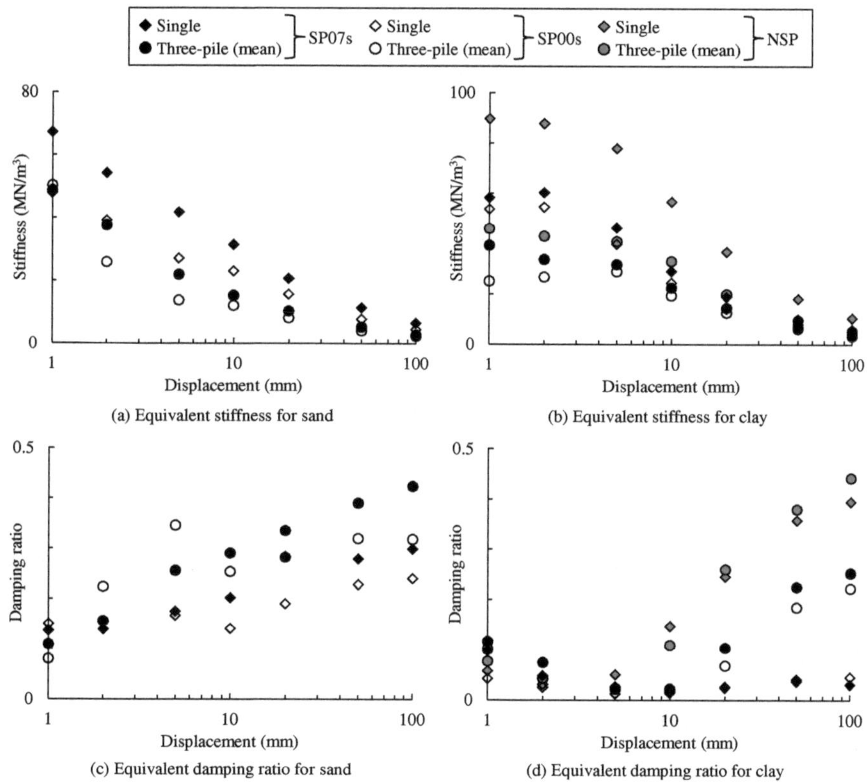

Fig. 10. Comparison of the pile arrangement and the contact conditions on the equivalent stiffness and the equivalent damping ratio at GL-1.5 m

Therefore, the equivalent stiffness and the equivalent damping ratio of the subgrade reaction were influenced by the pile arrangement and the contact conditions. Friction influenced both the stiffness and the damping ratio for sand. Both the stiffness and the damping ratio with separation for clay were smaller than those without separation because of the slip behavior.

5 Conclusions

We evaluated herein the subgrade reaction of piles using the finite element analysis to simulate the pile stress of a pile group through a shaking table test. We obtained the following conclusions from the analysis:

(1) The pile arrangement, soil types, and contact conditions influenced the hysteresis curve of the subgrade reaction. The friction coefficient at the interface influenced the maximum subgrade reaction for sand because the piles and the soil contacted because of the low cohesion.
(2) The slip displacement of the subgrade reaction was influenced by the pile arrangement because of the effect of the pile group. Friction reduced the slip displacement for sand.
(3) The equivalent stiffness and the equivalent damping ratio of the subgrade reaction were influenced by the pile arrangement. Friction influenced both the stiffness and the damping ratio for sand. The pile–soil separation reduced both the stiffness and the damping ratio for clay.

Therefore, we should properly consider the pile arrangements, soil types, and separation and friction at the interface to accurately evaluate the subgrade reaction of piles. These findings will contribute to realizing an improved dynamic soil–pile system model to better understand the seismic response of a pile-supported building.

In further research, we should obtain more experimental and analytical data on the subgrade reactions of pile groups. Moreover, we will model the subgrade reaction as discrete springs for a simplified model, such as a beam–spring model.

Acknowledgments. This work was supported by JSPS KAKENHI Grant Number JP17H06837.

References

Hijikata, K., Sugiyama, T., Itou, M., Fujiwara, K., Sako, Y., Miyamoto, Y.: Nonlinear behavior of pile-groups by experimental and analytical study. J. Struct. Constr. Eng. AIJ **596**, 249–260 (2007). (in Japanese)

Hirose, H., Nakano, T., Miyamoto, Y.: Lateral soil resistance of each pile in pile group by shaking table tests and its simulation analysis. In: 16th World Conference on Earthquake Engineering, Santiago, Chile (2017)

Kashiwa, H., Shouji, M., Hayashi, Y., Suita, K., Kurata, T., Inoue, W.: Influence of nonlinear behavior of pile–soil structure on displacement amplitude dependence for efficiency of pile group based on cyclic lateral loading tests subjected to large displacement amplitude. In: 14th World Conference on Earthquake Engineering, Beijing, China (2008)

Miyamoto, Y., Sako, Y, Miura, K.: Dynamic behavior of pile group in liquefied sand deposit. In: 10th World Conference on Earthquake Engineering, Rotterdam, Netherlands (1992)

Miyamoto, Y., Sako, Y., Kitamura, E., Miura, K.: Earthquake response of pile foundation in nonlinear liquefiable soil deposit. J. Struct. Constr. Eng. AIJ **471**, 41–50 (1995). (in Japanese)

Lemnitzer, A., Ahlberg, E.R., Khalili-Tehrani, P., Rha, C., Taciroglu, E., Stewart, J.P., Wallace, J.W.: Experimental testing of a full-scale pile group under lateral loading. In: 14th World Conference on Earthquake Engineering, Beijing, China (2008)

Tuzuki, M., Inada, O., Yamagishi, M., Yahata, K., Naito, Y., Kitamura, E.: Field testing and analysis of dynamic loaded pile group. In: 10th World Conference on Earthquake Engineering, Rotterdam, Netherlands (1992)

Suzuki, Y., Adachi, N.: Relation between subgrade reaction and displacement of model pile based on horizontal loading test. J. Struct. Constr. Eng. AIJ **570**, 115–122 (2003). (in Japanese)

Response of Four Pile Group Subjected to Eccentric Loading

M. Padmavathi[1(✉)], L. Koteswara Rao[1], V. Padmavathi[1], M. R. Madhav[1], and P. Sanjeeva[2]

[1] J.N.T. University, Hyderabad, India
mpadmace@gmail.com, koti0143@gmail.com, vpadma70@gmail.com, madhavmr@gmail.com
[2] Malla Reddy Engineering College, Hyderabad, India

Abstract. Pile foundations of engineering structures, such as bridges, offshore structures, and retaining walls are frequently subjected to eccentric loads. The eccentric loading on the foundation may cause the structure to tilt/rotate. Serviceability and safety of these structures depend mainly on responses of the piles and the soil while resisting moment loading. In case of structures resting on the soft soil, the foundations are often rigid and stability depends on moment-rotation relationship of the soil-foundation system. Several researchers have analysed the problem of pile group foundations under moment loading which are focused on ultimate bearing capacity of pile group foundation. But the response of pile group in terms of load-displacement and moment-rotation is not given much importance.

The aim of the present paper is to investigate the response of four pile group embedded in sand bed under vertical eccentric load. Model four pile groups made of mid steel with length to diameter ratios (L/d) of 5 and 7.5 were used for testing. Four pile groups with two spacing to diameter ratios (s/d) of 2 and 3 were casted. A total of 36 tests were performed with the pile groups subjected to normalized eccentricities (e/s) of 0 (central load), 0.1, 0.2 and 0.3. Tests have been conducted using an electronic load cell with four numbers of LVDTs and data-logger. The load-displacement and moment-rotation responses of a four pile group subjected to eccentric loading installed in sand bed at two different relative densities, 30% and 70% were studied. Load-settlement responses of the pile groups are analysed to quantify the effects of different L/d and s/d ratios with varying eccentricity ratios e/s. Results obtained also quantify the displacement of the pile group under the moment loading due to eccentricity. A new method is proposed to analyse the response of four pile group under vertical eccentric load. The predicted load-displacement response using the proposed nonlinear theory compares well with the measured response of eccentrically loaded pile groups.

1 Introduction

Piles are often grouped to carry relatively large loads originating from superstructure if the capacity of a single pile is not sufficient. The behavior of piles in a group is substantially different from that of a single pile due to the effect of pile to pile and pile

H. El-Naggar et al. (Eds.): GeoMEast 2018, SUCI, pp. 26–35, 2019.
https://doi.org/10.1007/978-3-030-01902-0_3

to soil interactions. Pile foundations of engineering structures, such as bridges, offshore structures, and retaining walls which are supported by pile foundations are frequently subjected to eccentric loads. The eccentric loading on the foundation may cause the structure to tilt/rotate. Serviceability and safety of these structures depend mainly on responses of the piles and the soil while resisting moment loading. Behavior of rigid piles and pile groups under eccentric vertical loads in sand and in clay has been studied by Kishida and Meyerhof (1965) and Saffery and Tate (1961) respectively. The ultimate bearing capacity of rigid model piles and pile groups in sand has been investigated for the general case of combined eccentricity and inclination of the load (Meyerhof et al. 1983). An analysis of the vertical deformation of pile groups based on superposition of individual pile displacement fields was proposed by Randolph and Wroth (1979).

Pile group behavior incorporating pile to pile and soil to pile interactions have been studied by several investigators experimentally and ana-lytically for vertical compression (Terzaghi and Peck 1967), lateral loading (Poulos 1973; Poulos and Davis 1980) and pullout (Das et al. 1976; Madhav 1986). But the pile group response under moment loading has received very limited attention in literature (Poulos and Madhav 1971). Komatsu et al. (2004) investigated the effect of the eccentricity on the ultimate capacity by model tests on pile groups. The effectiveness of using vertical short piles under an eccentrically loaded raft as structural members (connected piles) and as soil reinforcement (unconnected piles) were studied by El Sawwaf (2009). Test results indicate that the inclusion of short piles adjacent to the raft edges not only significantly improves the raft bearing pressures but also leads to a reduction in raft settlement and tilt leading to an economical design of the raft. Daniel et al. (2009, 2010) presented a forensic study on collapse of a three-span bridge supported by timber pile foundation. Capacity of eccentrically loaded piles is reduced by 60% compared with that of concentrically loaded specimens (Daniel et al. 2009, 2010). The present paper investigates the response of four pile group under vertical eccentric load in sand beds at 30% and 70% relative densities. The load-displacement and moment-rotation responses of a four pile group under eccentric loading are investigated by model tests. Load-displacement responses of the pile groups are analysed to quantify the effects of different length to diameter, L/d, and spacing to diameter, s/d, ratios with varying normalized eccentricity ratios, e/s and relative density (D_r). Results obtained quantify the tilt of the pile group under the moment loading due to eccentricity.

2 Methodology

Model tests were conducted on single piles and four pile groups under compression in a sand bed filled in a circular mild steel tank 111 cm in diameter and 55 cm deep. The test setup (Fig. 1) consists of the following components: (a) Test tank, (b) Model pile groups, (c) Foundation medium i.e. sand, (d) Arrangement for application of loads and (e) Measuring devices (electronic load cell, four numbers of LVDTs and data-logger). The pile cap made of steel is square in shape of size 9 cm. Two numbers of mild steel four pile group models of lengths 30 cm and 45 cm and 6 cm in diameter having length to diameter ratios, L/d = 5 and 7.5 (Fig. 2) were used. Spacing to diameter ratios (s/d) of four pile group models were 2 and 3. Sand in the tank was placed in five layers of uniform thickness up to 55 cm height, each layer being compacted well according to the required relative densities. The properties of sand used for the study are given in Table 1. It is classified as poorly graded sand (SP) as per IS classification system. The maximum and minimum densities are 17.7 kN/m^3 and 16.1 kN/m^3 respectively (Table 1).

(b)

(a)

Fig. 1. Schematic of test set up (a) load cell and LVDTs, (b) data logger, system and Hydraulic jack

Pile groups are made of mild steel. The length of the group is decided to get the required length to diameter ratio's (L/d) of 5 and 7.5 respectively with a diameter of 6 cm. Only two different L/d values are used as it is a small scale model study used to analyze the response of four pile group. The dimensions of the four pile group for the present experimental investigation are decided based on the previous research on the topic of eccentrically loaded pile groups. Use of reduced scale model tests on eccentrically loaded pile groups is indicated in the previous research (Meyerholf et al. 1983; Meyerhof 1960; Komatsu et al. 2004; Niket and Nitin 2015) on the response of eccentrically loaded pile groups. The basic limitation of the present study is that it is a qualitative approach to analyze the response of four pile group under eccentric load. And hence the confining stresses in the reduced scale model may not be in proportionate to any prototype or an actual pile group system.

2.1 Experimental Procedure

Sand was filled in 5 layers each of equal thickness. The total weights of the sand used for relative densities of 30% and 70% were respectively 880.3 kg and 914.3 kg. Thus the weight of sand for each layer was 176 kg and 182.8 kg respectively for 30% and 70% relative densities. The relative density achieved during the test was checked using small bins placed within the tank in all the five layers. The relative densities actually achieved after testing were 35% and 73% respectively. The pile model was pushed in to the second layer of the compacted sand bed vertically. The remaining sand was filled in

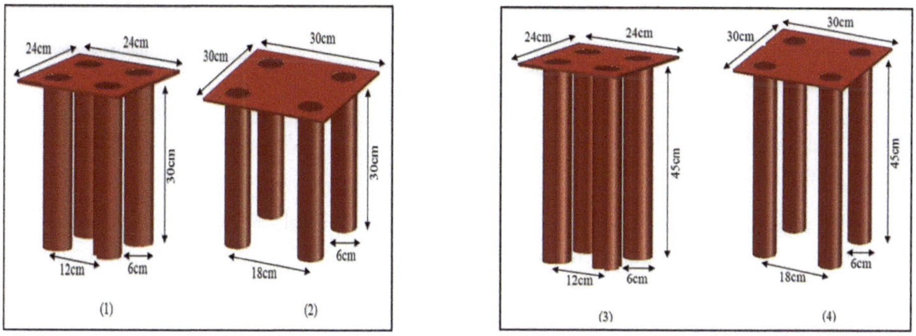

Fig. 2. Four pile group models-geometrical parameters

layers by maintaining proper compaction and without disturbing the pile group's verticality. The center of the pile group was properly placed below the jack as per the required eccentricity. The pile group verticality and eccentricity were checked before loading for each test. The surface of the top layer was leveled uniformly. Set of four LVDTs were placed on four corners of the pile cap (Fig. 1a). LVDTs have limiting displacements of 25 mm, 50 mm, 100 mm and 100 mm respectively. The LVDTs and the electronic load cell were connected to the data logger.

The data logger (Fig. 1b) is connected to a system from which the test data is retrieved. The four pile groups were subjected to normalized eccentricities of e/s = 0 (central load), 0.1, 0.2 and 0.3. Figure 2 shows the front view of the model pile group. The corresponding moment and rotation of the pile model were determined based on the measured values of loads and displacements.

2.2 Analytical Procedure

A nonlinear theory is proposed to analyze the response of four pile group under eccentric vertical load. Detailed methodology of the proposed theory is given in author's publication (Padmavathi et al. 2016). The analysis of a floating pile group under eccentric vertical load is carried out using elastic continuum theory considering pile to pile interaction. The analysis of four pile group is carried out using the concept of interaction factors and the principle of superposition (Poulos and Davis 1980). The method is based on the assumption that the soil is a homogeneous elastic medium, having a constant modulus with depth (Davis and Poulos 1968). Pile stiffness in compression, k_{pc} is predicted using elastic continuum approach. Interaction of an adjacent pile in causing additional displacement in a group is represented by an interaction factor, α. And the relationship between the load and displacement of the pile group is modeled using nonlinear load-displacement response (Winkler model). By the proposed method, applied load and displacement of pile are related nonlinearly (Eq. 1)

$$P_1 = \frac{k_c \delta_{11}}{1 + \frac{k_c \delta_{11}}{P_{ult,1}}} \tag{1}$$

Table 1. Material properties

Description	Coefficient of uniformity (C_u)	Coefficient of curvature (C_c)	Percentage of gravel	Percentage of coarse grained sand	Percentage of medium grained sand	Percentage of fine grained sand	Percentage of clay and silt	Specific gravity of sand	Maximum density of sand	Minimum density of sand
Value	2.44	0.94	1.5	5.5	71	21.5	0.5	2.65	1.79 g/cc	1.61 g/cc

where P_1 = load in pile 1 due to eccentric vertical load, P, k_{pc} = pile stiffness in compression, δ_{11} = displacement of pile, 1 due to load, P_1 and P_{ult} = ultimate pile capacity.

Equation 1 can be written as

$$\delta_{11}^* = \frac{\delta_{11}}{d} = \left(\frac{P_1^{**}}{1 - P_1^{**}}\right)\frac{1}{\mu}, \text{ where} = \frac{k_{pc}d}{P_{ult}}, P_1^{**} = \frac{P_1}{P_{ult}}. \tag{2}$$

Displacement of the single pile can be estimated using Eq. 2. This concept of nonlinear relationship between load and the displacement is extended for a four pile group with interaction among piles in the group. Displacement of the four pile group is predicted based on the Eq. 3 considering interaction.

$$\delta_1 = \delta_{11} + \alpha_a(\delta_{22} + \delta_{44}) + \alpha_d\delta_{33} = \text{Total displacement of pile 1} \tag{3}$$

where δ_{22} = displacement of pile, 2 due to load, P_2, δ_{33} = displacement of pile, 3 due to load, P_3, δ_{44} = displacement of pile, 4 due to load, P_4, α_a and α_d are interaction factors. P_1, P_2, P_3 and P_4 are loads induced in in each pile due to eccentric vertical load, P applied on the group. Total displacement of any pile in the can be estimated in a similar manner. Equation 3 can be expressed in terms of k_{pc} and μ like the one given in Eq. 2. Hence the total normalized displacement, δ_1/d can be estimated for the known values of k_{pc} and P_{ult}. Where k_{pc} and P_{ult} are namely initial stiffness, k_{pc} and ultimate load, P_{ult} called the Kondner's hyperbolic model parameters. These are estimated from measured load, P - displacement, δ response of a single pile with L/d = 5 and 7.5 in a poorly graded sand with relative density of 73%.

The load-displacement response is then predicted by the proposed nonlinear theory and is compared with the measured response of the four pile group.

3 Results and Discussion

Figure 3 depicts measured load, P, versus displacement, δ, responses of a four pile group with L/d = 5, s/d = 2 for e/S = 0, 0,1, 0.2 and 0.3 in a poorly graded sand at relative density of 35%. The ultimate load and pile group stiffness are estimated by Kondner's hyperbolic plot (Kondner 1963). The ratio of displacement, δ_{av} over load, P (on Y-axis) verses displacement, δ (on X-axis) is plotted to determine Kondner's hyperbolic model parameters namely initial stiffness, k_c and ultimate load, P_{ult} for each test. Inverse of the slope of the line fitted by δ_{av}/P verses δ gives ultimate load, P_{ult} and intersection point of fitted line on Y-axis gives pile stiffness, k_c.

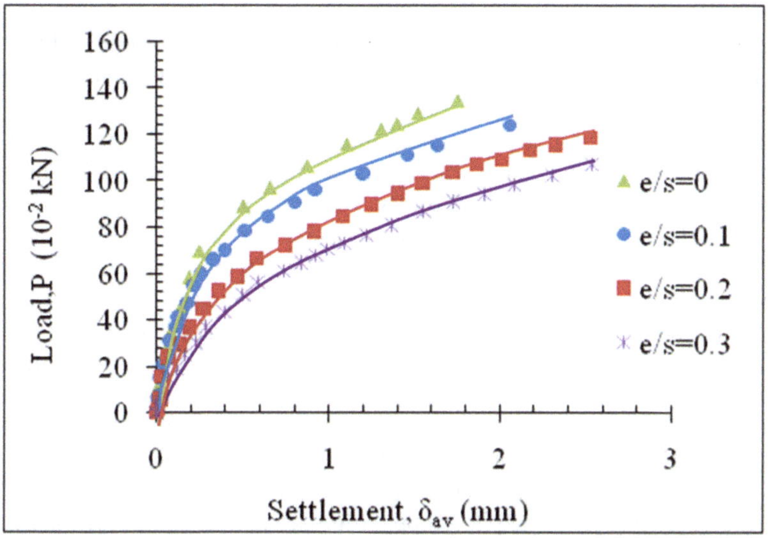

Fig. 3. Variation of load with average settlement – effect of e/s (L/d = 5, s/d = 2, D_r = 35%)

The ultimate load of the pile group decreases and the average displacement increases with increase in eccentricity of the applied load. Load for a constant displacement decreases with increase in normalized eccentricity. Loads for displacement of 1 mm are 0.108 kN, 0.98 kN, 0.8 kN and 0.7 kN for e/s of 0, 0.1, 0.2 and 0.3 respectively. The ultimate loads for e/s = 0 (central load), 0.1, 0.2 and 0.3 are 0.15 kN, 0.135 kN, 0.12 kN and 0.108 kN respectively. Figure 4 presents the variations of normalised ultimate load with normalised eccentricity for different s/d ratios for L/d ratio of 5 and relative density of 35%. The ultimate load ratio decreases with increase in e/s as is expected for all s/d. For s/d of 2, the ultimate load ratios with respect of that for vertical load (e/S = 0) are 0.9, 0.8 and 0.72 for e/s = 0.1, 0.2 and 0.3 respectively. Increase in s/d results in increase in ultimate load ratio. The normalised ultimate load

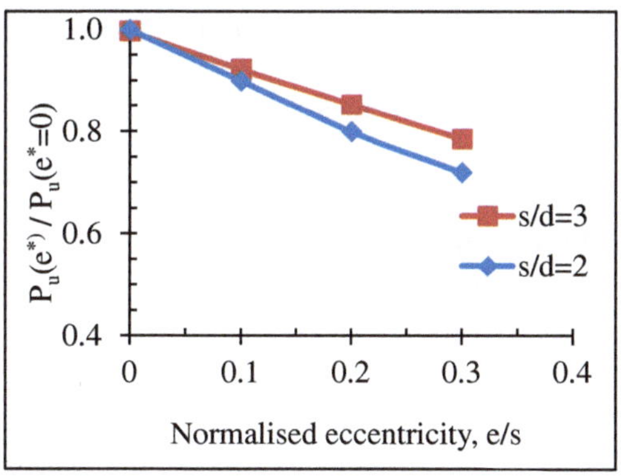

Fig. 4. Variation of normalised ultimate load with normalised eccentricity, e/s – effect of s/d (L/d = 5, D_r = 35%)

ratios for s/d of 3 are 0.92, 0.85 and 0.79 for e/s = 0.1, 0.2 and 0.3 respectively corresponding to 3%, 7% and 9% increase in ultimate load ratio compared to those for s/d = 2. For higher spacing, interaction between the piles is lesser and larger ultimate load ratios.

Figure 5 shows the variations of initial pile group stiffness defined as the ratio of load to displacement of pile group with normalised eccentricity for different s/d ratios for L/d ratio of 5 for 35% relative density. The pile group stiffness decreases with increase in e/s for all s/d. The pile group stiffnesses for e/s = 0, 0.1, 0.2 and 0.3 are 0.556 kN/mm, 0.4 kN/mm, 0.27 kN/mm and 0.167 kN/mm respectively for s/d = 2. Similarly for s/d ratio of 3, the pile group stiffness values are 0.67 kN/mm, 0.5 kN/mm, 0.33 kN/mm and 0.2 kN/mm for e/s = 0, 0.1, 0.2 and 0.3 respectively and correspond to 20%, 25%, 23% and 25% increases compared to those for s/d = 2. Increase in spacing increases pile group stiffness for any L/d. Higher the spacing higher is the pile group stiffness due to lesser interaction. Similar results are obtained for four pile group installed in sand with 73% relative density for s/d = 2 and 3. Table 2 presents the values for pile group stiffnesses and ultimate loads for relative density of 73% for different L/d and s/d ratios. The pile group stiffnesses and ultimate loads increase with increase in the relative density of the sand.

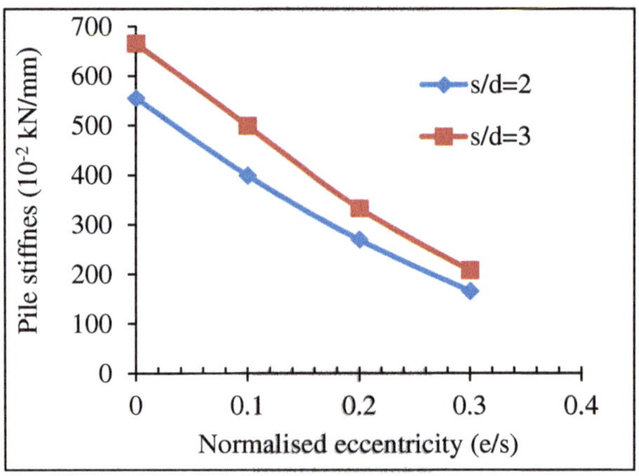

Fig. 5. Variation of Pile stiffness with normalised eccentricity, e/s for different s/d (L/d = 5, D_r = 35%)

3.1 Comparison of Predicted Response of Four Pile Group with Measured Response

Figure 6 shows measured load - displacement response of a single pile with L/d = 5 in a poorly graded sand with relative density of 73%. Variation of displacement load ratio with load of single pile is plotted with δ/Pas ordinate and δ as abscissa. the Kondner's hyperbolic model parameters namely initial stiffness, k_{pc} and ultimate load, P_{ult} are determined from the inverse of slope of the plot δ/P as ordinate and δ as abscissa gives the ultimate load, P_{ult}. The slope of the line fitted by δ/p verses δ is 0.0015. P_{ult} is about

Table 2. Pile group stiffness and ultimate load values for the four pile group for D_r = 73%

L/d	s/d	e/s	Pile Stiffness k_c (kN/mm)	Ultimate load (kN)
5	2	0	0.62	0.18
		0.1	0.50	0.17
		0.2	0.38	0.15
		0.3	0.26	0.14
5	3	0	0.71	0.2
		0.1	0.55	0.19
		0.2	0.41	0.18
		0.3	0.29	0.17
7.5	2	0	0.83	0.25
		0.1	0.62	0.24
		0.2	0.45	0.23
		0.3	0.32	0.21
7.5	3	0	0.90	0.28
		0.1	0.71	0.27
		0.2	0.55	0.26
		0.3	0.41	0.25

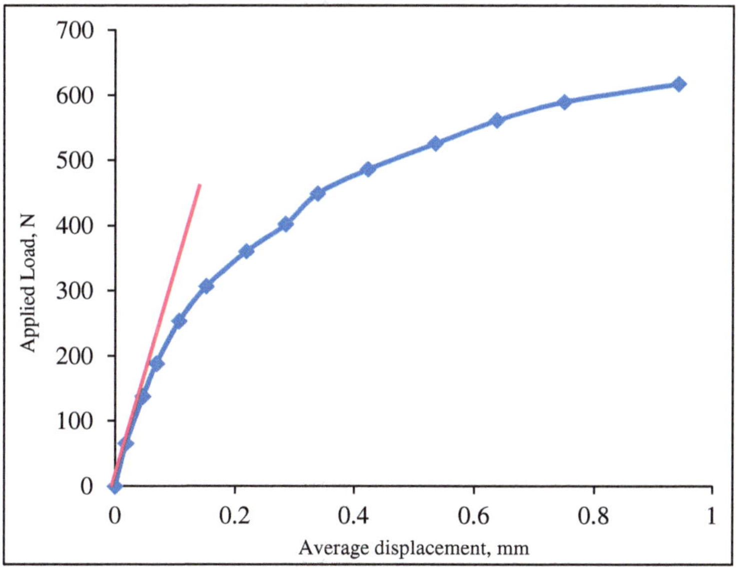

Fig. 6. Single pile measured response under compression for L/d = 5, D_r = 73%.

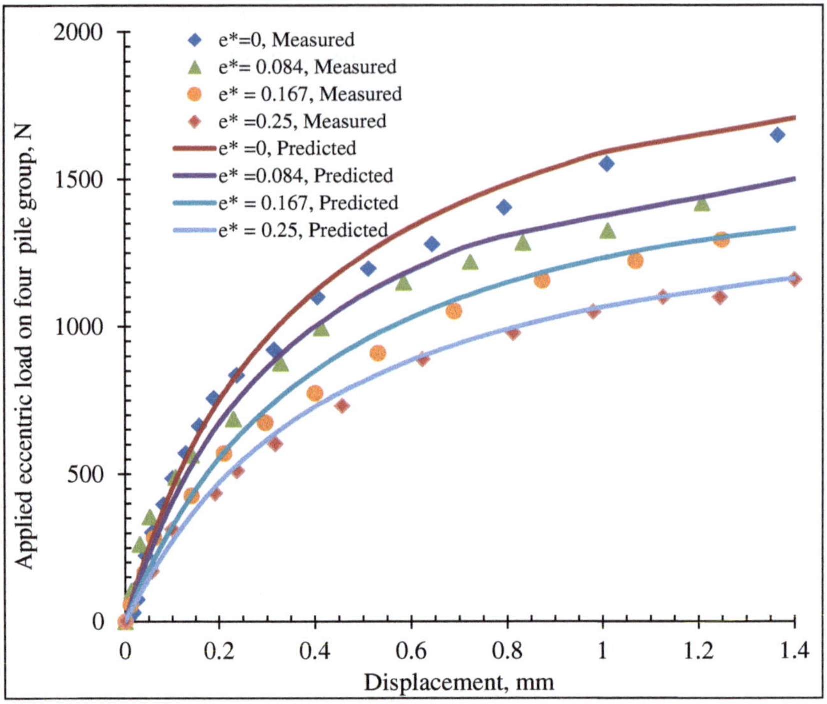

Fig. 7. Comparison of predicted load- displacement response of four pile group with measured response for different eccentricity ratios, e* (D_r = 73%, L/d = 5, s/d = 2.0)

660 N and the pile stiffness, k_{pc} is 400 N/mm. The estimated k_c and p_{ult} are used to validate the predicted response of four pile group with L/d = 5. Figure 7 shows the comparison of predicted load-displacement response of eccentrically loaded four pile group with measured response for four pile group with L/d = 5, s/d = 2 and relative stiffness factor, μ = 12. The predicted response of load-displacement is well in agreement with measured response for e = 0, 1, 2 and 3 cm.

4 Conclusions

1. The response of four pile group under eccentric loading is studied in terms of load versus displacement plots. Displacement of the pile group increases with increase in eccentricity for any s/d.
2. The response of the four pile group under different eccentric loading varies with L/d and s/d ratios of the piles.
3. The increase in relative density of the sand bed causes increase in stiffness of the pile group for all spacing to diameter (s/d) and length to diameter, (L/d) ratios.
4. The predicted response using proposed theory for eccentrically loaded pile groups based on Kondner's hyperbolic relation between load and displacement compares well with measured response.

References

Das, B.M., Seeley, G.R., Smith, E.J.: Uplift capacity of group piles in sand. J. Geotech. Eng. Div. ASCE **102**(3), 282–286 (1976)

Davis, E.H., Poulos, H.G.: The use of elastic theory for settlement prediction under three-dimensional conditions. Geotechnique **18**(1), 67–91 (1968)

Kondner, R.L.: Hyperbolic stress–strain response: cohesive soils. J. Soil Mech. Found. Div. ASCE **89**(SM1), 115–143 (1963)

Kishida, H., Meyerhof, G.G.: Bearing capacity of pile groups under eccentric loads in sand. In: 6th International Conference on Soil Mechanics and Foundation Engineering, Montreal, Canada, vol. 2, pp. 270–274 (1965)

Komatsu, A., Yoshito, M., Takahiro, S.: Study on seismic bearing capacity of grouped piles with battered piles. In: 13th World Conference on Earthquake Engineering, Vancouver, B.C., Canada, Paper No. 3294 (2004)

Saffery, M.R., Tate, A.P.K.: Model tests on pile groups in a clay soil with particular reference to the behavior of the group when it is loaded eccentrically. In: 5th International Conference on Soil Mechanics and Foundation Engineering, vol. 2, Paris, pp. 129–134 (1961)

Meyerholf, G.G., Yalcin, A.S., Mathur, S.K.: Ultimate pile capacity for eccentric inclined load. J. Geotech. Eng. **109**(3), 408–423 (1983)

Randolph, M.F., Wroth, C.P.: An analysis of the vertical deformation of pile groups. Geotechnique **29**, 423–439 (1979)

Terzaghi, K., Peck, R.B.: Soil Mechanics in Engineering Practice, 2nd edn, p. 729. Wiley, New York (1967)

Madhav, M.R.: Efficiency of pile groups in tension. Can. Geotech. J. **24**, 149–156 (1986)

Poulos, H.G.: Load deflection prediction for laterally loaded piles. J. Aust. Geomech. **FE3**(1), 1–8 (1973)

Poulos, H.G., Davis, E.H.: Pile Foundation Analysis and Design. Wiley, New York (1980)

Poulos, H.G., Madhav, M.R.: Analysis of the movements of battered pile. In: I Australia-New Zealand Conference on Geomechanics, Melbourne, July 1971

El Sawwaf, M.: Experimental and numerical study of eccentrically loaded strip footings resting on reinforced sand. J. Geotech. Geoenviron. Eng. **135**(10), 1509–1518 (2009)

Daniel, J.B., Andrawes, B., Hajjar, J.F., Scott, M.O., James, H., Buenker, J.: Forensic collapse investigation of a concrete bridge with timber piers. Research Report, ICT-R27-SP12, DeKalb County Bridge Collapse Investigation (Special Project) Illinois Center for Transportation (2009)

Daniel, J.B., Andrawes, B., Hajjar, J.F., Scott, M.O.: Experimental and analytical forensic investigation of bridge timber under piles eccentric loads. Eng. Struct. **30**(8), 2237–2246 (2010)

Padmavathi, M., Padmavathi, V., Madhav, M.R.: Load-displacement and moment-rotation responses of foundation subjected to moment loading. Int. J. Geomech. ASCE **16**(5) (2016). ISSN 1532-3641 (2016)

Niket, M.N., Nitin, J.H.: Carrying capacity of model piles under eccentric loading. In: 5th Young Indian Geotechnical Engineers Conference, Vadodara, India, pp. 9–12 (2015)

Negative Skin Friction Distribution on a Single Pile - Numerical Analysis

Talal Awwad[1,2(✉)], Salma Al Kodsi[3], and Alexey Shashkin[4]

[1] L.N. Gumilyov Eurasian National University, Astana, Kazakhstan
dr.awwad.gfce@gmail.com
[2] Damascus University, Damascus, Syria
[3] Geotechnical Division, Department of Civil Engineering, Graduate School of Osaka University 大阪大学, Osaka, Japan
[4] Emperor Alexander I St. Petersburg State Transport University, St. Petersburg, Russia

Abstract. Negative skin friction occurs when the settlement of the adjacent soil is larger than the pile settlement. Loading the surface next to the pile is one of the most common causes for negative friction to mobilize along the pile's shaft. In this paper, a numerical simulation will be carried out using Matsui-Abe soil model to study the case of a single pile driven in soft clay layer. Matsui-Abe soil constitutive model represents the Elasto-plastic relation between stress and strain. Negative skin friction (NSF) will be produced due to the soil consolidation. The model will be validated by comparing the results obtained from the primary consolidation with different field tests. A parametric study of the surcharge load will be conducted then to examine the change in negative skin friction (NSF) distribution. Graphs will be obtained showing the relation between loading and changing in the neutral plane location (NPL).

1 Introduction

NSF produces due to the settlements of the pile and the adjacent soil. Usually when the pile is driven into a clay layer, its settlement will be larger than the soil leading to produce a positive shear stresses along the pile's shaft. In case of applying a surcharge load on the surface next to the pile head, the settlements of the adjacent soil may be become larger than the pile's depending on the value of the surcharging and a negative shear stress distribution will occur along the pile's shaft. Terzaghi and Peck (1948) assumed full mobilization of shear strength along the pile-soil interface up to the pile toe for a single pile, or along the perimeter of pile group. Therefore, the neutral point (point of zero shear stress) is assumed to be located at the bearing stratum of the pile. Indraratna et al. (1992) suggested that in order to minimize the negative skin friction, piles may be driven few weeks, or a month later after the surcharge load is applied on the ground surface. Poorooshasb et al. (1996) studied the case of a single pile of circular cross-section considering an axisymmetric problem with a surcharge pressure

H. El-Naggar et al. (Eds.): GeoMEast 2018, SUCI, pp. 36–48, 2019.
https://doi.org/10.1007/978-3-030-01902-0_4

on the underlying clay layers. Hanna and Sharif (2006) conducted a study on piles driven into clay and subjected to indirect loading through the surcharge applied symmetrically on the surrounding area. The study was based on a numerical model using finite element technique and the soil was assumed to follow a linear elastic-perfectly plastic stress-strain relationship, which defined by Mohr-Coulomb failure criterion. They concluded that negative friction is developed on the pile's shaft during the undrained period and continues to exist until the consolidation of the surrounding soil is completed under surcharge loads. Awwad et al. (2017) and Awwad and Kodsi (2017) carried out a parametric study to examine the change in the neutral plane location by making a comparison of two different soil models. Arpan and Sujit (2017) studied the effect of surcharging on the soil consolidation using 3-D analysis.

The objective of this paper is to study the effect of surcharging on the distribution of shear stresses along the pile length beside the examination of change in the location of the neutral plane. A field test of a single pile in clay layer will be presented herein to compare the field measurements with the analytical modelling. By validating the model, a parametric study can be carried out.

2 Methodology

2.1 Model Validation

2.1.1 Field Loading Test of Suehiro Pile-Model Boundary, Mesh and Initial Conditions

Matsui (1993) carried out a loading test of a cast-in-place pile in Osaka Bay, Osaka, Japan. The soil profile consists of eight layers with the main clay layer with 14.9 m in height. The upper layers formed of loose, soft silts and clays. The bearing layer is a sandy gravel. The tested pile is 1.5 m in diameter and 28.5 m in length. Figure 1. shows a cross section in the field test soil profile. The parameters used for modelling are the elastic parameters Table 1, Elasto-plastic model defined by Matsui and Abe Table 2, bearing layer parameters represented by Yasufuku model Table 3, and the joint element parameters defined by Goodman et al. (1968) Table 4. All the soil parameters were determined from soil investigations (After Oda 2012).

The surface of the top layer is assumed to be permeable while the low and side boundaries of the model are supposed to be impermeable. The level of the groundwater is at 0.80 m down from the surface. The low model boundary is being fixed, while the boundaries at the axis of symmetry and sides are free to move vertically. The mesh was divided into 13 blocks. The first block -the pile- is divided into 360 elements behave as an elastic material. The main clay layer includes the block (2) and block (3) which both are divided into 234 elements. The abruption between pile shaft and surrounding soil is

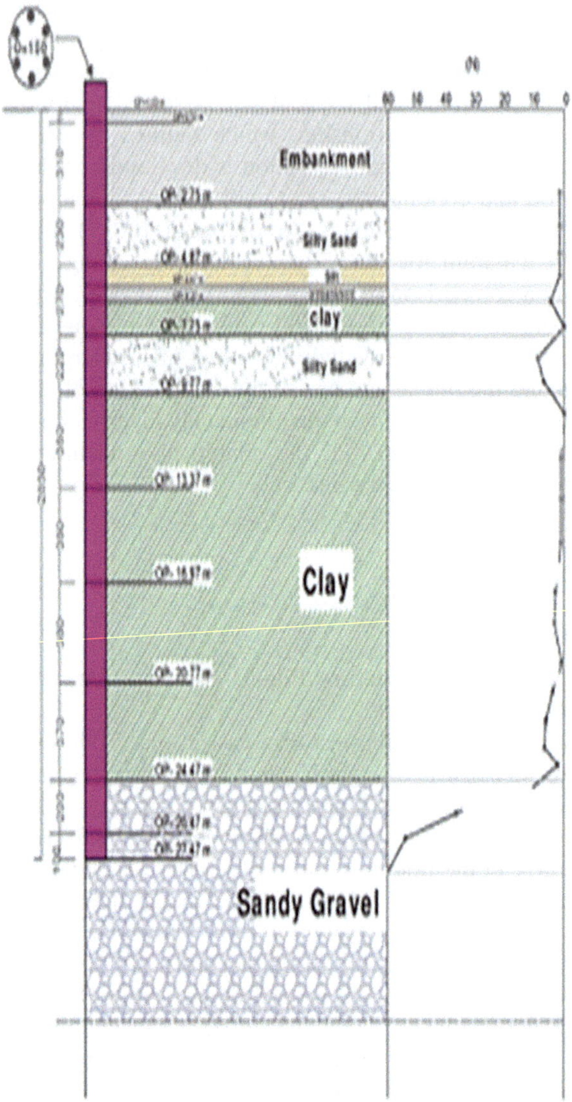

Fig. 1. Soil profile of Suehiro pile

represented by 36 joint elements (Goodman et al. 1968) in which the main clay layer includes 6 joint elements. The analytical study model and mesh boundary are shown in Figs. 2 and 3, respectively.

Table 1. Parameters of elasticity

No.	E (kN/m^2)	υ
E11	8.40E+04	0.47
E12	2.70E+04	0.47
E13	3.00E+04	0.35
E14	1.50E+04	0.30
E15	2.20E+07	0.22

Table 2. Parameters of Matsui-Abe model

(a)

No.	λ	κ	M	υ
Ma1	0.248	0.0124	1.41	0.333
Ma2	0.495	0.0248	1.26	0.333
Ma3	0.400	0.0248	1.26	0.333
Ma4	0.580	0.0243	1.26	0.333
Ma5	0.448	0.0224	1.32	0.333

(b)

No.	α_{a0}	α_{af}	α_{p0}	α_{pf}
Ma1	0.44	1.00	−1.44	1.00
Ma2	0.34	1.000	−1.34	1.00
Ma3	0.34	1.000	−1.34	1.00
Ma4	0.34	1.000	−1.34	1.00
Ma5	0.38	1.000	−1.38	1.00

2.1.2 Soil Constitutive Model

The main clay layer was modelled to behave as an elasto-plastic material to represent the relationship between stress and strain due to Matsui and Abe model (1981). This relation expressed by the following equations:

$$e_0 - e = \lambda \ln\left(\frac{p'}{p'_0}\right) + (\lambda - \kappa)\left(\frac{\alpha_a}{\alpha_a - 1}\right) \ln\left\{\frac{M_a + (\alpha_a - 1)\eta}{M_a + (\alpha_p - 1)\eta_{k0}}\right\} \quad \text{For the active state} \tag{1}$$

$$e_0 - e = \lambda \ln\left(\frac{p'}{p'_0}\right) + (\lambda - \kappa)\left(\frac{\alpha_p}{1 - \alpha_p}\right) \ln\left\{\frac{M_p + (1 - \alpha_p)\eta_{k0}}{M_p + (1 - \alpha_p)\eta}\right\} \quad \text{For the active state} \tag{2}$$

where e is the void ratio, e_0 is the initial value of e, λ is the compression index, κ is the swelling index and α_a, α_p are the strain increment ratio.

Table 3. Parameters of Yasufuku model

(a)

No.	h	ka	l	Pa
Yf1	0.4	1.90E−02	0.75	98.0
Yf2	0.4	2.82E−03	0.80	98.0
Yf3	0.4	2.82E−03	0.75	98.0
Yf4	0.4	2.82E−03	0.75	98.0
Yf5	0.4	3.27E−03	0.80	98.0
Yf6	0.4	2.82E−03	0.80	98.0
Yf7	0.4	3.27E−03	0.75	98.0

(b)

No.	α	N	M
Yf1	0.4	1.90E−02	0.75
Yf2	0.4	2.82E−03	0.80
Yf3	0.4	2.82E−03	0.75
Yf4	0.4	2.82E−03	0.75
Yf5	0.4	3.27E−03	0.80
Yf6	0.4	2.82E−03	0.8
Yf7	0.4	3.27E−03	0.75

(c)

No.	αh	αp	B	ν
Yf1	1.46	1.50	0.800	0.25
Yf2	1.95	1.85	0.915	0.25
Yf3	1.95	1.90	0.915	0.25
Yf4	1.95	1.90	0.915	0.25
Yf5	1.83	1.85	0.915	0.25
Yf6	1.80	1.85	0.915	0.25
Yf7	1.95	1.85	0.915	0.25

Note: Cg and Cf are constant.
Cg = 2.0, Cf = 1.5

A comparison of the numerical model and the field test measurements for the axial load distribution, settlements at pile head and settlements at pile toe are shown in Figs. 4, 5, and 6, respectively.

The model is valid and the parametric study can be carried out.

Table 4. Parameters of joint elements

No.	Kn (kN/m^3)	Ks (kN/m^3)	c0 (kN/m^2)	θ (°)
J01	9.8E+09	9.8E+09	15.55	0.0
J02	9.8E+09	9.8E+09	11.86	0.0
J03	9.8E+09	9.8E+09	19.58	0.0
J04	9.8E+09	9.8E+09	8.33	0.0
J05	9.8E+09	9.8E+09	9.07	0.0
J06	9.8E+09	9.8E+09	8.28	0.0
J07	9.8E+09	9.8E+09	5.64	0.0
J08	9.8E+09	9.8E+09	17.29	0.0
J09	9.8E+09	9.8E+09	15.90	0.0
J10	9.8E+09	9.8E+09	11.52	0.0
J11	9.8E+09	9.8E+09	15.01	0.0
J12	9.8E+09	9.8E+09	10.97	0.0
J13	9.8E+09	9.8E+09	10.40	0.0
J14	9.8E+09	9.8E+09	13.23	0.0
J15	9.8E+09	9.8E+09	12.87	0.0
J16	9.8E+09	9.8E+09	13.29	0.0
J17	9.8E+09	9.8E+09	9.09	0.0
J18	9.8E+09	9.8E+09	9.41	0.0
J19	9.8E+09	9.8E+09	5.89	0.0
J20	9.8E+09	9.8E+09	11.53	0.0
J21	9.8E+09	9.8E+09	34.56	0.0
J22	9.8E+09	9.8E+09	28.65	0.0
J23	9.8E+09	9.8E+09	43.03	0.0
J24	9.8E+09	9.8E+09	42.14	0.0
J25	9.8E+09	9.8E+09	36.43	0.0
J26	9.8E+09	9.8E+09	28.85	0.0
J27	9.8E+09	9.8E+09	33.26	0.0
J28	9.8E+09	9.8E+09	38.46	0.0
J29	9.8E+09	9.8E+09	32.61	0.0
J30	9.8E+09	9.8E+09	43.37	0.0
J31	9.8E+09	9.8E+09	37.00	0.0
J32	9.8E+09	9.8E+09	28.52	0.0
J33	9.8E+09	9.8E+09	42.45	0.0
J34	9.8E+09	9.8E+09	39.23	0.0
J35	9.8E+09	9.8E+09	37.01	0.0
J36	9.8E+09	9.8E+09	25.99	0.0

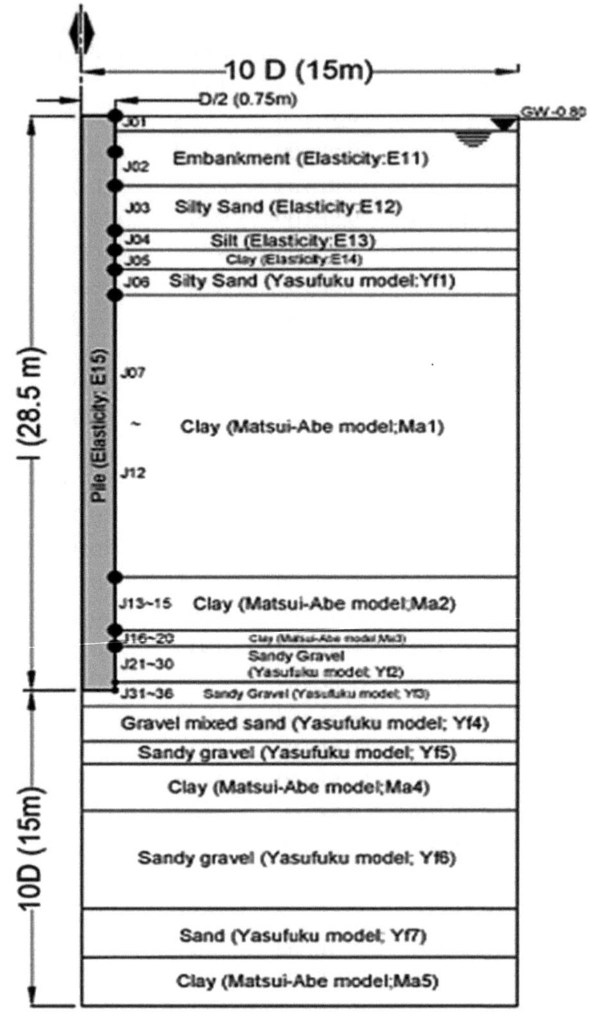

Fig. 2. Numerical model

3 Parametric Study

The main reason for the parametric study is to examine the effect of surcharging on the skin friction distribution.

The pile will be loaded first then the surcharge load will be applied next to the pile head. The distribution of skin friction will be shown in the figures at the end of primary consolidation. Changing the surcharge load will change the distribution of skin friction depending on the surcharging value. The distribution of skin friction defines the location of the neutral plane; the point of equilibrium.

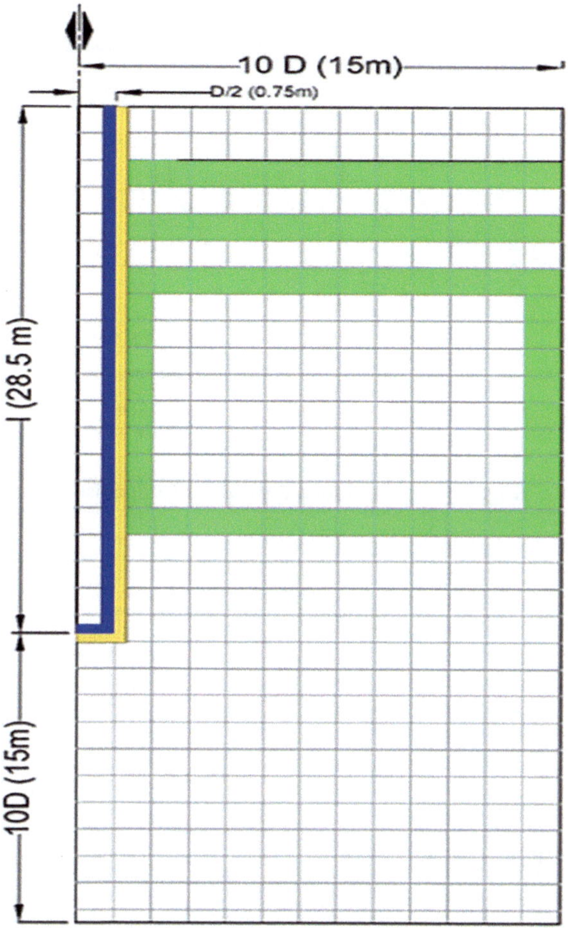

Fig. 3. The model mesh

The parametric study cases are shown in Table 5. The distribution of skin friction for each case is clarified in Figs. 7, 8, and 9, respectively. The change in the location of neutral plane according to the loading cases is shown in Fig. 10.

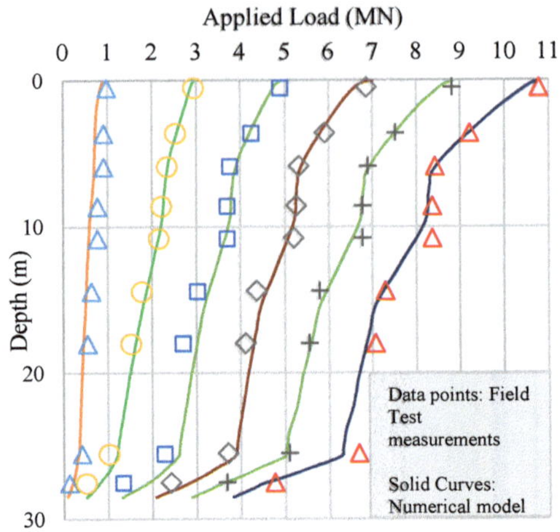

Fig. 4. Pile axial load distribution/comparison

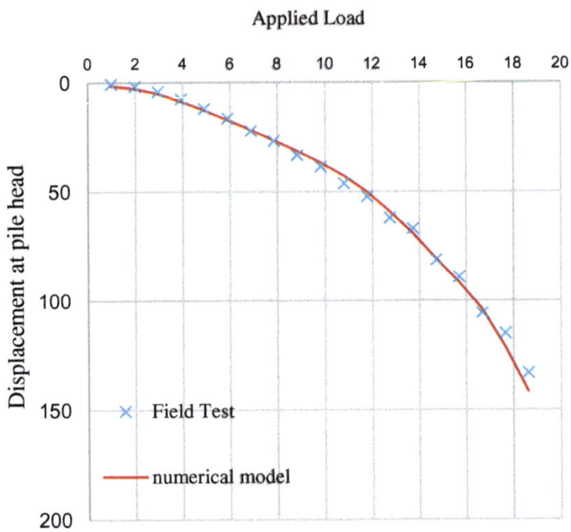

Fig. 5. Pile head settlements

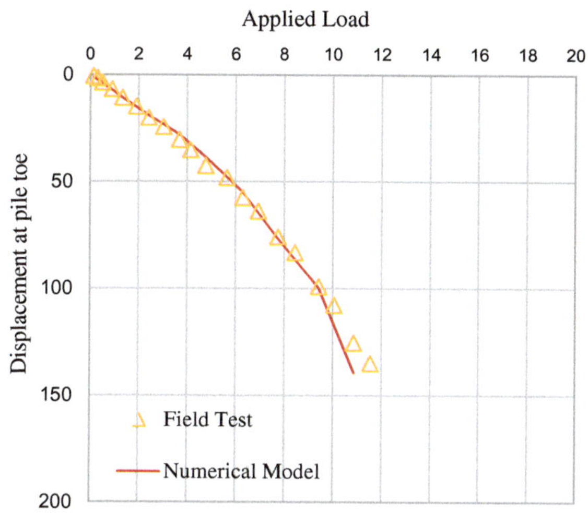

Fig. 6. Pile toe settlements

Table 5. Parametric study load cases

Case no.	Pile load	Surcharge load
Case 1	100 tf	2.00 tf/m^2
Case 2	100 tf	8.00 tf/m^2
Case 3	100 tf	16.00 tf/m^2

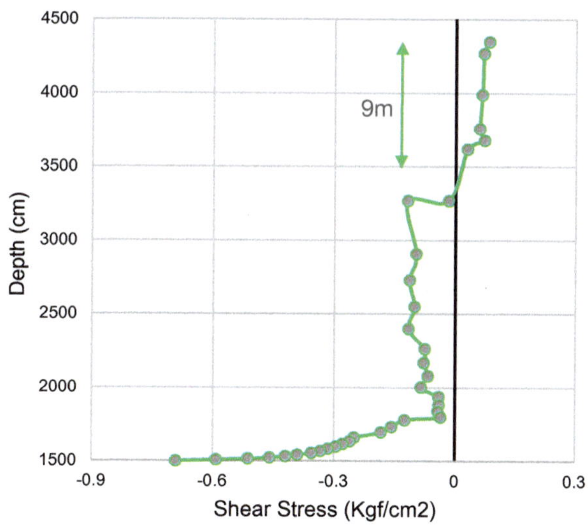

Fig. 7. Shear stress distribution along pile shaft/PL = 100 tf, SL = 2.00 tf/m^2

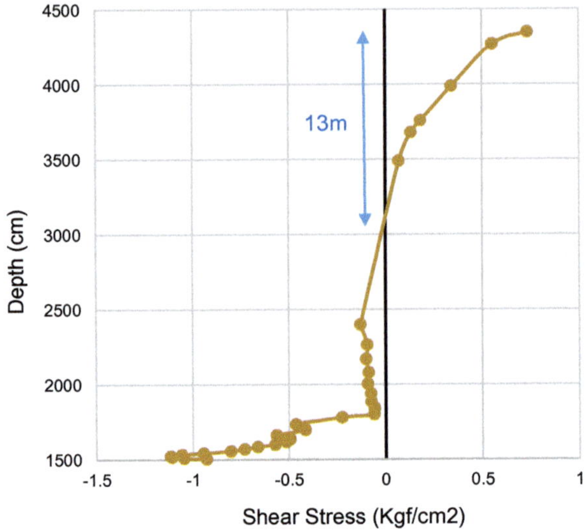

Fig. 8. Shear stress distribution along pile shaft/PL = 100 tf, SL = 8.00 tf/m^2

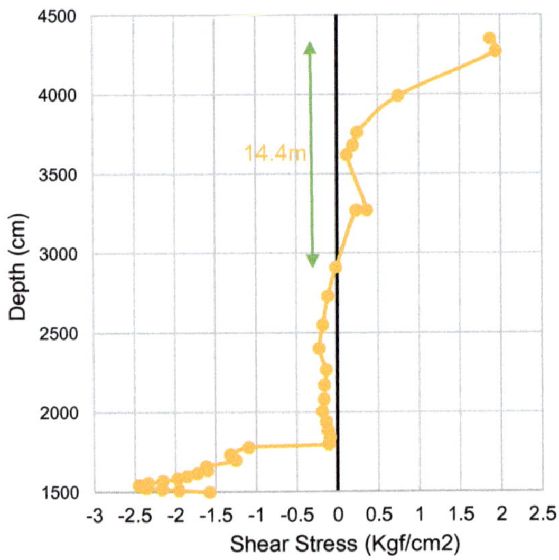

Fig. 9. Shear stress distribution along pile shaft/PL = 100 tf, SL = 16.00 tf/m^2

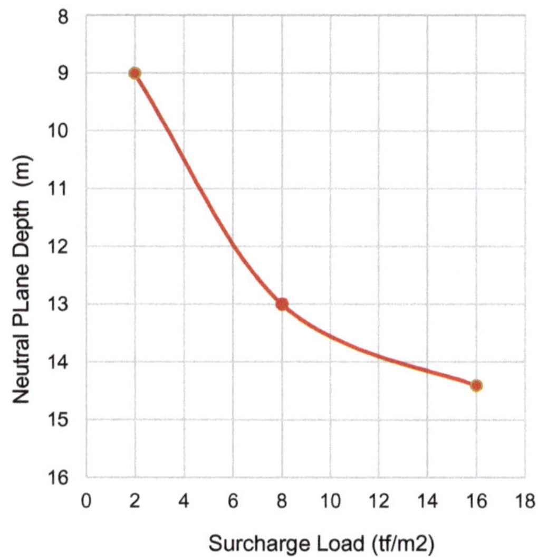

Fig. 10. Change in the location of the neutral plane

4 Conclusions

Applying surcharge load on the ground surface next to the pile leads the clay layer to settle due to the consolidation process. At this point, both pile and soil are moving. This relative movement produces shear stresses at the pile-soil interface. In this paper, different values of surcharge load were applied and the effect of loading was examined. it can be concluded that:

- The surcharge load plays the main role in mobilizing the shear stresses (negative and positive) whereas increasing the load increases the skin friction.
- Location of the neutral plane goes deeper by increasing the value of the surcharge load and therefore the arising problems due to negative skin friction phenomenon must resolve in case of applying a heavy load for a long-term.

References

Awwad, T., Donia, M., Awwad, L.: Effect of a stiff thin Foundation soil layer's depth on the dynamic response of an embankment dam. Proc. Eng. **189**(2017), 525–532 (2017)

Awwad, T., Kodsi S.A.: A comparison of numerical simulation models to determine the location of the neutral plane. In: Proceedings of the 19th International Conference on Soil Mechanics and Geotechnical Engineering, pp. 1947–1950 (2017)

Arpan, L., Sujit, K.P.: The effect of different surcharge pressure on 3-D consolidation of soil. Int. J. Appl. Eng. Res. **12**(8), 1610–1615 (2017). ISSN 0973-4562

Hanna, A., Sharif, A.: Negative skin friction on single piles in clay subjected to direct and indirect loading. Intl. J. Geomech. ASCE (2006) (in press)

Indraratna, B., Balasubramaniam, A.S., Phamvan, P., Wong, Y.K.: Development of negative skin friction on driven piles in soft Bangkok clay. Can. Geotech. J. **29**, 393–404 (1992)

Matsui, T., Abe, N.: Multi-dimensional elasto-plastic consolidation analysis by finite element method. Soils Found. **21**(1), 79–95 (1981)

Matsui, T.: Case studies on cast-in-place bored piles and some considerations for design. In: Deep Foundations on Bored and Auger Piles. Ghent, pp. 77–101 (1993)

Oda, K.: Numerical simulations of field loading tests of cast-in-place bored piles with a large diameter. In: International Conference (Proceedings) on Testing and Design Methods for Deep Foundations, pp. 859–866 (2012)

Poorooshasb, H.B., Alamgir, M., Miura, N.: Negative skin friction on rigid and deformable piles. Comput. Geotech. **18**(2), 109–126 (1996)

Terzaghi, K., Peck, R.B.: Mechanics in Engineering Practice. Wiley, New York (1948)

The Proposed Use of Generalized Regression Neural Network to Predict the Entire Static Load Test

A. K. Alzo'ubi[1(✉)] and Farid Ibrahim[2]

[1] Civil Engineering Department, Abu Dhabi University, Al Ain, UAE
abdel.alzoubi@adu.ac.ae
[2] Information Technology Department, Abu Dhabi University, Al Ain, UAE

Abstract. In the UAE, continuous flight auger piles (CFA) are the most commonly used type of foundations to construct high rise buildings, bridges, and other heavy structures due to the high groundwater table and the weak soil/rock layers near the ground surface. To minimize the risk of failure, of these CFA piles, mandatory expensive field tests need to be performed and the most important one is the Static Pile Load Test (SPLT). To minimize the number of tests required for a particular project in the field, this paper proposes using General Regression Neural Network (GRNN) to predict the pile performance ahead of any test. The data collected from thousands of loading points in over one hundred projects from Dubai, Abu Dhabi, and Al Ain cities are used to develop a GRNN capable of predicting SPLT curves with reasonable accuracy. The friction angle, unconfined compressive strength, depth, soil type, groundwater table, pile's diameter, and pile's length are the parameters that are input to predict the load-displacement curves of the SPLT. This approach can complement conventional SPLT and provide engineers with sufficient insight on the pile performance ahead of the actual test.

Keywords: Static load test · Piles · GRNN

1 Introduction

The upper geological formation in the United Arab Emirates (UAE) is soft ground consisting of sand, silty sand, or soft rocks with high groundwater table. This geotechnical environment is ideal for using CFA piles to safely transmit the load into the ground. These piles are easy and fast to construct, and in order to predict the pile capacity, two approaches are utilized; empirical method and/or field tests. The first approach is conservative due to many variables controlling the pile capacity while the latter is so expensive and it adds to the total cost of a project.

Many factors contribute to the most effective pile design; the size of the project, risk, and the significance of the project, to name few. Most analytical methods of determining the pile capacity are developed empirically by comparing the results of static load tests with measured soil properties (Coduto 2001). In UAE, conducting the static load tests is a vital part of designing these CFA piles and checking their performance to avoid the risk of foundation failure. According to the code of building in

H. El-Naggar et al. (Eds.): GeoMEast 2018, SUCI, pp. 49–60, 2019.
https://doi.org/10.1007/978-3-030-01902-0_5

UAE, a minimum of 5% of the piles in a project are selected and tested under the static load test after finishing the construction of all piles. The static pile test load is applied in increments up to 1.5 times the design load for piles, or more in rare cases, as required by the engineer. To accelerate the construction process in UAE, engineers started using CFA piles (Fig. 1). However, to minimize the risk of partial or complete failure of the foundation system in both excessive settlement and bearing capacity, over designing was the solution.

Fig. 1. CFA piles underneath the current Etisalat center in Abu Dhabi

In this paper, we propose to use General Regression Neural Network (GRNN) to predict the static load test prior to performing the actual field test. This approach would help engineers to minimize the risk of under testing piles failure and consequently reducing the overall cost. To reduce the variation of the parameters in the GRNN, we depend on three facts; the consistency of the geology, the dominant pile diameter, and the number of piles passing the tests. The tests in this paper are based in three major cities, the two coastal cities of Abu Dhabi and Dubai, and the city of Al Ain. The geological stratification in the three cities are very similar; Fig. 2 shows a typical soil profile that is used in this study. The diameter of the piles, however, was limited to half meters, while the pile's lengths range between 6 and 16 m. In the building code, a single pile that exceeds the limit of the acceptable downward movement of 1% of its diameter at 1.5 times the design load, is deemed as a failure, hence it was excluded from our study.

Depending on the previously mentioned factors, we assume that predicting the pile static load tests by means of GRNN is possible after training the model with the available data. This GRNN would allow engineers to predict the pile's performance prior to the actual test. This approach can reduce the cost of conducting multiple static load test, and therefore reducing the risk of failure while using a reasonable factor of safety. Coduto (2001), for example, stated that static load tests are much more expensive than analytic methods, and thus the latter is very attractive. However, engineers in UAE tend to over design the piles in order to minimize failure risk and thus, the calculated load capacities become so conservative that result in expensively constructed piles.

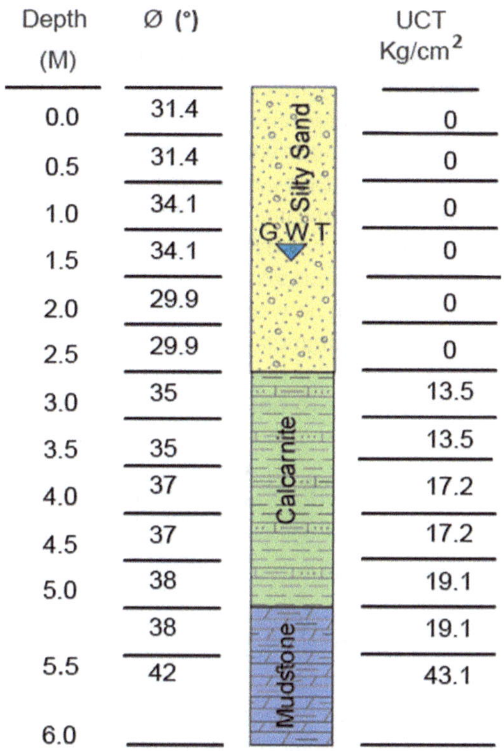

Fig. 2. General subsurface geotechnical profile in Abu Dhabi

2 Artificial Neural Networks

The artificial neural network is commonly used in many geotechnical engineering applications such as foundations, soil type, and liquefaction. One kind of such networks is the Backpropagation Neural Network (BPNN). In this learning model, a set of training records with known output are processed by an algorithm to learn the neural network. It consists of three layers: an input layer, hidden layer(s) and an output layer. A set of connected units are defined with each layer, and weights are associated with each connection between units. The input layer defines the attributes of each training record. Weights are calculated and passed to the next layer, known as the hidden layer. If the network is configured to have more than one hidden layer, then weights are recalculated and passed to the next hidden layer. The weighted output of the last hidden layer is passed to the output layer. This output is regarded as the network predicted value. This output is compared to the known output for this record and the network is said to back-propagate to adjust the weights and possible biases, and re-iterate again. The process continues until the error between the known output and the predicted value is minimized (Han and Kamber 2012).

Another kind of neural network models is the GRNN. This prediction model is based on function approximation. It is a feed forward network where regression is performed using a probability density function applied to the input data. It consists of four layers; input, pattern, summation and output layers. The input layer facilitates inputting all the attributes as a vector. These inputs are passed to the pattern layer which consists of units that store cluster centers that are determined using a particular clustering technique; such as the K-means. The objective of these centers is to group the sample inputs so that each group can be represented in one unit of the pattern layer that measures the distance of the input vector from the cluster center (Specht 1991). These units are used to subtract each new vector X from the stored cluster centers. The squares of all these differences are summed and input to an activation function which is an exponential function. The output of the pattern layer is stored as two parts in the summation layer which consists of two units. The first unit stores:

$$\sum_{j=1}^{n} yj\, exp\left(-\sum \frac{(xi - xij)2}{2\sigma 2}\right) \tag{1}$$

The second unit, however, stores

$$\sum_{j=1}^{n} exp\left(-\sum \frac{(xi - xij)2}{2\sigma 2}\right) \tag{2}$$

where (xi–xij) represents the distance between a new point xi and a training sample xij.

Finally, the output unit will compute the division of the values stored in the two summation units to produce the predicted value. The main advantage of GRNN, when compared to BPNN is that it is a simple and a fast learning machine.

3 Related Works

In this research, we are proposing to use GRNN to predict the static load test of CFA piles, while most of the work that was carried out was based on using BPNN to predict the friction and/or end bearing capacity of driven piles. Goh (1996) used a model to predict an estimate of the friction capacity of driven piles in clay soils. The results were promising when compared to the actual data and some empirical methods. Lee and Lee (1996) tried to predict the driven pile capacity. The error between the predicted and the actual pile test was around 20%. However, they did not attempt to predict the entire static load test. Abu Kiefa (1998a, b) developed three artificial neural network models to forecast driven piles capacity. He compared the results with four empirical formulas and found that the model that was devoted to forecasting the total pile capacity was more accurate than others with 0.95 coefficients of determination. Furthermore, Goh (1996) presented a new neural network model to predict pile capacity in sandy soils. The model's pile capacity was satisfactory when compared to other empirical formulas. Shahin (2001) discussed different applications of the artificial neural network in Geotechnical Engineering and mentioned different applications including predicting the pile capacity.

Benali and Ammar (2011) also used the artificial neural network to expect the pile capacity; they used a database of 80 cases collected from the literature and corresponding to different sites distributed all over the world. They reported that the networks are feasible for these kinds of problems but they did not attempt to predict the entire static load test. More recently, Maizir and Kassim (2013) used artificial neural networks for predicting the axial capacity of a driven pile; the data collected for this study consisted of 300 cases from several projects in Indonesia and Malaysia. He found that a good prediction was achieved if "both stress wave data and properties of both driven pile and driving system are considered in the input data". In another experiment (Tarawneh 2013) the pile setup was predicted using an artificial neural network model. The predicted values were compared with those produced by some empirical formulas. It showed that the model produced satisfactory results. A recent paper (Alzo'ubi et al. 2015) proposed a framework for a system model that combines the neural network technique to predict the pile capacity with the geographic information system (GIS) data. The proposal suggested building a new layer in the GIS containing the pile load tests.

GRNN, however, was used in predicting the ultimate capacity of driven piles in cohesionless soils (Abu Kiefa 1998a, b). It concluded that GRNN provided the best prediction for the pile capacity and its components. Another study (Sarkar et al. 2015) used GRNN to predict the soil type and the standard penetration test values at different depths. They found good between their GRNN results and the field results agreement.

As we can see in the above-mentioned studies, no serious attempt was made to predict the entire static load test, and no studies had been conducted on CFA piles. In this paper, we demonstrate that the entire static load test of CFA piles with two cycles of loading-unloading can be predicted by GRNN with enough training data. This can be done by associating the pile configuration with the geotechnical properties of the soil at the location of that pile.

4 General Geology

Fookes and Knill (1969) classified the coastline of UAE as "the base plain". The upper layer deposits of this zone consist of sand dunes, loess, and evaporate together with marine sand and silts. Wind-blown material predominates the movement and transportation of great quantities of silt and sand during periods of high wind. Water also contributes to this movement and deposition of high quantities of Silty Sand material. This soil layer depth varies from 3 to 10 m. This Silty Sand layer has a high friction angle and plays a major role in establishing the pile capacity; piles capacity depends on two components; friction at the sides and end bearing at the bottom of a pile.

Flash floods such as the one occurred in the last years are not so common, and any water stream actually reaching this zone is usually short-lived after the rainstorm. Despite that rain is very rare in the three cities, Dubai, Abu Dhabi, and Al Ain, the water table can be as close as 0.5 m from the ground surface. This is because the first two cities are coastal one while the latter is an oasis. This water table can dominate the desert processes by limiting the wind erosion to soils above the water table, as the capillary forces can limit wind erosion. This capillary pressure facilitates water movement from the water table to ground surface and evaporation occurs.

Usually, a thick salt crust can build and found as thick as twenty centimeters. These salt layers may affect the pile's capacity through friction reduction. These deposits are particularly extensive around the coastline of UAE, and hence a geotechnical engineer needs to be extra cautious for salt layers. Beneath the first layers of sand and silt, other soil and rock materials can also be identified The Sand and Silt deposits cover interbedded sandstones (see Fig. 2), conglomerates, calcisiltite, limestone, and silt-stones; clay deposits may also be found. In this research, the pile's length, in all of the projects that will be included, ranges from 6 to 16 m, so all the layers from 0 m to 20 m will be classified as shown in the next section.

5 Classification of Materials

To classify the soil layers and identify their mechanical properties, field and laboratory test results are conducted and summarized for each borehole in each project. The Unified Soil Classification System was used to classify the ground materials through conducting sieve analysis and Atterberg Limits tests. We used this classification as an attribute in the Neural Network input data to increase the model accuracy and precision. To do so, we coded the soil layer classification as a binary number, Table 1 shows samples of this coding which will be included in the input data of GRNN as a separate unit for each soil type at each respective depth.

Table 1. Coding of some soil layers as will be used in the GRNN

Soil description	Soil profile	Code
Silty SAND (SM): Dry to moist, light brown, fine non-plastic	1	10000000
Poorly graded SAND with silt (SP-SM): Wet, light brown, fine, and non-plastic	2	01000000
MUDSTONE: Very weak to weak, moderately to highly fractured, moderately weathered	3	00100000
MUDSTONE: Weak moderately fractured, moderately weathered	4	00010000
Crystalline Gypsum: Moderately weak moderately fractured, moderately weathered	5	00001000
Loose to medium dense SAND with trace silt	6	00000100
Very loose light brown clayey fine SAND	7	00000010
Dense, light brown fine silty SAND	8	00000001

6 Static Load Test

According to the field and laboratory investigations, subsurface conditions, engineering analysis, and practical experience, the recommended type of foundation were CFA piles in all of the projects in this research. Pile load tests are considered the most satisfactory method to assess the carrying capacity of a pile. Figure 3 shows the details of a static load test conducted in the field.

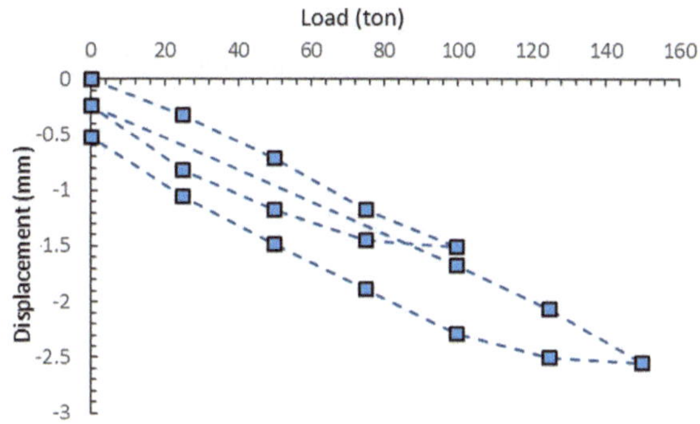

Fig. 3. Field data of a pile static load test

During the foundation construction period, it is recommended that such tests be performed according to British Standard Code of Practice 8004 on specially constructed piles that are installed before the start of the general construction works. The common practice in the UAE is to test the piles after construction for quality assurance and not for design; i.e. to check performance. As a result of this practice, CFA piles, in most cases, are over-designed to avoid failure in the testing process. However, if testing is carried out for design purposes, significant savings may result from a more economical pile design based upon the specific test data.

Since the theoretical design methods provide an approximate working load, the static load test should demonstrate the pile's performance and its load settlement characteristics. Figure 4, shows the static load test performed in the future Etisalat center (see Fig. 1). In the static pile load tests, the load-displacement curves were obtained through two cycles of loading-unloading.

Fig. 4. Running static load test in the current Etisalat center

As mentioned earlier, the pile load test is usually performed to verify the pile ultimate capacity. This value is based upon the allowable amount of settlements determined by the requirements of the engineer or a known standard specification. In all of the tests considered in this paper, the BS8004 (British Standard 2015) was used as the testing procedure. The test load is applied in increments up to 1.5 times the design load for working piles load. The load was applied by means of the hydraulic jack as shown in Fig. 4, four reaction piles were used to provide support for the loading frame. In addition, four settlement gauges were used to monitor the displacement.

The maintained load method was used in all of the projects. The load was increased in stages and maintained for at least 20 min or until the rate of settlement decreased to less than 0.10 mm/h, whichever is greater. As a matter of fact, in all of the projects in this research, the load was maintained until a settlement has seized. In the first loading cycle, increments of 25% of the design load-up to the working load- are imposed to the tested pile and maintained until settlement seized. Settlement readings were taken for each load increment every 5 min until the settlement stopped. In the first unloading cycle, the load was reduced in the inducement of 25% of the testing value until the zero load stage is reached, while the settlement was recorded as described above. Reloading was then applied to test the pile from zero to the test load (150% of the design load). First, the increment was from zero to the design load and kept for six hours at which the settlement seized. And then increased in 25% of the design load until the test load was reached. At last, the unloading occurred in 25% of the testing load until zero loads. In this model, the loading/unloading cycles were coded in binary numbers to reflect the status of the cycle. For example, the first loading cycle was coded as 1000, while the first unloading was coded as 0100. These codes were considered as 4 input units in the GRNN.

The above-mentioned testing procedure is adopted in the majority of tests in the UAE following the 2013 Abu Dhabi International Building Code. We rely on this procedure to verify the CFA piles performance ahead of construction. In case one of the tests failed at a particular site, the CFA piles must be extracted from the ground. This costly solution forced engineers to overdesign the piles to avoid failure in the test. The framework proposed in this paper aims at reducing this cost through predicting the static load test behavior based on the proposed pile configuration, groundwater table, and the soil profile mechanical properties.

7 The Network Model

The principle underlying this research is that it is possible to predict the displacement value using GRNN model that is trained over data identified by 28 input attributes constituting a combination of the soil profile (17 units), pile configuration (2 units), groundwater (1 unit), load (1 unit), loading status (4 units), strength properties (2 units) and depth (1 unit), while the displacement (1 unit) is to be the output predicted attribute. Coduto (2001) showed that the soil profile and the pile configuration are essential in determining the pile capacity and behavior under loading and unloading cycles.

Such network needs a substantial amount of data to be trained and verified to guarantee its performance and its ability to predict the displacement of new pile configuration and new soil profile. Actual pile test data were collected from 100 sites in the field and were implemented in the network. The reliability of the data was assured through three factors; the tests were performed on selected CFA piles, monitored by engineers at the sites, and has been approved by the municipality engineers. Moreover, we verified the data for completeness and filtered it to remove any outliers. As a result of this process, 6437 load-displacement points were used to train the model. The model should be able to generalize the mapping between input and output and should be capable of predicting the output for cases not included in the training set.

For this experiment, MATLAB Simulink 2016 software was utilized to construct the GRNN. The sample data was migrated to MATLAB worksheet at which they were transposed into a matrix P in the MATLAB command:

$$net = newgrnn(P, T) \tag{3}$$

This command generates the network: net, and it requires two parameters: the input transposed matrix P and the target predicted vector values T corresponding to the average displacement Two layers are created by newgrnn as shown in Fig. 5. The first layer has as many units as there are inputs in P. The weight W in this layer is set to P'. The unit's weighted input is computed as the distance between the input vector and its weight vector. The bias b is set to a column vector of 0.8326. The second layer has as many units as the target vector (the output) but with different weight setting W (MathWorks 2016).

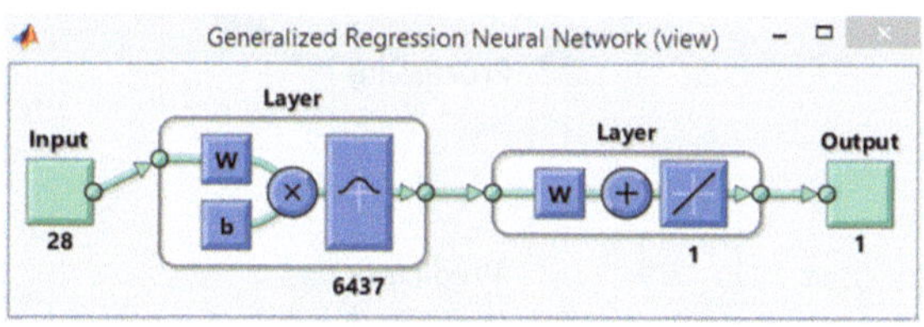

Fig. 5. The GRNN as produced by MATLAB

8 The Test Plan

Pile behavior is a complex process due to the many parameters controlling its performance. Predicting the pile capacity, let alone the entire load-displacement curve, has been a difficult task for geotechnical engineers and that is why the static load test is routinely conducted in the UAE. This paper introduces the GRNN model to predict not only the pile capacity but the two cycles of loading-unloading static load test curves for CFA piles under compression.

Although neural network models that have been used to predict the pile capacity, as discussed earlier in Sect. 3, no attempt has yet been made to predict the entire loading-unloading curves of the static load test. The test plan of our proposed model is that a site was randomly picked with a complete pile tests that was not part of the training data to guarantee the independence of the test. the geotechnical data, groundwater level, pile configuration, loading cycle, and load in the model were input leaving the displacement field to be predicted. Figure 6 shows the testing plan that will be adopted in this research using the MATLAB command:

$$V = sim(net, P) \tag{4}$$

Where P is the input transposed matrix of the tested pile, and V is the resulting output vector. Upon the completion of the testing plan, we will use different statistical methods to compare the predicted values against the actual field data values. The coefficient of determination, for example, will explain the relationship between the two sets of data. Figure 7 shows preliminary results of the test that was conducted using the setup discussed above. The results show impressive agreement between the actual and predicted displacement at each loading cycle.

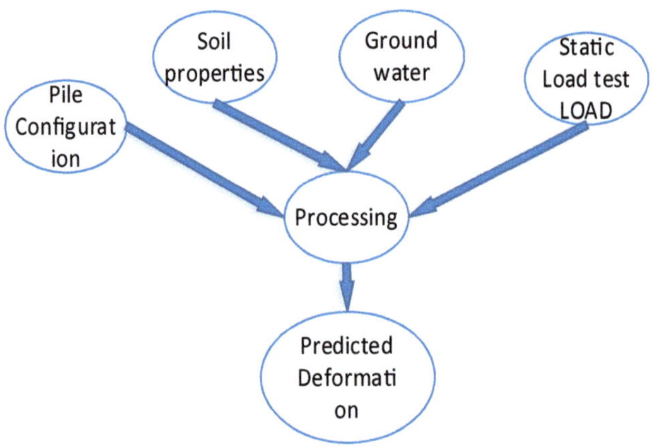

Fig. 6. Testing cycle of the artificial neural network model

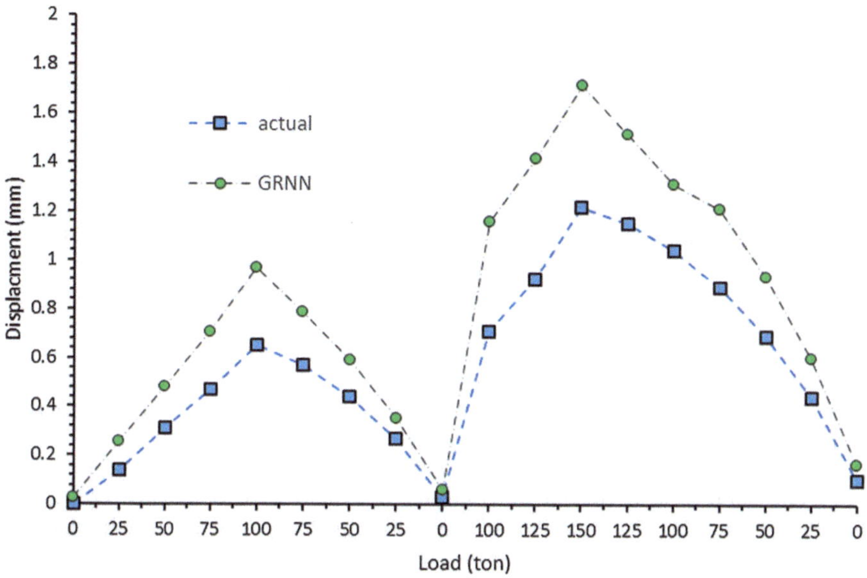

Fig. 7.

9 Conclusion

To minimize the reliance on physical static load test and reduce the risk of pile's failure, this paper introduced a general regression neural network approach to predict the static load tests curves with two cycles of loading-unloading. This prediction is attainable due to consistent geology, fixed pile diameter, and number of successful tests. The artificial network is based on general regression approach which is a one pass technique that is fast to conduct. Data such as material mechanical properties, soil classification, groundwater table height, pile length and diameter, pile load and displacement are used to train the network, and later test it to predict the displacement. This paper presents, for the first time, the usage of GRNN approach that is capable of modeling the test. The GRNN results were verified with existing field tests that were randomly excluded from the training stage. This approach would provide engineers with a tool to predict static load test results with a reasonable amount of accuracy, ahead of the actual field test. Consequently, engineers will be enabled to interfere at the right time to modify the design or the number of required static pile load field tests which reduces the overall cost and time while avoiding the risk of pile's underperformance.

Acknowledgment. The authors would like to extend their gratitude to the office of scholarship and sponsorship programs at Abu Dhabi University for their financial support of this project under grant number 19300093.

References

Abu Kiefa, M.: General regression neural networks for driven piles in cohesionless soils. J. Geotech. Geoenviron. Eng. **124**(12), 1177–1185 (1998a)

Abu Kiefa, M.A.: General regression neural networks for driven piles in cohesionless soils. J. Geotech. Geoenviron. Eng. ASCE **124**(12), 1177–1185 (1998b)

Alzo'ubi, A.K., Ati, M., Ibrahim, F.: Smart framework for predicting drilled shaft capacity based on data mining techniques and GIS data. In: Manzanal, D., Sfriso, A.O. (eds.) From Fundamentals to Applications in Geotechnics, The Pan American Conference on Soil Mechanics and Geotechnical Engineering, 15th PCSMGE/8th SCRM/6th IS-BA 2015, 15–18 November, pp. 1909–1915 (2015)

Benali, A., Nechnech, A.: Prediction of the pile capacity in purely coherent soils using the approach of the artificial neural networks. In: International Seminar, Innovation & Valorisation in Civil Engineering & Construction Materials, N° (2011): 5O-239, University of Sciences and Technology, Algiers (2011)

British Standard: BS 8004, Code of practice for foundations, BSI (2015)

Coduto: Foundation Design: Principles and Practices. Prentice Hall, New Jersey (2001)

Fookes, P.G., Knill, J.L.: The application of engineering geology in the regional development of northern and central Iran. Eng. Geol. **3**, 81–120 (1969)

Goh, A.T.C.: Pile driving records reanalyzed using neural networks. J. Geotech. Geoenviron. Eng. ASCE **122**(6), 492–495 (1996)

Han, J., Kamber, M.: Data mining, concepts and techniques, 3rd edn. Morgan Kaufman Publishers, Burlington (2012)

Lee, I.M., Lee, J.H.: Prediction of pile bearing capacity using artificial neural networks. Comput. Geotech. **18**(3), 189–200 (1996)

Maizir, H., Kassim, K.: Neural network application in prediction of axial bearing capacity of driven piles. In: Proceedings of the International MultiConference of Engineers and Computer Scientists, Hong Kong (2013)

MathWorks, MATLAB & SIMULINK: Generalized regression networks (2016). https://www.mathworks.com/help/nnet/ug/generalized-regression-neural-networks.html

Sarkar, G., Siddiqua, S., Banik, R., Rokonuzzaman, M.: Prediction of soil type and standard penetration test (SPT) value in Khulana city, Bangladesh using general regression neural networks. Q. J. Eng. Geol. Hydrol. **48**, 190–203 (2015)

Shahin, M.A., Jaska, M.B., Maier, H.R.: Artificial neural network applications in geotechnical engineering. Aust. Geomech. **36**, 49–62 (2001)

Specht, D.F.: A general regression neural network. IEEE Trans. Neural Netw. **2**(6), 568–576 (1991)

Tarawneh, B.: Pipe pile setup: database and prediction model using artificial neural network. Soils Found. **53**(4), 607–615 (2013)

Case Study: Micropiles as Integral Part of an Alternative Slope Stabilization System

Thomas Hangartner(✉) and Christophe Balg

Geobrugg AG, Romanshorn, Switzerland
thomas.hangartner@geobrugg.com

Abstract. The road widening project that connects Magelang and Selo - Boyolali in central Java Indonesia was foreseen to cut travel time and help to spur economic growth. In the past, the operation got massively constraint by the occurrence of a large slope failure, in which one road lane slipped down to the creek.

The site required a cost-effective protection system, one that actively and permanently stabilizes the problematic embankment. The applied protection system must further take the troublesome site conditions into account: almost vertical unstable slope with very limited accessibility, challenging geological condition, intense rainfalls, frequent occurrence of earth quakes and several installation time constraints.

The decision fell on a new technology, one that respects the challenging site conditions and on the same time blends with nature: the nailed flexible facing. An integral part of this system is the piled concrete beam at the top of the slope. Micro piles with a length of 8 meters are installed every two meters along the beam. The concrete beam absorbs the traffic load and stiffens the slope stabilization system. The high tensile steel wire mesh is attached at the concrete beam, rolled down the slope and pre-tensioned via the spike plates and the soil nails onto the slope.

Numerous implemented cases worldwide in soil as well as in rock conform that this technically and economically effective concept is suitable for practical applications. This case study gives an overview over the design and installation of this alternative slope stabilization method flexible facing in combination with a piled concrete beam.

Keywords: Micro piles · Flexible facing · Slope stabilization Landslide · Embankment

1 Introduction

The road widening project that connects Magelang and Selo - Boyolali in Central Java was foreseen to cut travel time and help to spur economic growth. In the past, the operation got massively constraint by the occurrence of a large slope failure, in which one road lane slipped down to the creek.

The site required an alternative and cost-effective protection system, one that actively and permanently stabilizes the problematic embankment. The applied protection system must further take the troublesome site conditions into account: almost

H. El-Naggar et al. (Eds.): GeoMEast 2018, SUCI, pp. 61–68, 2019.
https://doi.org/10.1007/978-3-030-01902-0_6

vertical unstable slope with very limited accessibility, challenging geological condition of volcanic origin, intense rainy seasons and a small river at the toe of the slope with varying water levels, frequent occurrence of earthquakes and severe installation time constraints.

The decision fell on a new technology that respects the challenging site conditions and one the same time blends with nature: the nailed flexible facing system. An integral part of this system is the piled concrete beam at the top of the slope. Micro piles with a length of 8 meters are installed every two meters along the beam. The concrete beam absorbs the traffic load and stiffens the slope stabilization system. The high tensile steel wire mesh is attached at the concrete beam, rolled down the slope and pre-tensioned via the spike plates and the soil nails onto the slope.

2 Project and Geology

Geotechnical investigations revealed that the release depth of the superficial slope failure measured approx. 1–2 m and caused one road line to collapse and slip down to the river. The landslide occurred right next to the strategic bridge around Selo Boyolali in Central Java. The prevailing geology consists of weathered Lahar of volcanic origin. The climatically condition is characterized through heavy and frequent rain falls, significantly increasing during rainy season. The water level and its variations of the creek is determined by the amount of precipitation. The inclination of the stabilized 80 m long and up to 45 m high slope increases from 50° at the toe of the slope towards vertical at the top. The total project area covers 2,500 m^2. The project area is exposed to the south and the sun ray incident is high. The area is surrounded by undisturbed slopes covered by a dense vegetation. This ecological system is in a fragile equilibrium and therefore reacts very sensitive to manmade cuttings, impacts and traffic loads (Fig. 1).

Fig. 1. Prevailing site condition before the installation of the flexible facing slope stabilization system. The release zone of the superficial slope failure is marked green.

3 Designing Flexible Facing Slope Stabilization Systems

Designing a nailed flexible facing slope stabilization system is based on the partial safety concept (EUROCODE 7, Cala et al. 2012). The nailing is adapted to the static requirements based on the investigations of the overall slope stability considerations. The dimensioning concept comprises two basic investigations and proofs which must be fulfilled: (1) Investigation of superficial slope parallel instabilities. (2) Investigations of local instabilities between the single nails (Cala et al. 2012) (Fig. 2).

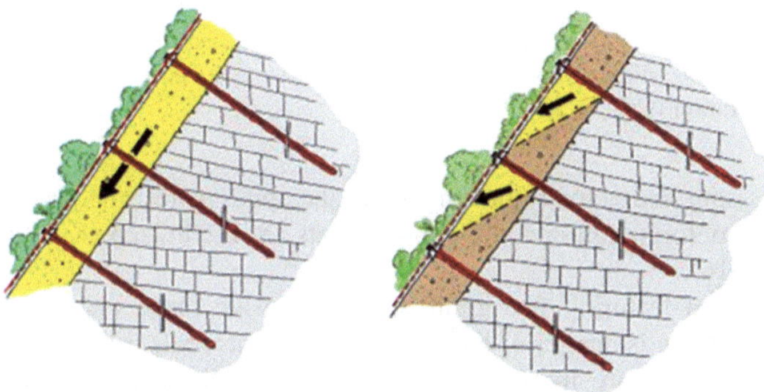

Fig. 2. Proofs required in order to successfully design a flexible slope stabilization system: slope parallel instabilities fig (a) and local instabilities between the individual nails fig (b, Flum et al. 2004, 2001).

Due to the active seismological situation in the Merapi region, effects from earthquakes are taken into account as additional forces. Those forces are included into the design as vertical and horizontal accelerations acting on the fracture bodies. Figure 3 shows the addition in the formulas resulting from the load case earthquake.

The influence of streaming pressure as a result of precipitation water, respectively inflowing ground or slope water is included the design considerations. It is assumed that streaming parallel to the slope occurs after saturation of the material.

4 Flexible Facing by the Means of High Tensile Steel Wire Mesh

In the standard layout, the key feature of flexible facing is the high tensile steel wire mesh. Every single wire has a minimal tensile strength of 1,770 N/mm^2 and a zinc-aluminum coating for the corrosion protection. The diamond-shaped high tensile steel wire meshes measure 83 mm × 143 mm and are produced by single twisting. The resulting mesh achieves a tensile strength of 150 kN/m. Special diamond-shaped spike plates matching the geometry of the high tensile steel wire mesh serve to fix and tension the nails. For the nails, steel bars with a diameter of d = 32 mm were used.

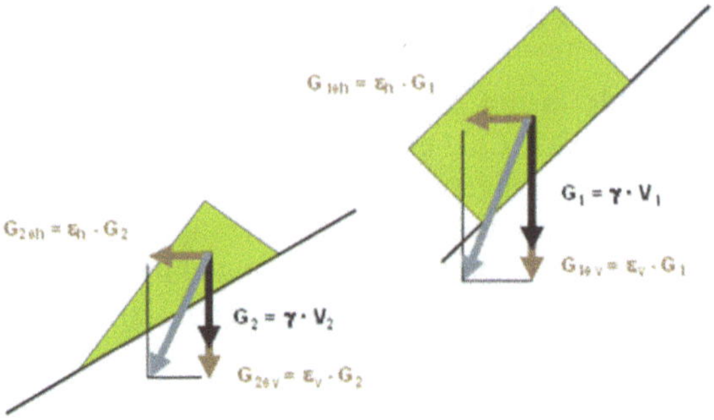

Fig. 3. Design components of the acceleration values induced by earthquakes whereas G = dead weight of sliding body, ε_h = coefficient vertical acceleration, ε_v = coefficient horizontal acceleration (Cala et al. 2012).

By tightly pressing and actively pretensioning the spike plates, the systems enable to minimize slope deformation and slope movements are minimized (Kühne et al. 2001; Kytzia et al. 2016).

In steep slopes featuring fine-grained, non-cohesive loose materials or severely weathered rocks there is danger of erosion. Such fine materials can be washed through the high tensile steel wire mesh and flushed away. Underlying a three-dimensional erosion control mat provides an optimal basis for such erosion protection and basis for successful revegetation (Rorem et al. 2003; Flum et al. 2004) (Fig. 4).

5 Flexible Facing and Micro Piles

The slope stabilization system is combined with a reinforced and anchored concrete beam. The 0.6 m high and 0.4 m wide concrete beam is placed along the 80 m long edge of the slope. Every two meters almost vertical 15° inclined micro piles and 60° inclined soil nail are installed. All boreholes for the micro piles as well as the nails have a diameter of d = 90 mm. The main function of the micro piles is to transfer the loads into the subsoil. The inclined soil nails keep the concrete beam in place and prevent any lateral displacement (Fig. 5).

6 Installation

The installation of the piled and anchored slope stabilization system could be completed within only three months. After cutting and lodging off the vegetation, the drilling works were done using a bamboo scaffold. Five drilling rigs were used simultaneously in order to match the project schedule. After completion of the drilling works, the steel bars were inserted and grouted from bottom to top (Fig. 6).

After drilling and grouting of the micro piles, the concrete beam is poured by using a wooden basket. The high tensile steel wire mesh is attached to the beam by the means

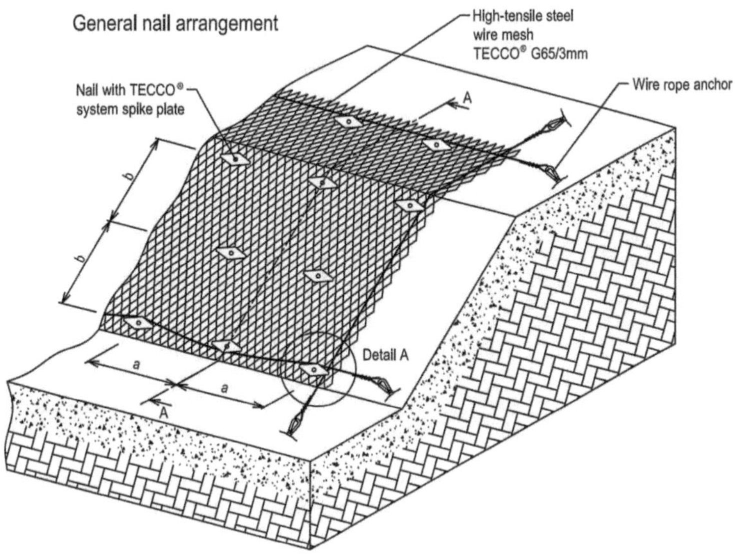

Fig. 4. Standard layout of the flexible facing slope stabilization system.

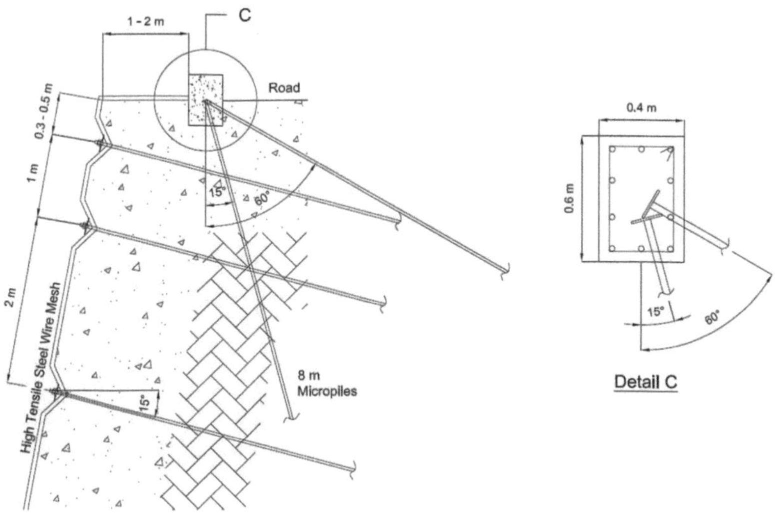

Fig. 5. Flexible facing by the means of high tensile steel wire mesh combined with around 40 micro piles along the anchored concrete beam in order to absorb the traffic loads. The inclined nails at the concrete beam keep the concrete beam in place.

of wire rope anchors and intermediate steel ring elements. A three-dimensional erosion control mat was installed below the mesh. The micro piles are entirely embedded in the concrete beam (Figs. 7 and 8).

Fig. 6. Impressive scaffold platform for drilling works. Small drillings rigs are sufficient and allow the installation of flexible slope stabilization system in challenging terrain with very limited access.

Fig. 7. Skeleton fig (a) and wooden basket fig (b) before pouring the concrete

Fig. 8. Piled concrete beam after pouring fig (a) and after installation of the flexible facing slope stabilization system in combination with a three dimensional erosion control mat fig (b).

Fig. 9. Already one month after installation, the vegetation starts to recover. The flexible facing slope stabilization system combined with a piled concrete beam will soon merge even more with the natural surroundings.

7 Revegetation

The presented slope stabilization system is a green solution. Blending with nature means that a vegetation layer is growing through the high tensile steel wire mesh. Besides great aesthetics, the vegetation layer prevents erosion induced by rainfall, water outflows or even winds.

The successful revegetation depends on the soil properties, the slope inclination, the slope exposition and also on the amount of water supply. Due to the steepness of the slope, hydro seeding was applied. Additionally, 700 pieces of Mucona Bratacea seed pockets were manually planted (Fig. 9).

8 Conclusions

The installation was completed within only three months despite the challenging accessibility. The pretensioned high tensile steel wire mesh guarantees the required system stiffness, whereas the Micro Piles absorb the traffic load. The visual impact of the slope is very appealing in comparison to massive concrete construction. Designed to be cost effective and maintenance free, the system will further grow into the surrounding landscape and contribute to save road operation. The application proves that flexible facing by the means of high tensile steel wire meshes in combination with a piled concrete beam can ideally combine slope stabilization with revegetation tailored to the actual climatic and environmental conditions.

References

Cala, M., et al.: TECCO® Slope Stabilization System and RUVOLUM® Dimensioning Method. AGH University of Science and Technology (2012)

Flum, D., Rüegger, R., Guasti, G.: Dimensionamento di sistemi di consolidamento flessibili superficiali costituiti da reti in acciaio ad alta resistenza in combinazione a elementi di ancoraggio in barra. GEAM – Associazione Georisorse e Ambiente Torino, Bonifica di versanti rocciosi per la protezione del territorio, Trento, Italia, 11–12 marzo (2004)

Kühne, M., Einstein, H.H., Krauter, E., Klapperich, H., Pöttler, R.: International Conference on Landslides, Causes, Impacts and Countermeasures, Davos, Switzerland, 17–21 June 2001

Kytzia, S., et al.: Stahldrahtgeflechte sind die beste Lösung. University of applied science Rapperswil, Switzerland, Magazine 2/2016, 11–14

Landslides, Investigation and Mitigation, Special Report 247. Transportation Research Board, National Research Council. National Academy Press, Washington, D.C. (1996)

Rorem, E., Flum, D.: TECCO high-tensile wire mesh & revegetation, system for slope stabilization. In: International Erosion Control Association, IECA's 35th Annual Conference, Philadelphia, USA, 16–20 February 2003

Rüegger, R.; Flum, D.: Slope stabilization with high-performance steel wire meshes in combination with nails and anchors. In: International Symposium, Earth Reinforcement, IS Kyushu, Fukuoka, Japan, 14–16 November 2001

Experimental Study on the Behavior of Pile and Soil at the Frost Condition

Talal Awwad[1,2(✉)], Zh. A. Shakhmov[1], R. E. Lukpanov[1], and S. B. Yenkebayev[1]

[1] L.N. Gumilyov, Eurasian National University, Astana, Kazakhstan
dr.awwad.gfce@gmail.com
[2] Damascus University, Damascus, Syria

Abstract. The paper including results of model tests and laboratory research of pile works in a condition of the seasonally frozen soil. It is very important during geotechnical design and construction to make a definition for the frost-hazardous properties of soils that could negatively affect to foundations. Frost-susceptibility of soil ground of foundations should be found during experimental, laboratory testing, modelling and so on. This paper present model pile testing on frozen conditions of the silty soil. Kazakhstan territory is located on seasonally freezing-thawing soil ground conditions where an influence of negative temperatures is a very sensitive topic to geotechnical conditions of the region. Geotechnical conditions of soil are very variable and different on the territory of Kazakhstan. There are presented salty soils and loess soils in the West and South Kazakhstan, while silty and clay soils are usually spreading in North and East Kazakhstan. The moisture content of a soil is too much by percentage in the North and East Kazakhstan compared with West and South Kazakhstan geotechnical conditions. The frozen soil depth in East Kazakhstan has maximum 2.5–3.0 m while the minimum frozen soil depth is 0.5 m, in South Kazakhstan. Therefore, investigation and inspection of freezing soil influence to soil ground and foundations ought to be a very important research area among geotechnical professionals in Kazakhstan. Unfortunately, the frozen soil ground testing and research is very slow and usually not provided during the designing of buildings and structures in Kazakhstan. The main reason for this situation is an absence of researchers, equipped laboratories, special devices and interests from construction companies. Historically Kazakhstan was under the rule of Soviet Union before 1991, so a lot of research institutes and centres located in Russia. After Kazakhstan got independence in 1991 it wasn't researched and investigated problems of an influence of freezing to soil ground and foundation. Thus, this paper has high importance for investigation and study the problem of frost susceptibility of soil ground.

1 Introduction

Freezing problem of soil and its influence is researched many times in Russia, USA, Canada, Finland, Japan. Scientists such as Tsytovich, Karlov, Kudryavcev from Russia study formation of soil freezing, a cryogenic structure of soil during freezing, frost susceptibility, capillarity process during freezing and so on. There are providing a lot of

H. El-Naggar et al. (Eds.): GeoMEast 2018, SUCI, pp. 69–76, 2019.
https://doi.org/10.1007/978-3-030-01902-0_7

laboratory works and in-situ testing in frozen condition. The frozen process also is very important because of the permanently frozen soil in large part of Russia. Researchers such as Taber, Bescow, Morgenstern from West countries also deeply investigate frozen soil characteristics during negative temperatures.

Usually, the problem of frozen soil is related to frost-susceptible soils, which has pore sizes between particles and particle surface area that provide capillary flows. The materials could be classified as frost susceptible if 10% or more constituent particles pass through 0.075 mm (# 200) sieve or 3% or more pass through 0.02 mm (# 635) sieve. While non-frost-susceptible soils should be too dense to promote water through soil body or too open by porosity to promote fast capillary flow. There are examples such as dense clay with small pore size and low hydraulic conductivity, or clean sands and gravels with contain small amounts of fine particles and therefore its pore sizes are too open for promoting capillary flow.

Construction of pile foundations is very much spreading type foundation in Kazakhstan (Zhussupbekov et al. 2013). Especially many interesting buildings are built by using piles in foundation construction of Kazakhstan. Kazakhstan is located on a seasonally climatic region where hot and freezing temperatures present during of year. The temperature of freezing could reach sometimes −45 °C and hot temperatures +45 °C in summer (Zhussupbekov et al. 2012). So the harsh climatic condition like this creates a lot of problems related to the freezing and thawing of the soil ground (Shakhmov and Zhussupbekov 2015; Zhussupbekov et al. 2017a, b). Therefore a lot of deformations of foundations occur in construction area by the lack of frost susceptibility research, investigation of soil ground. This paper presents research about the influence of freezing to pile foundations. Model testing of a pile in normal and freezing condition is presented in this paper.

2 Methods to Provide Experiment

2.1 Instruments Usage in the Experiment

In the construction, it is very important to take into account the frost-hazardous properties of soils that can affect the foundations from the negative side (Beskow 1935). Many works are written about the effect of the soil's unprotected properties of foundations on natural grounds. Frost heaving, the strength of the whipping is especially problematic during construction (Zhussupbekov and Shakhmov 2015; Zhussupbekov et al. 2017a, b).

The experiment is assisted by special devices which help to get results. The list of equipment and devices such as next.

- device for pile loading (weight, load);
- supporting structure for the perception of reactive forces (beam system);
- device for measuring the movement of the pile during the test (manometer);
- thermometer;
- model pile;
- mechanism of piling and extraction of model piles;
- fishing line.

The device for loading piles should provide coaxial and central transfer of loads to the pile, the possibility of transferring the load in steps, the constant pressure at each loading stage is given.

2.2 The Process of Providing the Experiment

The experiments were carried out in a tray, which is a frame construction. The capacity of the box for backfilling of the equivalent mixture has dimensions of 0.9 * 0.5 * 0.9 m. The frame is intended for attachment of devices, which in these experiments belonged to a loading device and a system of dial indicators, which allows recording movements and settlements.

In the experiments, the work of a single pile was investigated, as well as the work of fragments of pile foundations with 18 m long piles.

The materials for the model are selected on the basis of the general law of dynamic similarity, taking into account the simultaneous action of gravity and internal stresses:

$$m_C = c_M/c_H \cdot \gamma_H/\gamma_M = 0.9/38 \cdot 19/17{,}7 = 1/40$$

Preparation for the experiment was conducted in the following order and steps:

- the equivalent mixture to the ground condition was laid with layers of 10 cm thickness. The stacked portions of soil were preliminarily weighed, which provided the required specific gravity corresponding to the total deformation modulus E = 1.628–2.035 MPa.
- After laying the equivalent mixture, the piles were dipped. The immersion was carried out using a conductor, which was set on the level to give piles in the ground a strictly vertical position.
- After each experiment, the models were dug out, and the results of only those experiments in which the piles did not deviate from the vertical were taken into account. The final penetration of the piles to the required depth was carried out after the removal of the conductor.

Load on the foundation model was applied with levers with a ratio of the arms 1:10. The load stages were taken equal to 1/10 ÷ 1/15 of the value of the maximum loads determined by preliminary methodological experiments. Each stage was maintained until complete stabilization of the deformation, after which indicators were taken to determine the movement of the piles.

Soil material is compacted by special manual compactor to its optimum compaction unit weight. Knowing dimensions of the box which mentioned above it is possible to provide by definition of soil mass for optimal compaction condition of the soil.

The freezing process for the modelled pile in fulfilled soil box was done in special freezing chamber (camera) Controls. The range of the temperature in the camera could reach about −40 °C. The first period was the stabilization of the experimental model in temperature −4 °C for 8 h, after that temperature dropped to −12 °C to 16 h. Totally after 24 h frozen soil with pile is tasted to static load step by step.

2.3 Content and Types of the Testing

The experiment consists of two parts. There is first part with a normal condition and second part with a frozen condition. The features of the 1st part presented in Fig. 1 and below:

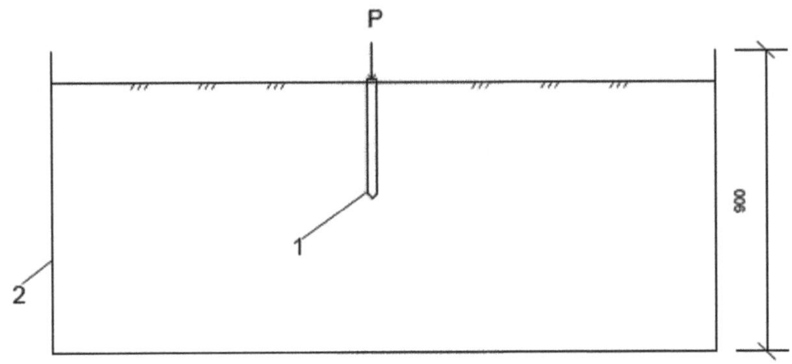

Fig. 1. The scheme of the model pile in ordinary condition (1-pile, 2-tray, 3-equivalent material).

1. Study of the stability of piles of various designs in non-heaped soils with a repeated freezing-thawing process
2. Investigation of the interaction of the ground base (Determination of the bearing capacity of piles under normal conditions)
3. Calculation for the II group of the limiting state on the compressive thickness
4. Develop a pile model on a scale.
5. Install a crate for installing thermometers - 2 pcs.
6. Calculation of soil freezing depth on a scale.

Results of experiments with models of piles (Fig. 2):

1. Analysis of the results of experiments on the stability of piles in non-heaped soils
2. Draw up "sludge-load" graphs based on the results of static pile tests.

The features of 2nd part presented in Fig. 3 and below:

1. Study of the stability of piles of various designs in ragged soils with repeated freezing-thawing
2. Investigation of the interaction of subsoil base (Determination of the bearing capacity of piles in conditions which same to seasonally freezing soils)
3. Calculation for the II group of the limiting state on the compressive thickness
4. Developing a pile model on a scale
5. Install the crate for installing thermometers
6. Calculate the depth of soil freezing on a scale.

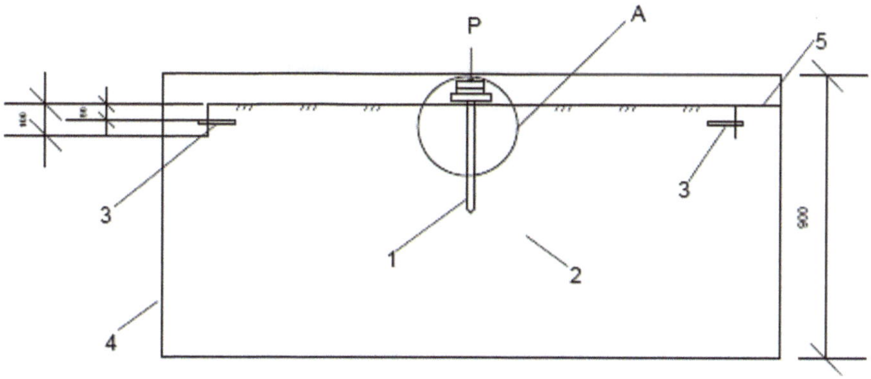

Fig. 2. The scheme of the model pile in frozen condition (1-model pile, 2-equivalent material, 3-thermometer, 4-tray, 5-technological pits).

Fig. 3. Preparation of the soil mass in the track

Results of experiments with models of piles:

1. Analysis of the results of experiments on the stability of piles in the subsoil soils
2. Compile the "sludge-load" graphs from the static pile tests.

General test procedure and visual provision of testing preparations also presented in Figs. 3 and 4.

3 Results and Discussion

Model tests of piles in ordinary and frozen ground conditions showed different results. In particular, piles with a length of 3 m in frozen ground conditions show an increase in the bearing capacity of almost 3 times more than in the same length under ordinary

Fig. 4. Loaded pile model in track during the test

conditions. This is due to the compression of the piles along the perimeter and the freezing of the soils along the sides because of the expansion and the pore pressure that occurs during freezing. These tests are presented in a comparative form in Table 1.

Table 1. Results of the model testing pile

Pile L (m)		Ordinary soil	Frozen soil
Length	Penetration	F_O, kg	F_M, kg
3	2.5	4	11
5	4.5	7	13.75
6	5.4	8.75	15.25
10	9.4	14	19.5
12	11.5	17.5	22
14	13.5	20.25	25.25

The results of the test for clarity are also presented below in the form of two curvilinear graphs in Fig. 5, which show more clearly the difference in bearing capacity of piles in ordinary soils and in frozen ground conditions. I would like to note that the effect of freezing on the bearing capacity decreases in direct proportion to the increase in the length of the piles.

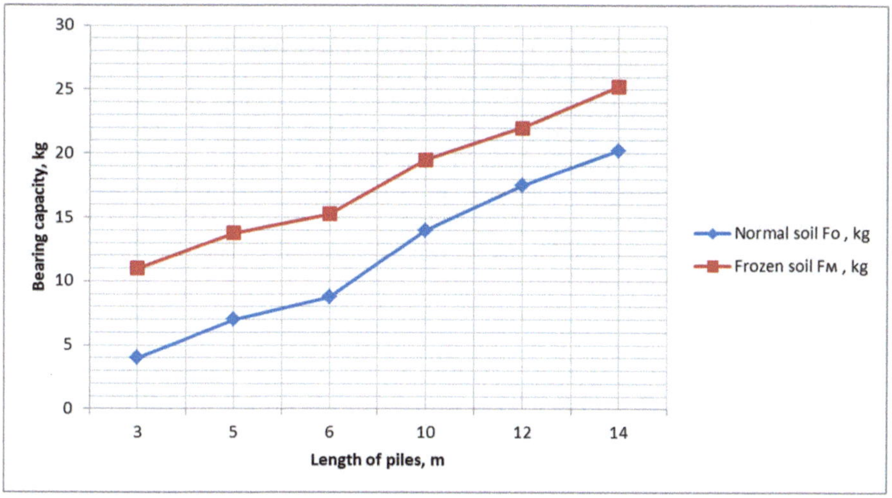

Fig. 5. Pile testing results in ordinary and frozen conditions

4 Conclusion

The tests which given above shows how the freezing process can influence the pile foundations, which is very important to consider during testing and determining the bearing capacity of piles. In particular, short piles can show the bearing capacity in the frozen ground state increased by several times, which is very important to take into account when designing pile foundations in winter conditions. As the length increases, this tendency is reduced. We believe that in the future it is necessary to study in more details the problems of shearing stress during the freezing of soil bases.

Acknowledgements. Deeply thanks to master students and doctoral students such as Tleulenova Gulshat of L.N. Gumilyov Eurasian National University for their assistance during providing testing and also to Head of Civil Engineering Department of Temirtau University Prof. Bazarov Baurzhan in Temirtau city.

References

Beskow, G.: Soil freezing and frost heaving with special application to roads and railroads. Swed. Geol. Soc. Ser. C. **375**(3), 67–75 (1935). (translated by J.O. Osterberg, Northwestern University, 1947)

Shakhmov, Zh., Zhussupbekov, A.: Frost susceptibility of soil and in-situ monitoring of frost depth in construction. In: The 15th Asian Regional Conference on Soil Mechanics and Geotechnical Engineering, Fukuoka, Japan, p. 558 (2015)

Zhussupbekov, A., Shakhmov, Z., Shin, E.C., Krasnikov, S.: Challenges for transportation geotechnics in extreme climates of Kazakhstan and Korea. In: 2nd International Conference on Transportation Geotechnics, Advances in Transportation Geotechnics II, pp. 655–660. CRC Press, Taylor & Francis Group, Hokkaido (2012)

Zhussupbekov, A., Shakhmov, Zh.: Experimental investigations of freezing soils at ground conditions of Astana, Kazakhstan. Sci. Cold Arid Reg. **7**(4), 399–406 (2015)

Zhussupbekov, A., Shakhmov, Zh., Tleulenova, G.: Geotechnical problems on freezing ground soil and experimental investigation in Kazakhstan. Sci. Cold Arid Reg. **9**(3), 331–334 (2017a)

Zhussupbekov, A., Ye, Utepov, Shakhmov, Z., Ling, H.: Model testing of piles in a centrifuge for prediction of their in-situ performance. J. Soil Mech. Found. Eng. **50**(3), 92–96 (2013)

Zhussupbekov, A., Shakhmov, Z., Lukpanov, R., Tleulenova, G.: Frost depth monitoring of pavement and evaluation of frost susceptibility at soil ground of Kazakhstan. In: 19th International Conference on Soil Mechanics and Geotechnical Engineering, Seoul, Republic of Korea, pp. 1455–1458 (2017b)

Analysis of Driven Pile Bearing Capacity Results by Static and Dynamic Load Tests

Talal Awwad(✉), S. B. Yenkebayev, D. V. Tsigulyov, and R. E. Lukpanov

Eurasian National University of L.N. Gumilyov, Astana, Kazakhstan
dr.awwad@ymail.com

Abstract. Traditionally, practical definition of bearing capacity of piles is stipulated by pile load tests performance by using static and dynamic load test methods. Field load test (namely Static and Dynamic load tests) is a very important part of pile design and accuracy of its performance reflecting on the future safety of building and construction.

There are two sustainable notions: static load tests of piles are more expensive, but results are more accurate; dynamic load tests of piles are less expensive, and the results are less accurate. Fields tests of driven piles were carried out on the different construction sites in Astana. The possibility to analyze static and dynamic test data is stipulated by the requirement of local regulations that demand to conduct of both tests on construction site regardless of the category of responsibility.

Moreover, it has been noticed the results of bearing capacity of pile depend on the type of hammer. The paper also provides an analysis of bearing capacities obtained by dynamic tests using a diesel and hydraulic hammer. As practice has shown, there is a difference between the bearing capabilities obtained with the use of a diesel and hydraulic hammer. The reason for that is the empirical equation (regulated by the local standard) for calculating the bearing capacity. The equation does not provide for the use of a hydraulic hammer with high impact energy and requires revision and input of correction factors. In order to analyze the latter, the impact energy of different hammers was measured by using pile dynamic analyzer sensors.

The paper presents the results of a comparative analysis of data obtained by static and dynamic tests using different types of hammers. The results of the research are presenting both in tabular and diagram interpretation.

1 Introduction

Traditionally, practical definition of bearing capacity of piles is stipulated by pile load tests performance by using static and dynamic load test methods. Field load test (namely Static and Dynamic load tests) is a very important part of pile design and accuracy of its performance reflecting on the future safety of building and construction. A lot of studies have been done by using numerical modelling (Awwad and Kodsi 2017; Ulitsky et al. 2017; Awwad et al. 2017; Awwad and Donia 2016).

H. El-Naggar et al. (Eds.): GeoMEast 2018, SUCI, pp. 77–84, 2019.
https://doi.org/10.1007/978-3-030-01902-0_8

There are two sustainable notions: static load tests of piles are more expensive, but results are more accurate; dynamic load tests of piles are less expensive, and the results are less accurate.

Nowadays many megaprojects are emerging in the new capital of Kazakhstan – Astana. The high rates of construction and appearance of high-rise buildings having modern architecture, and engineering megaprojects, led to a wide use of pile foundations.

During the last 25 years of capital construction, many high-rise buildings built upon the driven pile. Modern megaprojects put forward modern requirements to engineers. Application of pile foundation is explained by the necessity of ensuring a high bearing capacity for high-rise buildings. This led to refuse from traditional out-of-dates technologies and use new more economical and reliable technologies like CFA (continuous flight auger), DDS (drilling displacement system) and the others.

In any case, driven pile technology still stay most available alternative for the pile foundation choosing.

The general concept for the pile foundation design, installation and testing are presented in Fig. 1. In this case, field load test (Static and Dynamic load tests) is a very important part of pile design and accuracy of its performance reflecting on the future safety condition of building and construction.

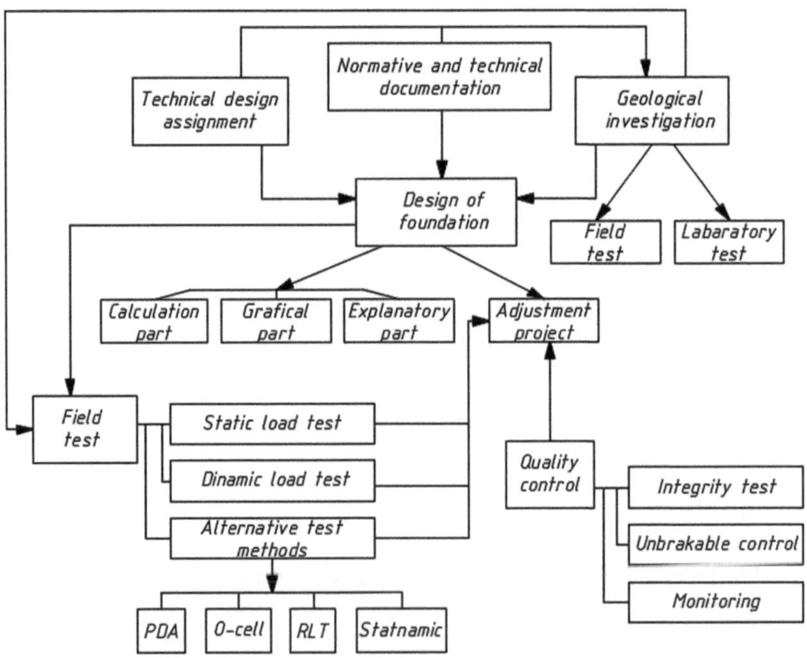

Fig. 1. Pile foundation design concept

2 Dynamic Pile Load Test (DLT)

DLT is express bearing capacity field test (Fig. 2). Before the driving, it is necessary to mark the pile through every 1 m of the length. However, the last meter of the pile should be marked for each 0.1 m. During the driving, the numbers of blows are recording for each meter of pile penetration and last one meter for each 0.1 m. At the same time, it is necessary to fix average drop height of the hammer. For a definition of the bearing capacities of piles, it is required to use maximum refusal, which is obtaining during re-driving of the piles after the "rest". The rest time depends on soil condition: for clay up to 6 days, for sand and gravel up to 3 days.

Fig. 2. Dynamic load test performance

Bearing capacity of the pile F_u could be fined by following equation (SNiP RK 5.01-03-2002, Kazakhstan Standard):

$$F_u = \frac{\eta AM}{2}\left[\sqrt{1 + \frac{4E_d}{\eta AS} \cdot \frac{m_1 + \varepsilon^2(m_2 + m_3)}{m_1 + m_2 + m_3}} - 1\right] \quad (1)$$

Where η = coefficient depending on pile material, A = cross-sectional area of pile, M = coefficient depending on hammer, E_d = effective energy of hammer, S = set per blow, ε = coefficient of restitution, m_1 = total mass of hammer device, m_2 = total mass of pile and cap, and m_3 = mass of spacer placed between the pile head and pile cap.

Effective energy of blows of the hammer E_d could be fined by the following equation:

$$E_d = G \cdot H \quad (2)$$

Where G = weight of blowing part of the hammer, kN; H = drop height of blowing part of the hammer.

According to Kazakhstan Standard (GOST 5686, Kazakhstan Standard), 1% of working piles on construction site must be tested by DLT, but at least 6 DLT must be done.

3 Static Pile Load Test (SLT)

SLT is the most reliable field tests to an analysis of pile bearing capacity. SLT should be carried out for driving piles after the "rest" and for bored piles after achievements of the concrete strength more than 80%. For SLT the following equipment is used: hydraulic jack; pump station; caving in-measurers; benchmark system; beam system for supporting; anchor piles (Fig. 3).

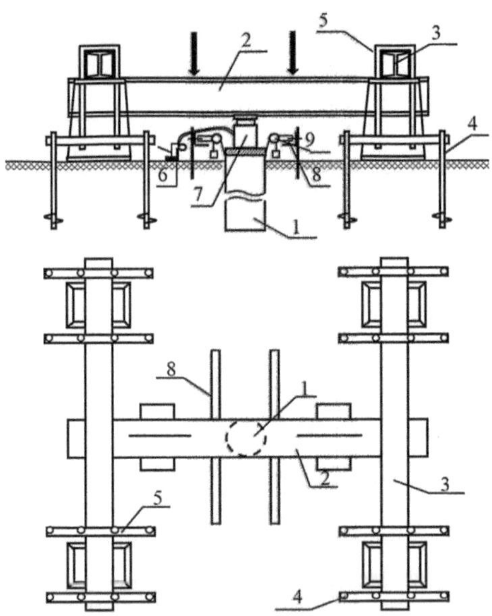

1 – bored pile; 2 – main beam; 3 – auxiliary beam; 4 – anchor screw pile; 5 – auxiliary device; 6 – pump with a manometer; 7 – lifting jack;
8 – benchmark system; 9 – deflectometer.

Fig. 3. Static load testing equipment

The first record was performed just after putting the loading, then consequently 4 records with an interval of 15 min, then 2 records with an interval of 30 min and further for every hour until the conditional stabilization of pile settlement. For the criterion of conditional stabilization of pile, a settlement was taken when the speed of settlement of piles on the given stage of loading did not exceed 0.1 mm during the last 1–2 h of observations.

According to requirements of Kazakhstan Standard - SNiP RK 5.01-03-2002 – ultimate value of settlement of the tested pile is determined by the following equation:

$$S = \zeta S_{u,mt} \tag{3}$$

where ζ = coefficient for conversion factor of the limit value of mean settlement of foundation of the building or structure $S_{u,mt}$ into pile settlement obtained while static tests at conventional settlement stabilization; $S_{u,mt}$ is a maximum permissible value of foundation settlement of the designed building or structure as stated either in the project statement or Kazakhstan Standards requirements.

According to Kazakhstan Standard (GOST 5686, Kazakhstan Standard), 0.5% of working piles on construction site must be tested by SLT, but at least 2 SLT must be done.

4 Comparison of DLT and SLT Results

DLT and SLT were carried out for precast concrete piles with a cross-section of 30 × 30 cm, a length of 12 m. Figure 4 shows the driving records of the tested piles: the fall height of the hammer, the number of blows the number of blow per 0.1 m penetration and set per blow, S. The re-driving of the tested piles were carried out 32 days after the end of the driving.

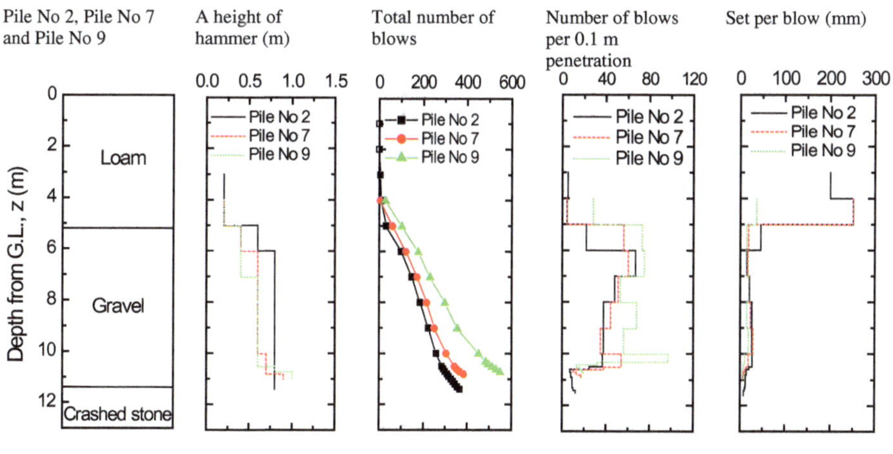

Fig. 4. Driving records of piles No. 2, No. 7 and No. 9 during initial driving

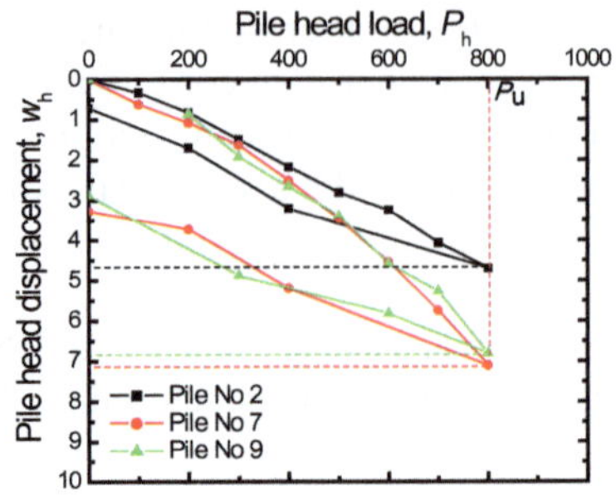

Fig. 5. Measured pile head load versus pile head displacement of piles No. 2, No. 7 and No. 9

The measured static relationships between the pile head load (P_h) and the pile head displacement (w_h) of the test piles are shown in Fig. 5. It is seen from Fig. 5 that the load-displacement curves of Piles No. 7 and Pile No. 9 are almost identical compared to Pile No. 2, having the maximum bearing capacity of 800 kN.

The ultimate bearing capacity of piles have not been identified due to a limitation of load and maximum load of 800 kN obtained from the static load tests are used as the design value of the ultimate capacity.

Table 1. Results of dynamic load tests and static load tests

Pile no	DLT			SLT	
	H (m)	S (mm)	Fu (kN)	S_{max} (mm)	P_u (kN)
2	0.8	8.3	670	4.7	800
7	0.9	5.8	680	7.1	800
9	1.0	5.2	670	6.79	800

The results of dynamic and static pile load tests are presents in Table 1. Where H = drop height of the hammer, m; S = set per blow, mm; F_u = DLT bearing capacity of the pile, kN; S_{max} = SLT pile settlement, mm; P_u= SLT bearing capacity of the pile, kN.

According to experience on construction sites of Astana, some difference exists between SLT and DLT results (Seidmarova 2009). Moreover, results of the bearing capacity of pile depend on the type of hammer. Thus, DLT results obtained by using hydro-hammer are approximate to the SLT results, namely more reliable than results obtained by using diesel-hammer. The safety factor as defined by comparative analysis of many DLT and SLT data is presented in Fig. 6.

5 Alternative Load Test Methods

Nowadays alternative load test method called Rapid Load Test (RLT) in process of adaption in Kazakhstan. RLT allow to testing up to 10 piles per day and much cost effective than SLT (Matsuzawa et al. 2008). The comparison of SLT and RLT as

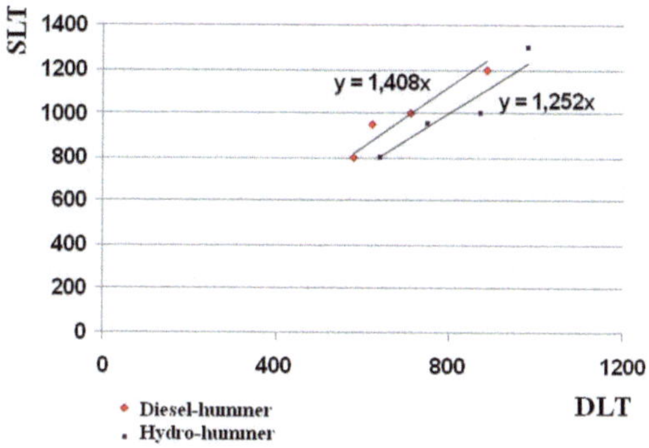

Fig. 6. Comparison results of DLT and SLT

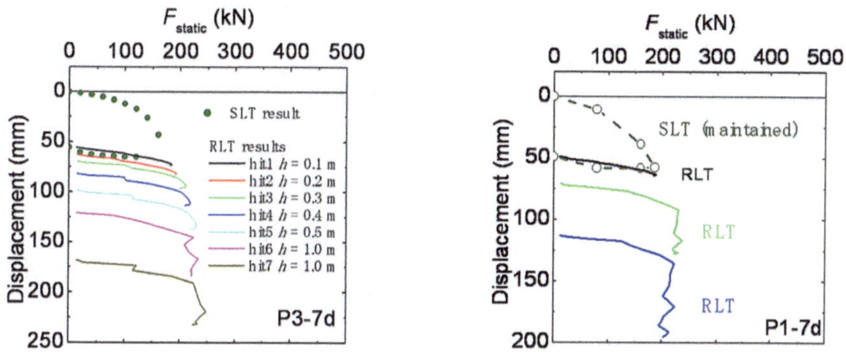

Fig. 7. Comparison of results of SLT and RLT

obtained by Matsumoto are presented in Fig. 7 that shows the reliability of RLT. Satisfied convergence of the SLT and RLT curves had noticed.

From the aforementioned, it follows that SLT and DLT both have features: SLT required a lot of time and cost, and prescribed by Standard quantity of SLT is not enough to realize bearing capacity of all the piles (2 SLT for 200 piles only); DLT is comparatively much faster but less reliable and restricted by testing driving piles only. Therefore, RLT could be used as a reliable test method compatible with SLT. However, for successful usage of RLT, it is necessary to research this method in follow.

6 Conclusions

Existing pile foundation Standards practised in Kazakhstan are out-of-date and are in urgent need of modernization. This paper presented very short descriptions of coming changes to the concept of Kazakhstan pile foundation design.

According to experience on construction sites of Astana, some difference exists between SLT and DLT results. Moreover, results of the bearing capacity of pile depend on the type of hammer. Thus, DLT results obtained by using hydro-hammer are more approximate to the SLT results, namely more reliable than results obtained by using diesel-hammer.

From the aforementioned is follow that it is necessary to inculcate a more reliable method of analysis of bearing capacity of pile such as RLT. RLT allows performing up to 10 piles per day, much cost effective than SLT, and more authentic than DLT. RLT is a type of DLT and is appropriate for any type of pile, but cannot be used to full extent on construction sites of Kazakhstan due to the absence of Standard.

References

Awwad, T., Kodsi, S.A.: A comparison of numerical simulation models to determine the location of the neutral plane. In: Proceedings of the 19th International Conference on Soil Mechanics and Geotechnical Engineering, Seoul, Korea, September 2017, pp. 1947–1950 (2017)

GOST 5686. Soil. Field load test of piles. Kazakhstan Standard

Matsuzawa, K., Nakashima, Y., Matsumoto, T.: Spring hammer rapid load test method and its validation. In: Proceedings of the BGA International Conference on Foundations, Dundee, Scotland (2008)

Seidmarova, T.: Bearing capacity evaluation by results of static and dynamic load tests. A Dissertation Submitted in Partial Fulfilment of the Requirements for the Degree Ph.D., 2009. Astana, Kazakhstan (2009)

SNiP RK 5.01-03-2002. Pile foundation. Kazakhstan Standard

Awwad, T., Donia, M.: The efficiency of using a seismic base isolation system for a 2D concrete frame founded upon improved soft soil with rigid inclusions. Earthq. Eng. Eng. **15**(1), 49–60 (2016)

Awwad, T., Donia, M., Awwad, L.: Effect of a stiff thin Foundation soil layer's depth on dynamic response of an embankment dam. Procedia Eng. **189**(2017), 525–532 (2017)

Ulitsky, V., Shashkin, A., Shashkin, K., Lisyuk, M., Awwad, T.: Numerical simulation of new construction projects and existing buildings and structures taking into account their deformation scheme. In: Proceedings of the 19th International Conference on Soil Mechanics and Geotechnical Engineering. Seoul, Korea, September 2017, pp. 2061–2065 (2017)

Bearing Capacity of an Incompletely End-Supported Pile Based on Rigid Plastic Finite Element Method (RPFEM)

Hamadoum Hamidou(✉) and Koichi Isobe

Faculty of Engineering, Hokkaido University, Sapporo, Japan
hamidou@eis.hokudai.ac.jp

Abstract. The Incompletely End Supported Piles (IESP) are piles which the tips do not reach the bearing stratum because of complex geologic soil layers. These piles have a deficiency of bearing capacity and cause the tilting of the structure they support. It is, therefore, necessary to assess the end-bearing capacity of IESP appropriately.

This study estimates the end bearing capacity of IESP by using two-dimensional RPFEM, considering material non-linearity and soil-foundation interaction.

Bearing capacities of footing on a soft ground were simulated and the results were compared with the Terzaghi and Prandtl formula's results. The Cohesion and the Shear Resistance angle were the main parameters and have been varied.

The influence of a hard ground layer below the surface soft ground on the bearing capacity was investigated. The relative thickness H/B (or Spacing) which is the thickness H of the soft layer normalized by the footing width B has been varied.

The end-bearing capacity of IESP was estimated without consideration of the skin friction resistance between the pile and the surrounding ground for simplicity. A parametric analysis was conducted to investigate the relationship between the relative thickness and the End-bearing capacity of IESP. The relative thicknesses of 0; 0.25; 0.5; 0.75; 1.5; 3 and 6 are considered in this series of simulation.

The results show that the RPFEM simulates the soil bearing capacity accurately and the bearing capacity of a soft ground is lower than the bearing capacity of a ground layer of the soft ground overlying a hard ground layer. The end-bearing capacity of IESP decreases inversely with the relative thickness up to a certain relative thickness and then becomes a constant value. A relationship between the relative thickness and the end-bearing capacity of IESP is proposed based on the results.

1 Introduction

One of the reasons why pile foundation is used in foundation engineering is to transfer loads to a deeper and competent soil layer, in other words, to reach a hard support layer because the ground of the surface layer is soft and has bad conditions. Recently, in a practical case of pile foundation, it is reported that some pile tips of a structure do not

H. El-Naggar et al. (Eds.): GeoMEast 2018, SUCI, pp. 85–103, 2019.
https://doi.org/10.1007/978-3-030-01902-0_9

reach the bearing stratum because of a complex geologic soil layer (Fig. 1). These piles called Incompletely End Supported Piles (IESP) cause the tilting of the structure they support. The tilting is a direct result of a differential settlement due to a difference in the bearing capacity of the piles which reached the hard support layer and those which did not reach this layer (IESP). Although it is the most important for geotechnical engineers to install piles to bearing stratum so as to have enough bearing capacity, it is necessary to assess the end-bearing capacity of IESP appropriately.

Fig. 1. IESP caused by a non-horizontal bearing bud stratum

There are many methods for determining the bearing capacity of deep foundations, we can mention among others, application of in situ test records (Meyerhof 1976; Eslami and Fellenius 1995, 1997); interpretation of full-scale pile load tests (Fellenius 1990); theoretical formula Terzaghi (1943) and static analysis (Poulos 1989). The theoretical approaches, namely static analysis, are often used in designing. But these theoretical approaches have been developed based on some assumptions and simplifications, their applications are therefore limited in complex situations.

Finite element method (FEM) is widely used as a convenient technique to solve complex problems in various fields of engineering.

Rigid Plastic Finite Element Method (RPFEM) has been applied to geotechnical engineering by Tamura et al. (1984) and developed for frictional material as Tamura et al. (1987). Du et al. (2016) introduced the non-linear shear strength property against confining pressure into RPFEM in order to assess the ultimate bearing capacity of sandy soils by taking account of the size effect of the footing. By using the non-linear shear strength property of soils and introducing the rigid plastic constitutive equation of the parabolic yield function of the confining pressure (non linear RPFEM model), Du et al. (2016) has improved RPFEM.

This study estimates the end bearing capacity of IESP subjected to a uniform vertical load without consideration of the skin friction resistance between the pile and the surrounding ground for simplicity by using two-dimensional RPFEM, considering material non-linearity and soil-foundation interaction.

2 Constitutive Equation for Rigid Plastic Finite Element Method

The rigid-plastic finite element method (RPFEM) was developed for geotechnical engineering by Tamura et al. (1984, 1990). In this method, the limit load is calculated without any assumption on potential failure modes. The method is effective in calculating the ultimate bearing capacity of shallow foundation and pile foundation where the soil conditions are varied such as a multi-layered ground. The advantage of the rigid plastic constitutive equation is that only few material parameters such as unit weight, cohesion and shear resistance are necessary.

In this study, the rigid plastic constitutive equation for the Drucker–Prager yield function as shown below is employed although Du et al. (2016) improved RPFEM by introducing the non-linear shear strength property against confining pressure as so to consider the influence of particle breakage.

$$f(\boldsymbol{\sigma}) = \alpha I_1 + \sqrt{J_2} - \kappa \tag{1}$$

Where, I_1 is tr (σ_{ij}) (the first invariant of stress), J_2 is 1/2 s_{ij} s_{ij} (the second invariant of deviator stress s_{ij}), α and κ are soil parameters for plane strain condition expressed as follows;

$$\alpha = \frac{\tan\phi}{\sqrt{9 + 12\tan^2\phi}} \tag{2}$$

$$\kappa = \frac{3c}{\sqrt{9 + 12\tan^2\phi}} \tag{3}$$

where, c is cohesion, ϕ is shear resistance angle.

The volumetric strain rate is expressed as follows;

$$\dot{\varepsilon}_v = \mathrm{tr}(\dot{\boldsymbol{\varepsilon}}) = \mathrm{tr}\left(\lambda \frac{\partial f(\boldsymbol{\sigma})}{\partial \boldsymbol{\sigma}}\right) = \mathrm{tr}\left(\lambda\left(\alpha \mathbf{I} + \frac{\mathbf{s}}{2\sqrt{J_2}}\right)\right) = \frac{3\alpha}{\sqrt{3\alpha^2 + 1/2}}\dot{e} \tag{4}$$

Where, λ is the plastic multiplier and $\dot{e}$ is the norm of the strain rate. $\mathbf{I}$ and $\mathbf{s}$ express the unit and the deviator stress tensors, respectively. Strain rate $\dot{\boldsymbol{\varepsilon}}$, which is a purely plastic component, should satisfy the volumetric constraint condition which is derived by Eq. (5), as follows;

$$h(\dot{\boldsymbol{\varepsilon}}) = \dot{\varepsilon}_v - \frac{3\alpha}{\sqrt{3\alpha^2 + 1/2}}\dot{e} = \dot{\varepsilon}_v - \eta\dot{e} = 0 \tag{5}$$

Any strain rate which is compatible with the Drucker–Prager yield criterion must satisfy the kinematical constraint conditions of Eq. (5). η is a coefficient determined by Eq. (5) which is one of the dilation characteristics. The rigid plastic constitutive equation is expressed by the Lagrangian method after Tamura et al. (1987), as follows:

$$\boldsymbol{\sigma} = \frac{\kappa}{\sqrt{3\alpha^2 + 1/2}} \frac{\dot{\boldsymbol{\varepsilon}}}{\dot{e}} + \beta\left(\mathbf{I} - \eta\frac{\dot{\boldsymbol{\varepsilon}}}{\dot{e}}\right) \tag{6}$$

The first term expresses the stress component, uniquely determined for the yield function, and the second term expresses the indeterminate stress component along the yield function. The indeterminate stress parameter β remains unknown until the boundary value problem with Eq. (5) is solved. Du et al. (2016) directly introduced the constraint condition on the strain rate into the constitutive equation using the penalty method.

$$\boldsymbol{\sigma} = \frac{\kappa}{\sqrt{3\alpha^2 + 1/2}} \frac{\dot{\boldsymbol{\varepsilon}}}{\dot{e}} + P(\dot{\varepsilon}_v - \eta\dot{e})\left(\mathbf{I} - \eta\frac{\dot{\boldsymbol{\varepsilon}}}{\dot{e}}\right) \tag{7}$$

Where, P is a penalty constant. This technique makes the computation faster and more stable.

3 Finite Element Mesh and Boundary Conditions

The preliminary analyses of this study focus on the bearing capacities of shallow foundation on a uniform soft ground and on a layered ground (a soft ground overlying a hard ground layer). The finite element mesh and the boundary conditions used are shown in Fig. 2. A mesh of approximately 2000 rectangular elements have been used to model the footing and the soil. The modeled soil has 10 m of height and 21 m of width. Finer mesh near the footing base were employed. The bottom is fixed in both direction and the sides are fixed in x direction and free in y direction. An increasing load is applied at the footing bottom. For the case of the layered ground, the relative thickness H/B (thickness of soft layer H normalized by the footing width B) is varied, the colors in Fig. 2(b) show the boundary between the layers for cases of H/B = 0.00, 0.25, 0.50, 0.75, 1.50, 3.00, 6.00, respectively. The footing has the following properties; width B = 1 m, height h = 0.5 m, cohesion c = 50000 kPa, shear resistance angle ϕ = 0°. The hard layer has a cohesion soil with c = 10000 kPa, shear resistance angle ϕ = 0° and unit weight γ = 22 kN/m^3. The characteristics of the soft layer are changed in each series. As mentioned above, the aim of this study is to estimate the end bearing capacity of IESP without consideration of the skin friction resistance between the pile and the surrounding ground. In this study, this is simply simulated as bearing capacity of a footing with a uniform constant load distribution corresponding to a surcharge of an embedment (D_f) of the pile length applied on the sides of the footing as shown in Fig. 2(b).

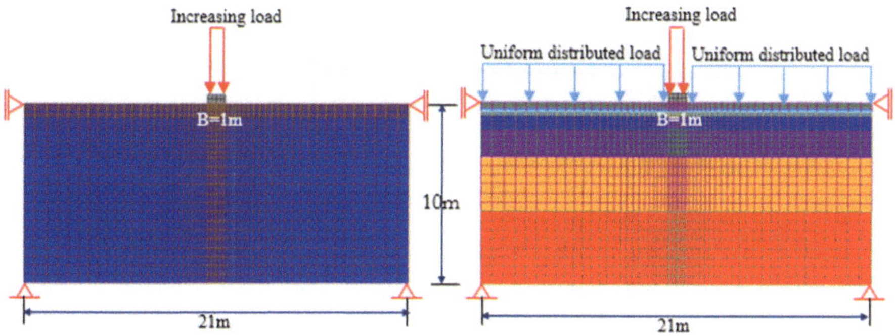

(a) Uniform ground (b) Layered ground

Fig. 2. Finite element mesh and boundary condition used in the simulation.

4 Analysis Cases

4.1 Bearing Capacity of a Footing on a Uniform Soft Ground

The purpose of this simulation is to verify the validity of the simulation method. Bearing capacities of shallow foundation on the uniform ground was simulated and the results were compared with the Terzaghi formula's and Prandtl formula's results. Tables 1 and 2 show the characteristics of the soft ground for different cases simulated.

Table 1. Variation of ϕ

Cases	ϕ (°)	c (kPa)	γ (kN/m^3)	D_f (m)
1-1	0	1	17.5	0
1-2	10	1	17.5	0
1-3	20	1	17.5	0
1-4	30	1	17.5	0
1-5	32	1	17.5	0
1-6	34	1	17.5	0
1-7	36	1	17.5	0
1-8	38	1	17.5	0
1-9	40	1	17.5	0

The results from the cases of Table 1 are compared to the results from the Terzaghi's formula (Eq. 8) and the results from the cases of Table 2 are compared to the results from Prandtl's formula (Eq. 9). Table 3 shows the used load factors N_c, N_q and N_γ and the shape factors α and β for the Terzaghi's formula.

Table 2. Variation of c

Cases	ϕ (°)	c (kPa)	γ (kN/m^3)	D_f (m)
2-1	0	5	0	0
2-2	0	10	0	0
2-3	0	15	0	0
2-4	0	20	0	0
2-5	0	25	0	0
2-6	0	30	0	0
2-7	0	40	0	0
2-8	0	50	0	0
2-9	0	60	0	0
2-10	0	80	0	0
2-11	0	100	0	0

Table 3. Coefficient for Terzaghi's formula

(a) Bearing capacity factors

ϕ (deg.)	N_c	N_q	N_γ
0	5.1	1.0	0.0
10	8.3	2.5	0.4
20	14.8	6.4	2.9
30	30.1	18.4	15.7
32	35.5	23.2	22.0
34	42.2	29.4	31.1
36	50.6	37.8	44.4
38	61.4	48.9	64.1
40	75.3	64.2	93.7

(b) Shape factors

coefficients	square	circular	continuous	rectangular
α	1.0	1.2	1.0+0.2*(B/L)	1.2
β	0.5	0.3	0.5-0.2*(B/L)	0.3

$$q_u = \alpha c N_c + \beta \gamma B N_\gamma + q N_q \quad \text{(Terzaghi's formula)} \tag{8}$$

$$q_u = (2 + \pi)c \quad \text{(Prandtl's formula)} \tag{9}$$

4.2 Bearing Capacity of Footing on Layered Ground

In this simulation, the influence of a hard ground layer below the surface soft ground on the bearing capacity is investigated. The main parameter is the relative thickness *H*/*B*. The Case 1–4 in Table 1 is taken as characteristics of the softer layer and the hard layer has cohesion of c = 10000 kPa, shear resistance angle of ϕ = 0 degree and unit weight γ = 22 kN/m^3 as mentioned above. Table 4 shows the different cases of the simulation.

Table 4. Case study for the influence of a hard layer

Cases	*H*/*B*	ϕ (°)	c (kPa)	γ (kN/m^3)	D_f (m)
3-1	0.00	30	1	17.5	0
3-2	0.25	30	1	17.5	0
3-3	0.50	30	1	17.5	0
3-4	0.75	30	1	17.5	0
3-5	1.50	30	1	17.5	0
3-6	3.00	30	1	17.5	0
3-7	6.00	30	1	17.5	0

4.3 Simulation of the End Bearing Capacity of IESP

The end bearing capacity of IESP is simulated as bearing capacity of a footing with a uniform constant load distribution corresponding to a surcharge of an embedment (D_f) of the pile length applied on the sides of the footing.

This simulation is done in the same way of the previous one, but by changing the equivalent surcharge of the embedment D_f. The embedment is equal to the pile length in this simulation. In that way, the computed bearing capacity corresponds to the end bearing capacity of the pile. The influence of two parameters of the soft layer which are the shear resistance angle ϕ and the cohesion c are investigated. And for each soil parameter the influence of the relative thickness (*H*/*B*), and the embedment (D_f) are investigated. The characteristics of the hard layer are same as the previous simulation and are constant. Tables 5 and 6 show the characteristics of the soft layers for the investigation of the influence of the soil parameters such as shear resistance angle ϕ (named ϕ materials) and Cohesion c (named c materials), respectively.

4.4 Influence of the Strength of the Lower Hard Layer

In this section the way that the variation of the strength of the lower hard layer affects the bearing capacity is investigated for both cohesive material (c material) and

Table 5. Soil parameters for ϕ material series

Cases	H/B	ϕ (°)	c (kPa)	γ (kN/m^3)	D_f (m)
3-1	0	30	1	17.5	0
3-2	0.25	30	1	17.5	0
3-3	0.50	30	1	17.5	0
3-4	0.75	30	1	17.5	0
3-5	1.5	30	1	17.5	0
3-6	3	30	1	17.5	0
3-7	6	30	1	17.5	0
9-1	0	30	1	17.5	20
9-2	0.25	30	1	17.5	20
9-3	0.50	30	1	17.5	20
9-4	0.75	30	1	17.5	20
9-5	1.5	30	1	17.5	20
9-6	3	30	1	17.5	20
9-7	6	30	1	17.5	20

Table 6. Soil parameters for c material series

Cases	H/B	ϕ (°)	c (kPa)	γ (kN/m^3)	D_f (m)
10-1	0	0	100	17.5	0
10-2	0.25	0	100	17.5	0
10-3	0.50	0	100	17.5	0
10-4	0.75	0	100	17.5	0
10-5	1.5	0	100	17.5	0
10-6	3	0	100	17.5	0
10-7	6	0	100	17.5	0
16-1	0	0	100	17.5	20
16-2	0.25	0	100	17.5	20
16-3	0.50	0	100	17.5	20
16-4	0.75	0	100	17.5	20
16-5	1.5	0	100	17.5	20
16-6	3	0	100	17.5	20
16-7	6	0	100	17.5	20

cohesionless material (ϕ material). The case of the surcharge equivalent to an embedment of 10 m was considered for both types of material.

For c material, the upper layer has a cohesion of c = 100 kPa and an internal frictional angle $\phi = 0°$. The lower hard layer has a shear resistance angle $\phi = 0$ and the cohesion was varied from 200 to 10000 kPa.

For ϕ material, the upper layer has a cohesion of c = kPa and a shear resistance angle $\phi = 30°$. The lower hard layer has a shear resistance angle $\phi = 0$ but in that condition, a cohesion of at least 6000 kPa is necessary to get strength higher than the one of the upper layer. For that reason, the cohesion was varied from 6000 to 10000 kPa.

5 Results Analysis

5.1 Bearing Capacity of Footing on Uniform Ground

Figure 3 shows the comparison of the result from the cases of Table 1 and the results from the Terzaghi's formula, Fig. 4 shows the comparison of the results from the cases of Table 2 and the results from Prandtl's formula. And Fig. 5 shows the strain rate distribution style in the two cases. The norm of the strain rate presented contour lines in the range of $\dot{e}_{\max} \sim \dot{e}_{\min}$ (= 0). From these results, it can be concluded that the simulation method can estimate accurately the bearing capacity of a shallow foundation.

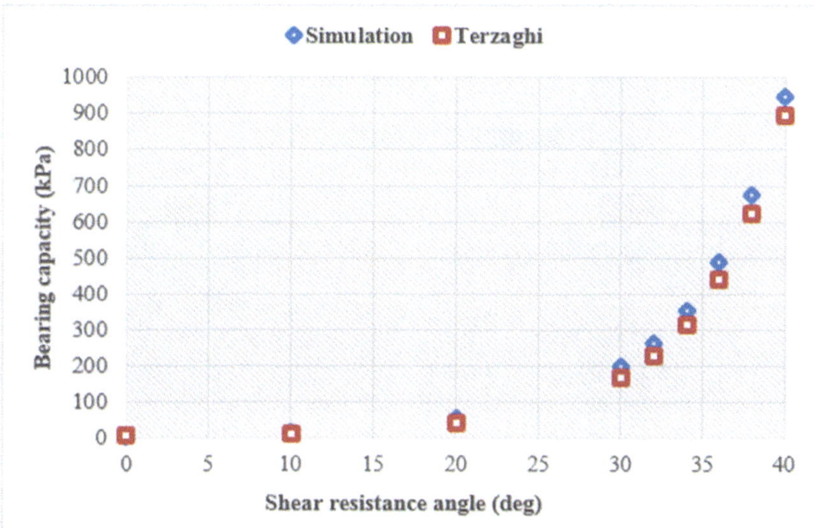

Fig. 3. Comparison with Terzaghi's formula

5.2 Bearing Capacity of Footing on Layered Ground

The results of the simulation with the cases of Table 4 are plotted on Fig. 6. Figure 7 shows the strain rate distribution style of some cases. The norm of the strain rate was presented by contour lines in the range of $\dot{e}_{\max} \sim \dot{e}_{\min}$ (= 0). The bearing capacity decreases inversely with the relative thickness as claimed in Fig. 7, but it becomes almost steady after $H/B = 0.75$ so that from $H/B = 1.50$ to $H/B = 6.00$ the bearing capacity is almost the same. This bearing capacity is the same with the bearing capacity obtained with the uniform soft ground for $\phi = 30°$ as claimed in Fig. 3. It means that the bearing capacity decreases from the bearing capacity of the hard layer to the bearing

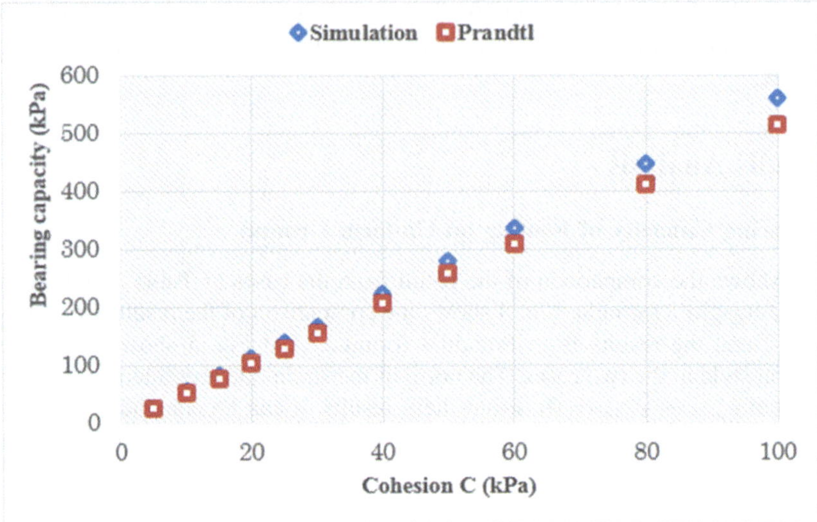

Fig. 4. Comparison with Prandtl's formula

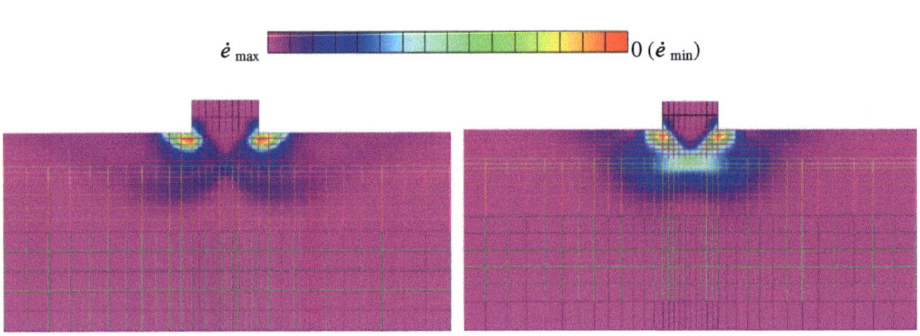

(a) Case 1-4 (b) Case 2-11

Fig. 5. Strain rate distribution

capacity of the soft layer. Figure 7 shows the influence of the relative thickness on the strain rate distribution style. It can be seen that the shear band is limited by the surface of the hard ground. The band expands by increasing H/B until $H/B = 1.50$, confirming that from that value of H/B the hard ground has no longer effect on the bearing capacity.

5.3 End Bearing Capacity of IESP

The obtained results are plotted on Figs. 8 and 9. The figures show that the capacity in case of $H/B = 0$ (uniform hard ground) is too high compared to the other cases, this can be because the strength of the hard layer is too high compared to that of the soft layer. It also can be seen that bearing capacity increases when D_f increases.

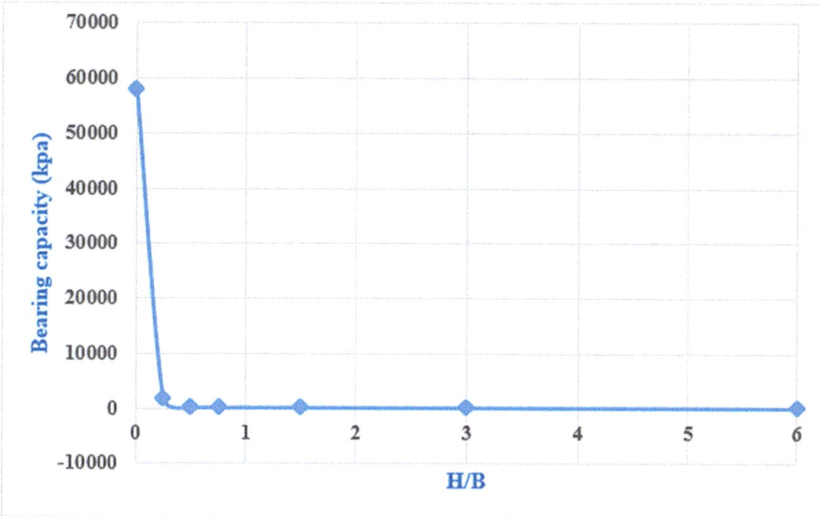

Fig. 6. Variation of the bearing capacity according to the relative thickness

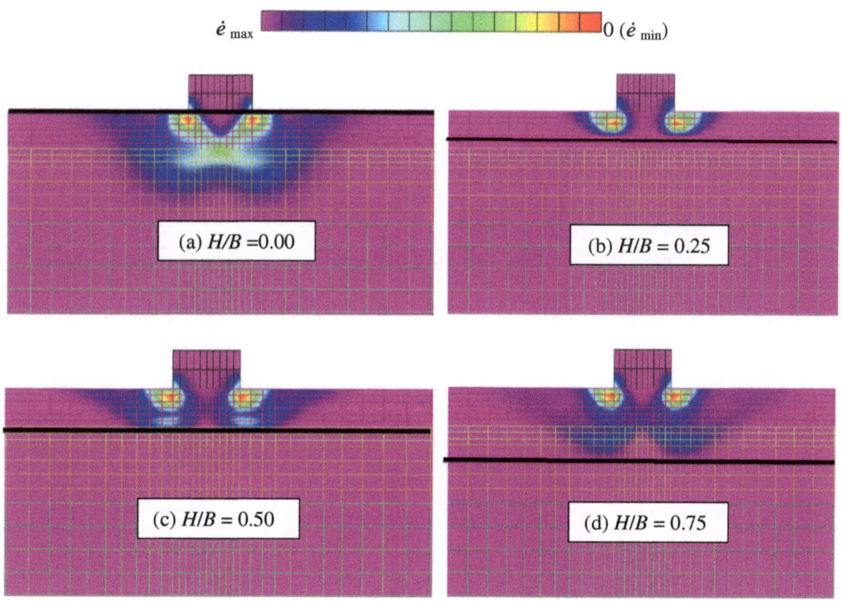

Fig. 7. Influence of the hard layer on the strain rate distribution

5.4 The Influence of the Strength of the Hard Layer

For both *c* material and *ϕ* material, the influence of the variation of the strength of the hard layer can be seen only for the case of a relative thickness H/B = 0 (completely

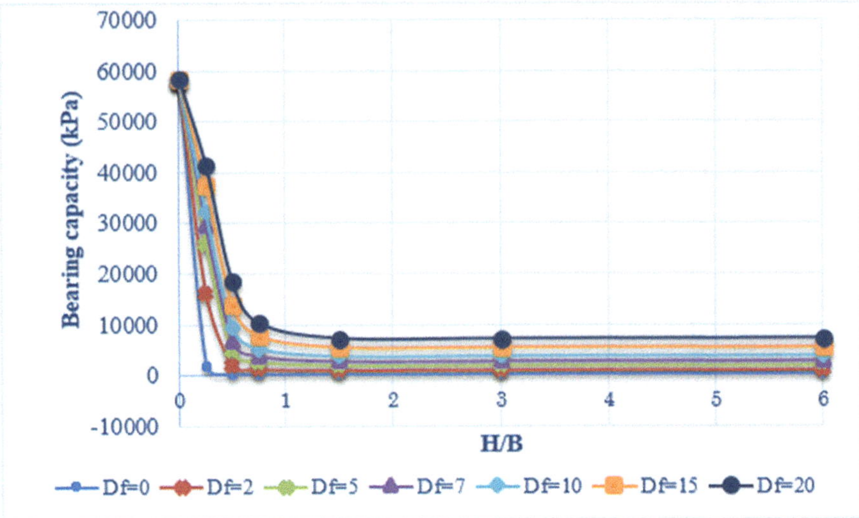

Fig. 8. Variation of the bearing capacity according to the relative thickness and embedment for $\phi = 30°$

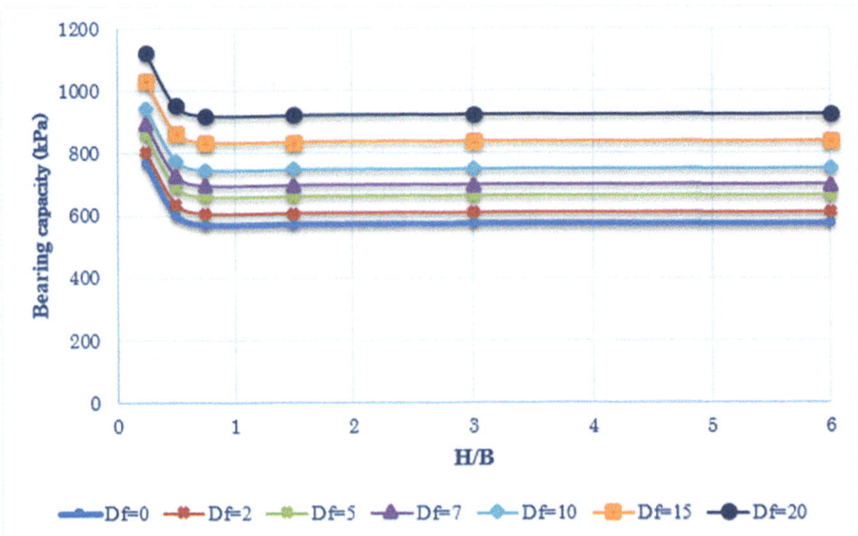

Fig. 9. Variation of the bearing capacity according to the relative thickness and embedment for c = 100 kPa

hard ground). From the H/B = 0.25, the variation of the strength has no longer any influence on the bearing capacity. But still, the bearing capacity inversely decreases with the relative thickness up to the relative thickness H/B = 1.5 to become constant. The Fig. 10 shows that in the case of ϕ material and the Figs. 11(a) and (b) show it for *c* material. Meaning that in a real case of IESP (H/B greater than 0) the variation of the strength of hard layer has no much influence on the end bearing capacity. This can be because the strength of the hard layer is too much higher than the one of the soft layer so that any strain cannot happen in the hard layer, the Figs. 12(a) and (b) illustrate it for ϕ and *c* material respectively.

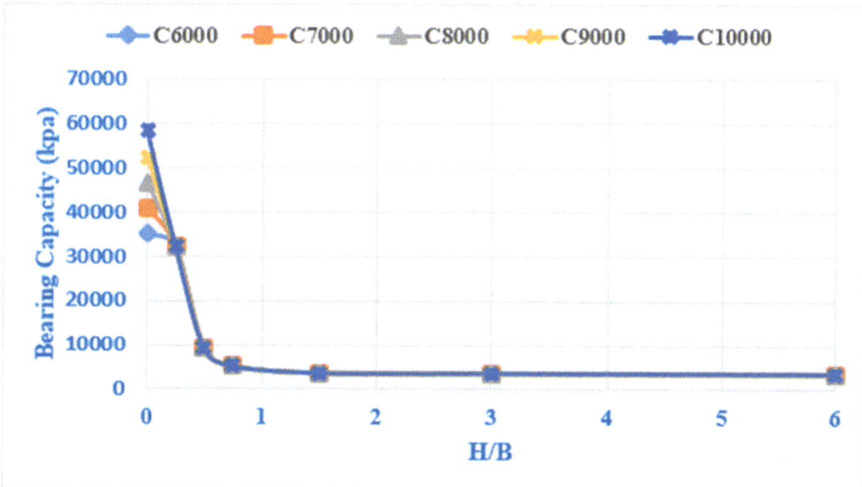

Fig. 10. Influence of the variation of the strength of the hard layer on the bearing capacity of IESP for ϕ materials (D_f = 10 m)

6 Estimation of the End Bearing Capacity of Iesp

6.1 Proposed Formula for Estimation of End Bearing Capacity of IESP

Based upon the above results a relationship between the bearing capacities normalized by the bearing capacity of *H/B* = 0.00 is proposed with Eqs. (10) and (11)

$$\xi_x = \frac{q_x - q_6}{q_0 - q_6} \tag{10}$$

$$q_x = \xi_x q_0 + (1 - \xi_x) q_6 = \xi_x (q_0 - q_6) + q_6 \tag{11}$$

where, q_x is the bearing capacity with a relative thickness of *x*, q_0 is the bearing capacity with a relative thickness of 0.00, q_6 is the bearing capacity with a relative thickness of 6.00. ξ_x is a coefficient depending on the relative thickness *x* and the embedment D_f. q_0 is the bearing capacity of the pile tip on the uniform hard ground called End Supported pile (ESP). And q_6 can be considered as the bearing capacity on the uniform soft ground according to Figs. 3 and 6. Figures 13 and 14 show the values of the coefficient ξ_x which depend on *H/B* and D_f for the shear resistance angle ϕ materials and the cohesion *c* materials, respectively. For *c* material, one unique line of ξ_x is obtained independently of D_f

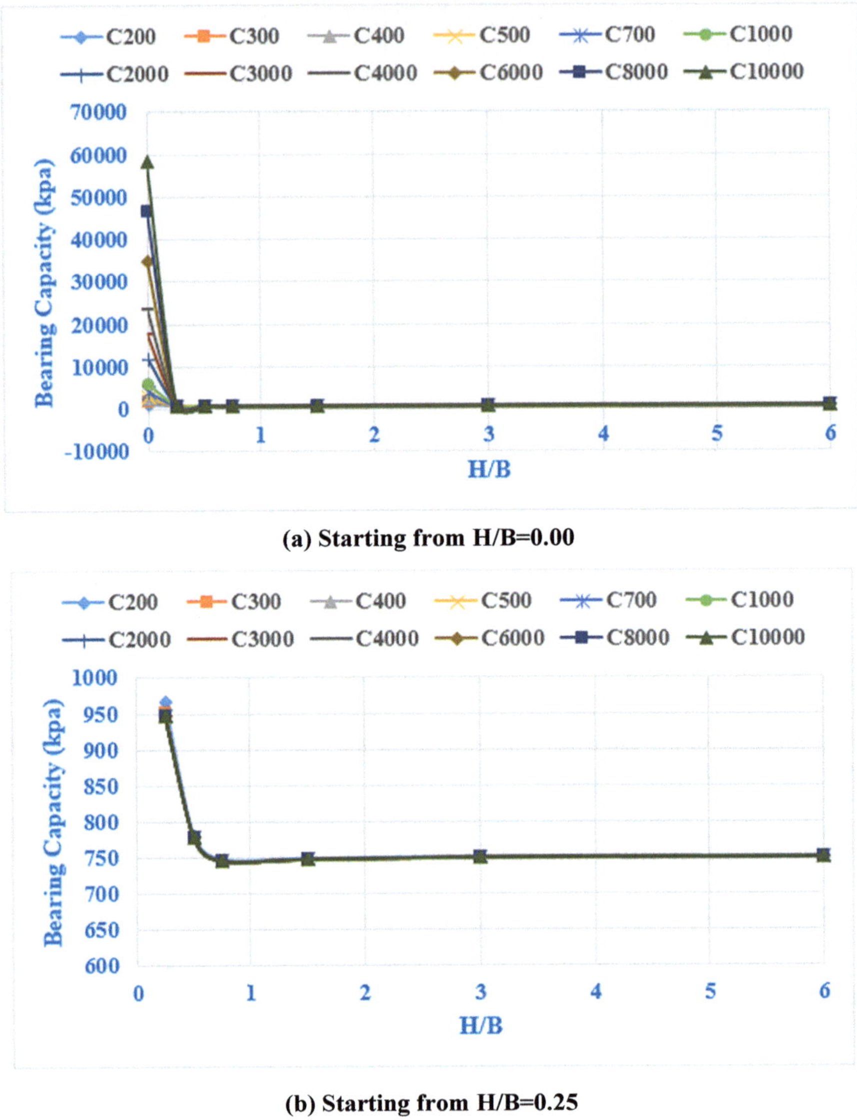

Fig. 11. Influence of the variation of the strength of the hard layer on the bearing capacity of IESP for *c* material (D_f = 10 m)

as claimed in Fig. 14. But for the ϕ materials, different curves of ξ_x depending on D_f are drawn. And there is a big difference among them as D_f increases. It can be justified by the increase of the confining pressure as D_f increases for ϕ material. Figure 13 also shows that for the shear resistance angle ϕ materials, when D_f (Pile length) is small the value of ξ_x decreases rapidly for a small increase of *H*/*B*. In the interval of *B*/*H* = 0.00 ~ 1.50, for the same value of *H*/*B* the value of ξ_x increase with the increase of D_f.

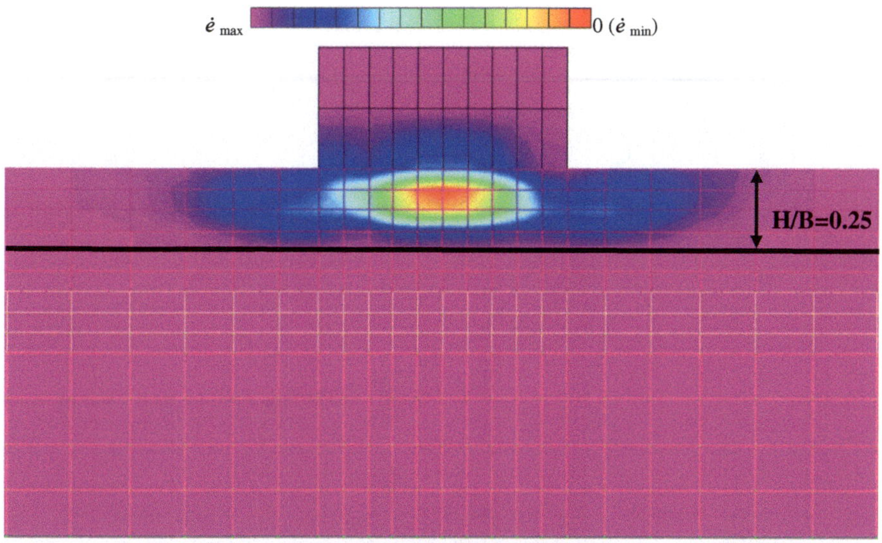

(a) For Phi material with a hard layer of a cohesion c = 6000 kPa

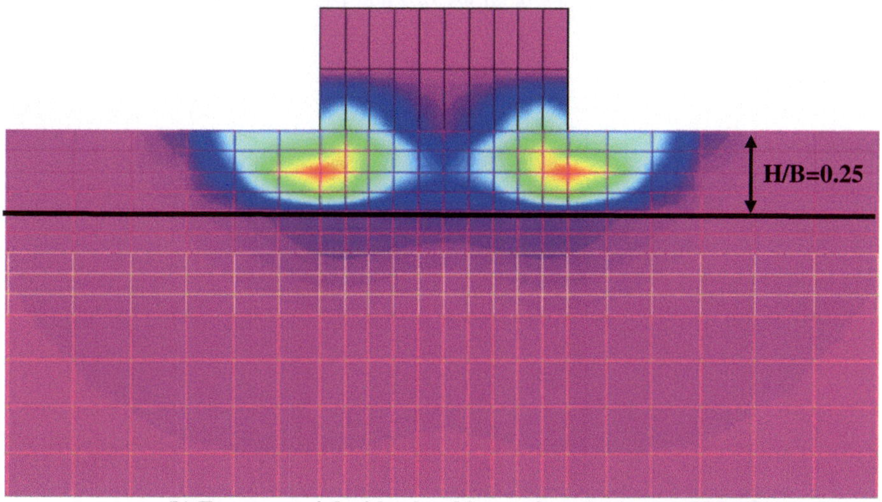

(b) For c material with a hard layer of a cohesion c = 200 kPa

Fig. 12. Strain distribution for the case H/B = 0.25

6.2 Influence of the Variation of the Strength of the Hard Layer on the Values of ξ_x

The variation of the strength of the hard layer has an effect on the bearing capacity in the case of H/B = 0 corresponding to the bearing capacity of the hard layer. And that bearing capacity of the hard layer is included in the formula. So, the variation of the strength of the hard layer has also an effect on the formula particularly on the values of the coefficient ξ_x. For the same D_f (D_f = 10 in this case) the value of ξ_x decreases when

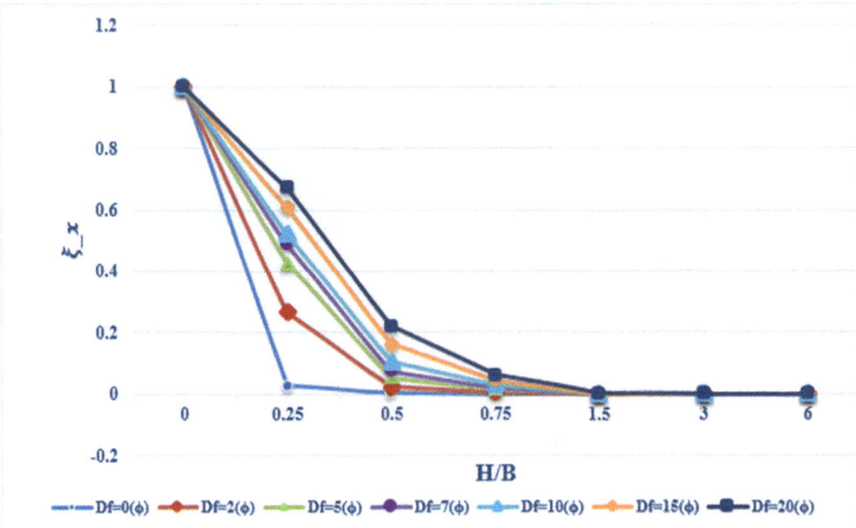

Fig. 13. ξ_x depending on the *H*/*B* and D_f (for ϕ materials)

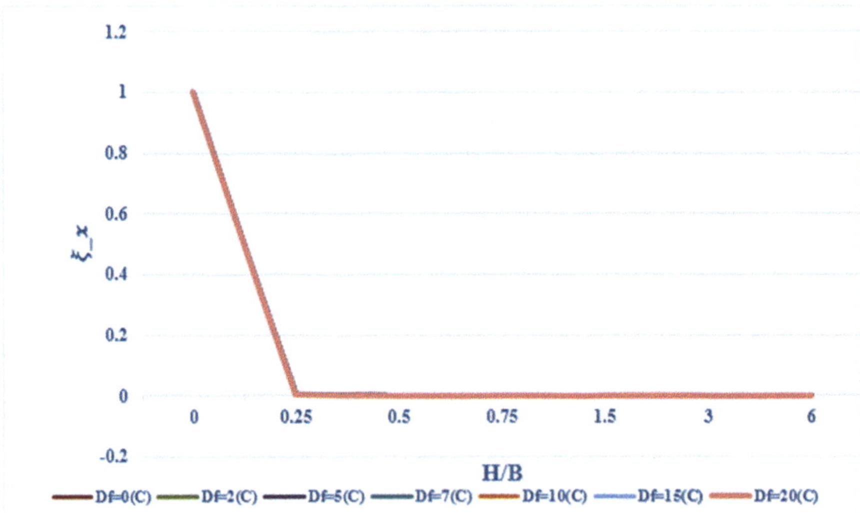

Fig. 14. ξ_x depending on the *H*/*B* and D_f (for *c* materials)

the strength of the hard layer gets greater and greater than the one of the upper layer. That trend can be noticed in Fig. 15 for Phi material and Fig. 16 for *c* material. Those Figures show also that the values of ξ_x inversely decrease with the relative thickness up H/B = 1.5 to be constant.

In a future study, an acceptable approximate curve of trend of the values of ξ_x will be proposed.

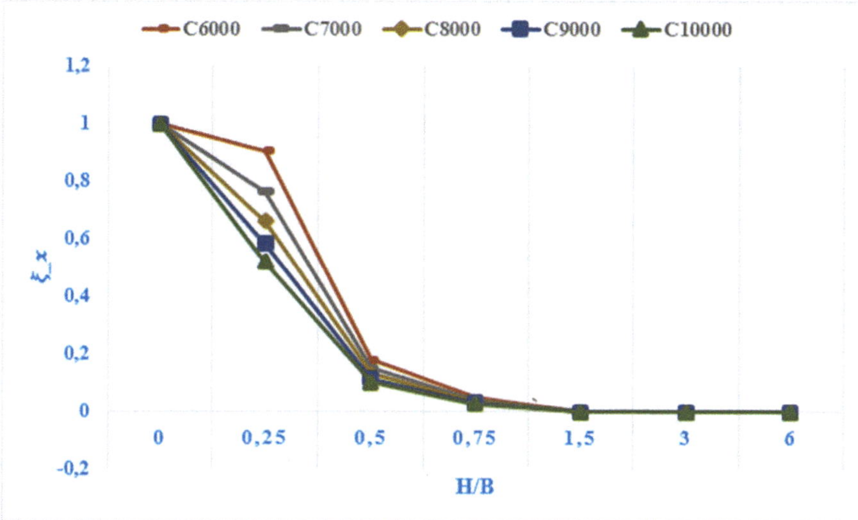

Fig. 15. Influence of the variation of the strength of the hard layer on the coefficient ξ_x (for ϕ materials)

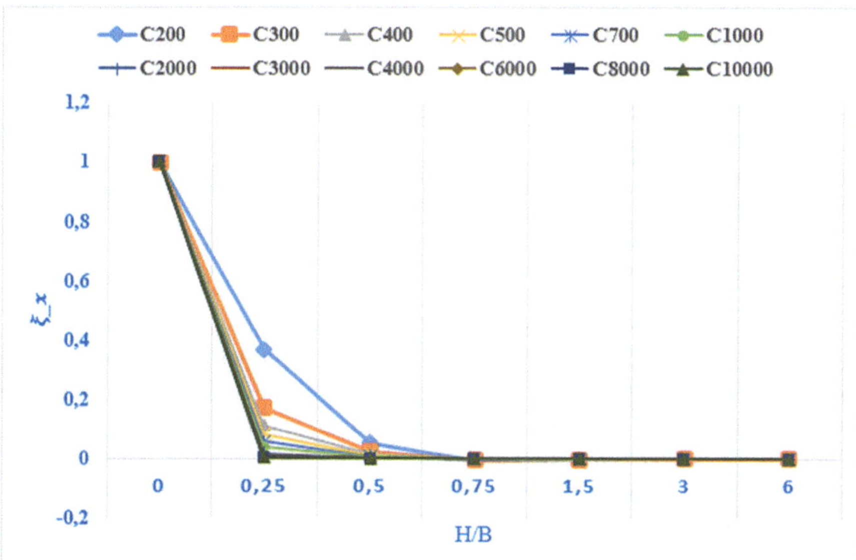

Fig. 16. Influence of the variation of the strength of the hard layer on the coefficient ξ_x (for c materials)

7 Conclusions and Future Studies

7.1 Conclusion

This study carried out two-dimensional RPFEM analyses to estimate the end bearing capacity of IESP without consideration of the skin friction resistance between the pile and the surrounding ground. After verifying the validity of the simulation method, the effect of the following parameters was investigated; the relative thickness of the soft layer by the foundation width (*H*/*B*), the embedment D_f (pile length), the strength parameters of the soft ground. The following conclusions were drawn;

(1) The bearing capacity of the footing decrease inversely with the relative thickness (*H*/*B*), but it becomes steady from $H/B = 1.5$. The bearing capacity decreases from the bearing capacity of the hard layer corresponding to $H/B = 0.00$ to the bearing capacity of the soft layer corresponding to $H/B \geq 1.50$. From $H/B = 1.50$ the hard ground has no longer influence on the bearing capacity in these simulation conditions.

(2) The variation of the strength of the hard layer has influence only on the bearing capacity of the case of H/B = 0 but does not have a significant effect for the cases of a relative thickness higher than H/B = 0.

(3) A formula to estimate the end bearing capacity of IESP is proposed. This formula includes the bearing capacities of both layers and a coefficient ξ_x which is highly influenced by the materials properties. For *c* material, ξ_x is unique regardless of D_f, but for the shear resistance angle ϕ materials, in the interval of $H/B = 0.00 \sim 1.50$, for the same value of *H*/*B* the value of ξ_x increase with the increase of D_f. It is caused by the difference of shear band size in *c* and ϕ material.

(4) The values of the coefficient ξ_x decrease when the strength of the hard layer gets higher than the one of the upper layer. And the values inversely decrease with the relative thickness from 1 for H/B = 0 to 0 for H/B = 1.5.

7.2 Future Studies

From this study, it has been confirmed that RPFEM can simulate soil bearing capacity. The end bearing capacity of IESP decreases inversely with the space between the pile tip and the hard layer. But the method used in this simulation to estimate the end bearing capacity of IESP is a very simplified method, it does not take account of the real effect of the surrounding ground of the pile and therefore does not simulated the real behavior of the ground. Future studies will be done to estimate the end bearing capacity of IESP by using a hollow at the place of the pile and by using joint elements between the pile and the surrounding ground based on RPFEM.

References

Tamura, T., Kobayashi, S., Sumi, T.: Limit analysis of soil structure by rigid plastic finite element method. Soils Found. **24**(1), 34–42 (1984)

Tamura, T., Kobayashi, S., Sumi, T.: Rigid plastic finite element method for frictional materials. Soils Found. **27**(3), 1–12 (1987)

Du, N.L., Ohtsuka, S., Hoshina, T., Isobe, K.: Discussion on size effect of footing in ultimate bearing capacity of sandy soil using rigid plastic finite element method. Soils Found. **56**(1), 93–103 (2016)

Terzaghi, K.: Theoretical Soil Mechanics. Wiley, New York (1943)

Meyerhof, G.G.: Bearing capacity and settlement of pile foundations. J. Geotech. Eng. Div. ASCE **102**(3), 195–228 (1976)

Poulos, H.G.: Pile behavior—theory and application. Géotechnique **39**(3), 365–415 (1989). https://doi.org/10.1680/geot.1989.39.3.365

Eslami, A., Fellenius, B.H.: Toe bearing capacity of piles from cone penetration test (CPT) data. In: Proceedings of the International Symposium on Cone Penetrometer Testing, CPT 1995, Linkoping, Sweden, 4–5 October 1995. Swedish Geotechnical Society, Gothenburg (1995)

Eslami, A., Fellenius, B.H.: Pile capacity by direct CPT and CPTu methods applied to 102 case histories. Can. Geotech. J. **34**(6), 886–904 (1997). https://doi.org/10.1139/cgj-34-6-886

Influence of Cap Rigidity on the Performance of Piled-Cap Foundation

Lan Lin(✉), Anup Sinha, and Adel Hanna

Department of Building, Civil and Environmental Engineering, Concordia University, Montreal, Canada
lan.lin@concordia.ca, hanna@civil.concordia.ca

Abstract. Piled-cap foundations are recognized for sharing the total load by the piles and the cap. In practice, it is assumed that the cap will distribute uniformly the load to the piles, which is often used in choosing the thickness of the cap. In the literature, research in this field remained lagging behind due to the difficulties associated in modeling. This is due to the fact that the load sharing mechanism of piled-cap foundation is a complex soil-structure problem governed by pile/cap geometry, soil condition, etc. It is of interest to note that a flexible cap will lead to uneven load distribution among the piles, which offsets the principle of design of pile foundation. It may, in turn, lead to failure of some piles in the group or even a catastrophic failure of the foundation. While, a non-flexible (i.e., rigid) cap may lead to uniform load distribution among the piles in the group, which will minimize or eliminate differential settlements within the cap. However, a thick cap will impose additional load due to its own weight and accordingly, it may in increase the total settlement of the foundation.

This paper presents the results from a three-dimension finite element model to examine the role of the cap geometry on the load sharing mechanism of the piled-cap foundation. The effect of the cap rigidity in terms of its size, thickness and pile spacing were examined. The modified Drucker-Prager cap plasticity model provides an excellent basis for analyzing the stress-strain of soil continuum constitutive law, and the software "ABAQUS", which is based on pure solid mechanics principles has been used successfully to simulate the problem stated.

1 Introduction

The function of the foundation is to transmit the loads of the building to the underlying soil safely and economically, and to insure reliability and serviceability of the structure during its lifespan. In practice, the design of pile foundation is based on the notion that the cap will distribute uniformly the load on the piles within the group. Accordingly, the cap must be a rigid enough to satisfy this condition. Nevertheless, excessively rigid cap may add additional load on the piles, due to its own weight. On the other hand, flexible caps may lead to uneven distribution of the column load on the piles within the cap. While extensive research on pile foundations can be found in the literature, studies on the optimization of the cap rigidity is lagging, which believed to have significant influence on the performance of the system.

H. El-Naggar et al. (Eds.): GeoMEast 2018, SUCI, pp. 104–115, 2019.
https://doi.org/10.1007/978-3-030-01902-0_10

Sabnis and Gogate (1984) investigated the behavior of thick slabs (pile cap) by comparing different design methods available at that time. Adebar et al. (1990) preformed an experimental study on piled-cap to study their behavior. He used the results to validate the ACI Building code procedures for the design of pile caps. Adebar and Zhou (1996) proposed a simple design procedure to calculate the shear strength of deep piled-caps using the maximum bearing stress. The simplified methods proposed by Randolph (1983), Clancy and Randolph (1996), Poulos (2005), and Ai et al. (2005) were based on a single piled-cap unit, while the interactions between the piles and the cap were not considered. Griffiths et al. (1991) and Kim et al. (2001) ignored the interactions between piles, soil and the cap. Other forms of hybrid methods include the combination of finite element method with finite layer method as proposed in Small and Zhang (2002). Recently, Jiu and Huang (2014) introduced a finite difference method to reduce the computational time. Notwithstanding the above, the cap rigidity represented by its thickness, size and pile spacing was ignored in all these studies, which believed to have significant influence on the performance of this system.

2 Numerical Modeling

A three-dimensional finite element model was developed using ABAQUS to examine the performance of the cap in a pile foundation as illustrated in Fig. 1. The model consists of a solid block (AA′-BB′-CC′-DD′), supporting a cap (PQRS) and the pile group embedded inside the block. Due to the geometrical symmetry and material isotropy, one quarter of the block was analyzed (AA′-HH′-FF′-GG′). Figure 2 present a summary of the boundary conditions of the numerical model. Due to the planner symmetry, the boundary condition along the mid-axis was set as XSIMM (symmetry about plane x = constant) and YSIMM (symmetry about plane y = constant). No translation was allowed in the YZ and XZ directions. Similarly, the rotational degree of freedom R_x and R_z in XZ plane were not allowed. Furthermore, the nodal restraints were applied on R_y and R_z in YZ plane, all other translations were allowed. In order to account for the interaction between the soil and the foundation, the master-slave concept was used, where the pile and the cap were treated as master surface and soil surface as slave surface.

In this investigation, the 'surface to surface' contact option was considered due to its high level of accuracy as reported in Hibbitt et al. (2007). Furthermore, the penalty type frictional constraint enforcement method is used to formulate the stiffness of the contact surfaces. In addition, the soil was treated as an isotropic homogeneous single-phase medium in modelling. In order to reduce the computational cost and time, a number of stress-strain analyses on elastic, elasto-plastic soil continuum of different length, breadth and depth were examined. Based on the analysis results, the horizontal length of the continuum was taken as 30 times the pile's diameter, and the vertical depth as two times the pile's length, beyond which, no appreciable stress and strain were observed at the boundaries. Preliminary analyses were performed and the results confirmed the continuum configuration proposed.

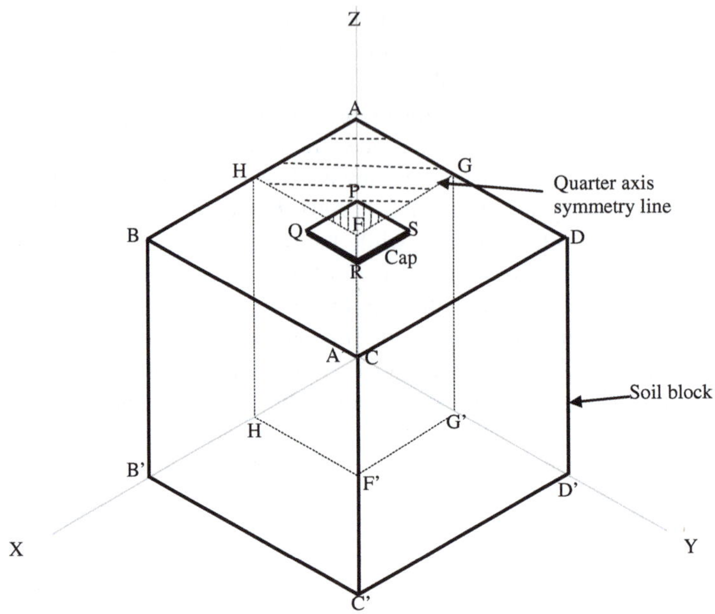

Fig. 1. Schematic sketch of the numerical model.

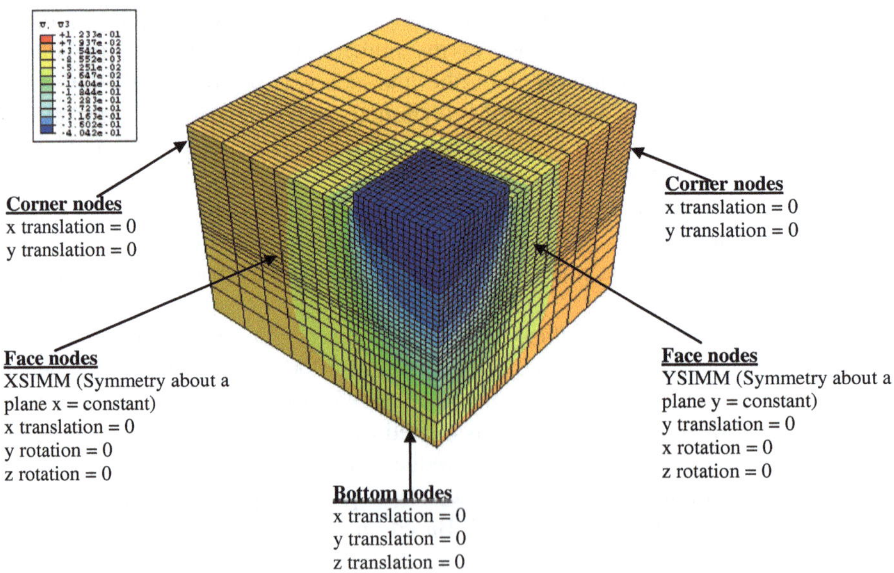

Fig. 2. Boundary condition of the 3-D axisymmetric model.

Given this, a quarter cut of the soil continuum of 80 × 80 × 30 m, with 16 piles of 15 m in length and 1 m in diameter, and a cap of 24 × 24 × 2 m subjected to a uniformly distributed load of 0.5 MPa was analyzed. Figure 3 presents the produced pressure pulb (in grey) when the load was applied slowly to reach 0.5 MPa. In order to make a smooth transition of elements from fine to coarse, special partition technique was used. During this phase the bias and the aspect ratio were kept within a reasonable limit to run the program successfully.

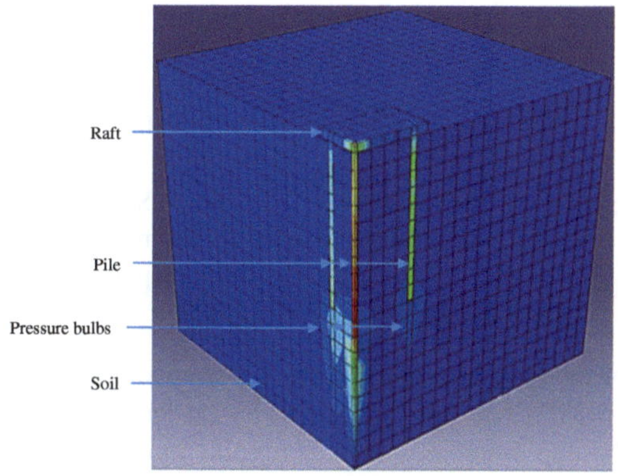

Fig. 3. Pressure bulb of quarter cut of the pile cap.

3 Model Validation

The results of the present numerical investigation were compared with the available results of pile foundation summarized by Poulos (2001). Table 1 summarizes the geometry and the material property used in Poulos (2001), and Fig. 4 presents a summary of these results, where a good agreement was noted with the results of Poulos-Davis-Randolph (PDR), GARP5, GASP, while the results produced by FLAC 2-D and 3-D were regarded as less satisfactory. This can be attributed to the fact that FLAC software is mainly used in practice as fast tool for producing results, while ABAQUS is widely used as strong research tools known for its accurate results.

4 Analysis Results

4.1 Effect of Cap Size

To examine the influence of the cap size on the rigidity of the cap, three groups of pile-cap foundation were examined. The caps were 16 × 16 m, 24 × 24 m and 32 × 32 m and having a thickness of 2 m supported by piles of 15.0 m in length and 1.0 m in diameter with a center to center spacing of 4 m. The properties of the material used in this analysis were taken from Katzenbach et al. (2000) and the parameters used for the modified Drucker-Prager constitutive law are given in Table 2. The volumetric plastic

Table 1. Geometry and material properties of the data in ASCE TC 18.

	Soil	Pile	Cap
Size (m)	20 × 20 × 20	0.5(D) × 10(L) S = 1m (laterally) 2m (longitudinally)	10 × 6 × 0.5
Unit weight, γ (kg/m^3)	1700	2500	2500
Elastic modulus, E (MPa)	20	30,000	30,000
Poisson's ratio, ν	0.3	0.2	0.2
Load (MN)	Self-Weight	Self-Weight	Self-Weight + 2MN on middle row and 1MN on sides row of piles

hardening property of the clay was taken from Katzenbach and Reul (1997). A uniform distributed load of 0.5 MPa was applied gradually on the top of the cap. The responses from the piles and the cap were recorded at regular intervals.

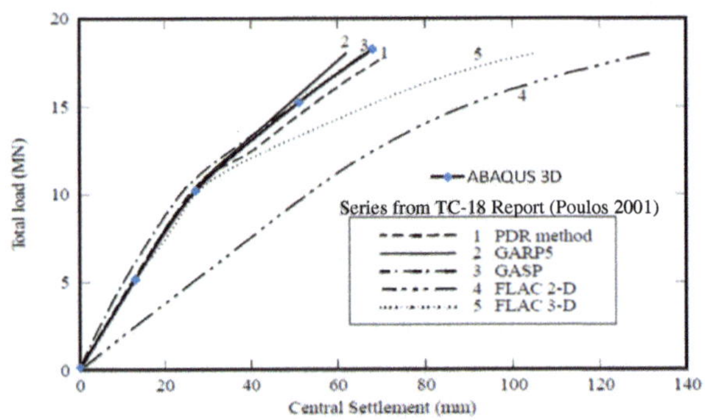

Fig. 4. Settlement at the center point of cap measured by different methods.

In this study, the settlement profile of cap was examined along the center line of the cap, diagonal line of the cap, and central line of piles as shown in Fig. 5. For illustration, Fig. 6 shows the settlements along the three paths mentioned above, for the cap with 24 × 24 m. It can be seen in Fig. 6 that the deflected cap was formed in a dish or bowl shape, with the maximum settlement occurred at the center. Furthermore, it is noted that the settlements measured along these three paths are almost identical. It is necessary to mention that the results for the other two caps (i.e., 16 × 16 m and 32 × 32 m) are similar to those illustrated in Fig. 6, and are not presented in the paper.

Figure 7 presents the settlement along the diagonal line for all the three tested caps. It can be noted that the larger (32 × 32 m) the cap size, the higher the settlement at the

Table 2. Material properties used in the numerical model.

Parameter	Unit	Clay	Pile	Raft
Young modulus, E	MPa	54	25000	34000
Poisson's ratio, ν		0.15	0.2	0.2
Unit weight, γ	kN/m^3	19	25	25
Submerged unit weight, γ'	kN/m^3	9	15	15
Angle of internal friction, φ	degree	20		
Cohesion, c′	kPa	20		
Cap parameter				
d	kPa	42.42		
β	degree	37.67		
α		0		
K		0.795		
R		0.1		

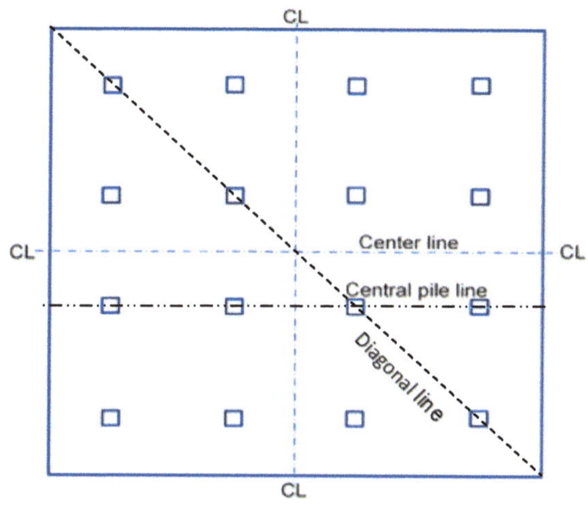

Fig. 5. Path definition to investigate the maximum settlement.

center (0.36 m) as compared to the edges (0.31 m). The maximum settlement in the cap of 16 × 16 m is about half in the cap of 32 × 32 m. In addition, the settlement in the cap of 16 × 16 m is almost the same. Accordingly, the cap of 16 × 16 m is quite rigid while the 32 × 32 m is very much flexible.

4.2 Effect of Cap Thickness

To examine the effect of the cap thickness on the rigidity of the cap and the performance of the pile-cap system, the cap thickness (t) was tested at 1.0 m, 1.5 m, 2.0 m and 3.0 m under a constant load of 0.5 MPa, on the three caps as described above. Figure 8 presents the measured settlement along the diagonal line of the cap for the cap

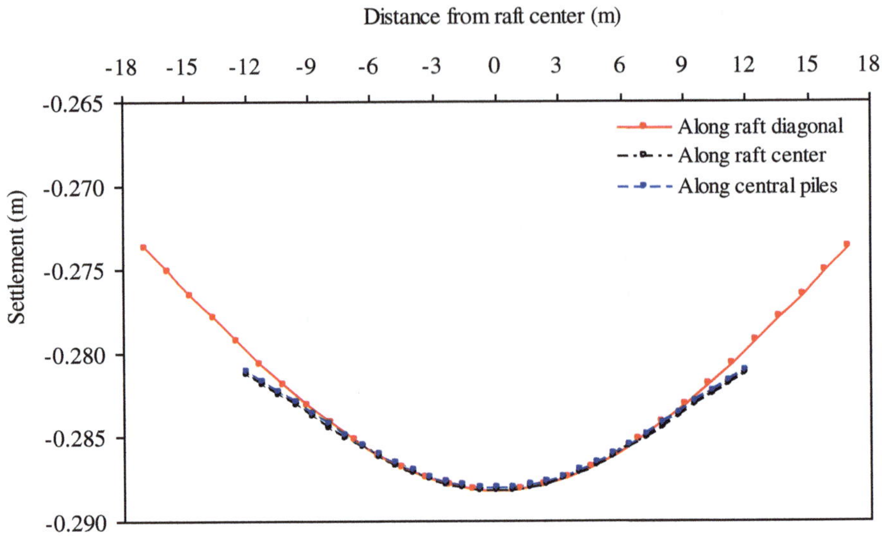

Fig. 6. Settlement along different path for the cap of 24 m × 24m.

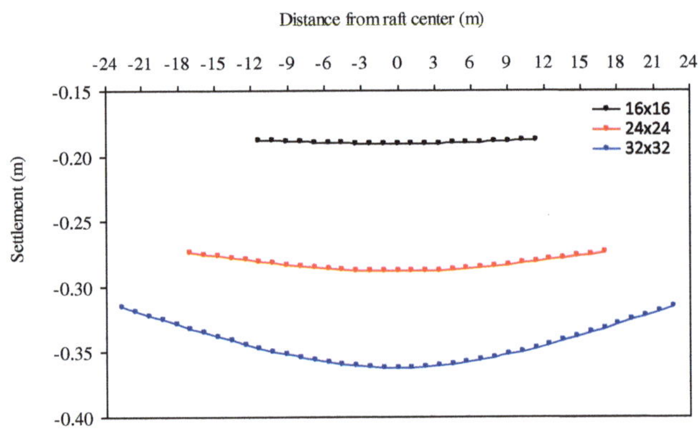

Fig. 7. Settlement along the cap diagonal.

size of 16 m × 16m for the thickness ranging from 1.0 m to 3.0 m. It can be seen in the figure that, for thinner caps (t = 1.0 m and 1.5 m) the measured settlement was in a parabolic shape, which indicates that these caps behave as a flexible structure. It may lead to overloading on some piles within the group. Furthermore, piles may separate from their cap if the deflection in the cap is excessive. On the other hand, for thicker caps having thickness of 2.0 m, 2.5 m and 3.0 m, the measured settlement was uniform. It confirms they perform as a rigid structure. It is also necessary to mention that under the same load, the 3.0 m thick cap settled more than the other two rigid ones (2.0 m and 2.5 m). This can be explained by the fact that thicker caps may impose additional load on the system. Figure 9 shows the results for the cap of 32 × 32 m.

It is of interest to note that the thickness of 2.0 m leads to rigid performance for the cap of 16 × 16 m, but it leads to flexible performance for the cap of 32 × 32 m. This observation indicates the effect of the cap size on its rigidity. Accordingly, in the design, the cap thickness should be thick enough to uniformly distribute the load on the plies, but thin enough for not to add additional load on the system. Therefore, optimization of the cap thickness is required to achieve a safe and economical design.

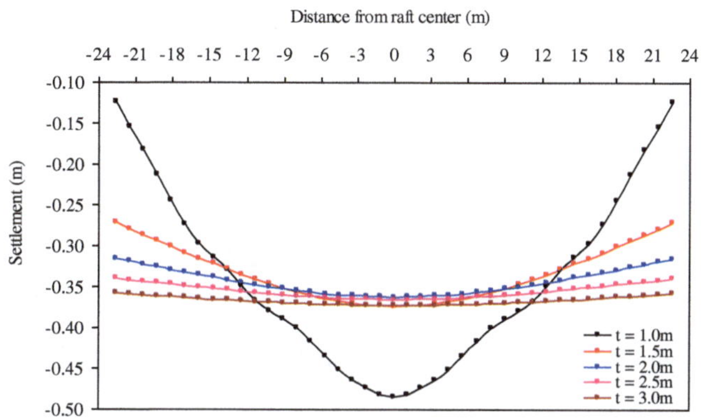

Fig. 8. Cap top settlement for various cap thickness (Cap area = 16 × 16 m).

Figure 10 illustrates the results of the settlement *versus* cap thickness. It can be seen in the figure that the settlement decreases by increasing the cap thickness up to a minimum value (i.e., 1.5 m), at which the cap becomes rigid, beyond which the settlement increases due to the own weight of the cap.

4.3 Effect of Pile Spacing

To examine the effect of the pile spacing on the rigidity of the cap, the pile spacing was tested at the range from 3D to 8D (D is the pile diameter). The reason of selecting this range is that, for pile spacing smaller than 3D the piles carry almost the entire column load, while higher than 8D the cap takes over the entire load as reported in Sinha and Hanna (2016) and Yamashita et al. (2011). The cap of 24 × 24 m was considered on this subject and the results are presented in Figs. 11 and 12 for the spacing of 4D and 8D, respectively. The results in these two figures clearly show the effects of pile spacing on the rigidity of the cap. More specifically, it can be noted in Figs. 11 and 12 that the settlement is uniform for thicker caps up to 2.5 m, where the cap performs as a rigid structure, beyond which the settlement takes parabolic shape, as the cap performs as a flexible structure.

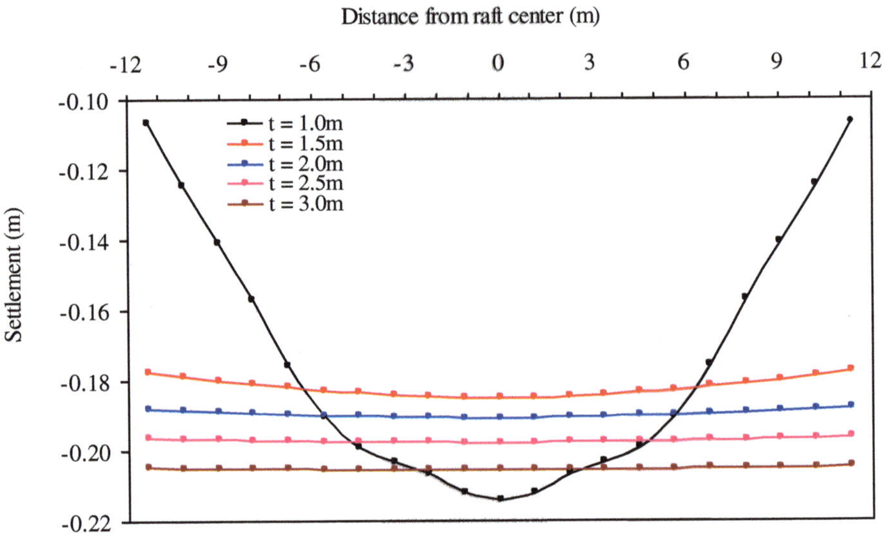

Fig. 9. Cap top settlement for various cap thickness (Cap area = 32 × 32 m).

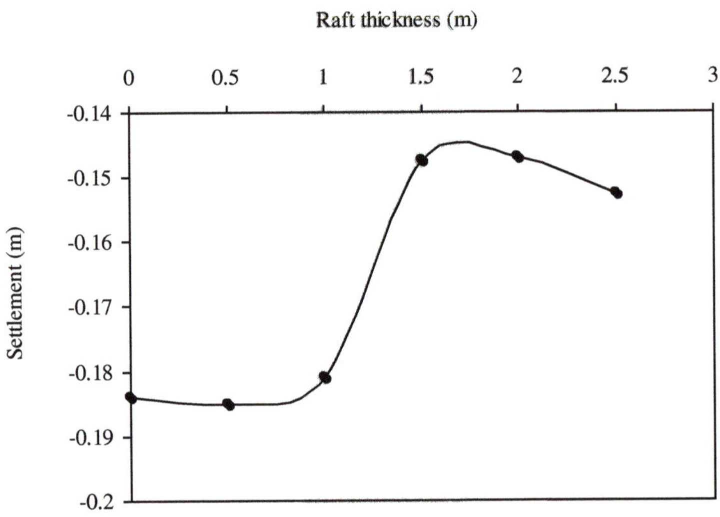

Fig. 10. Results of settlement *vs* cap thickness.

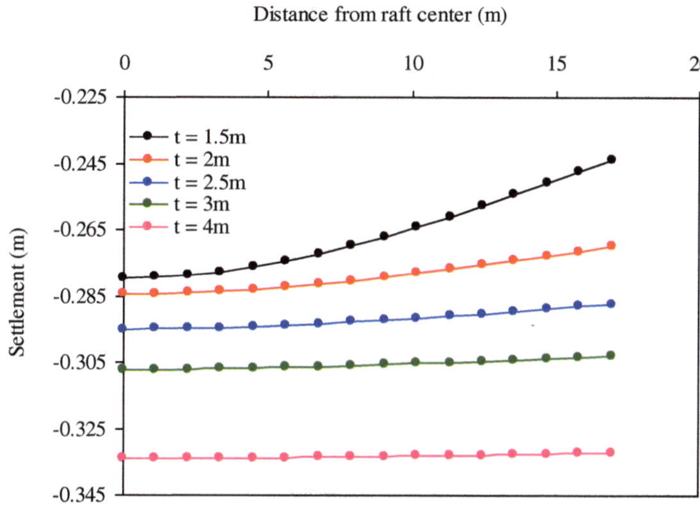

Fig. 11. Settlement along diagonal path of the cap, pile spacing 4D.

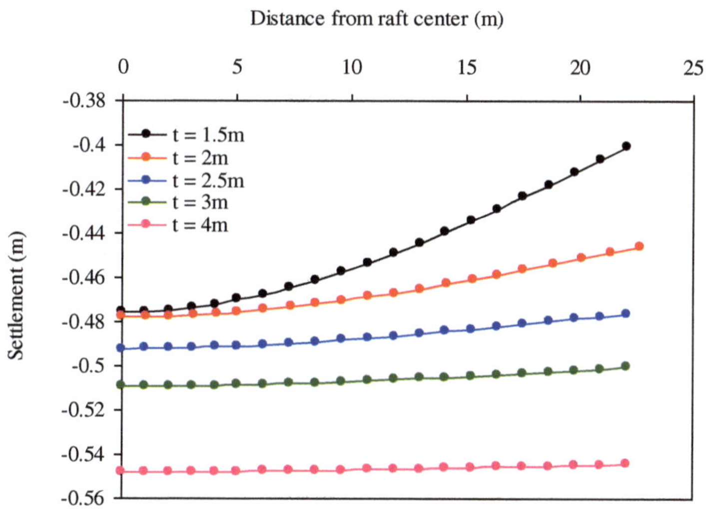

Fig. 12. Settlement along diagonal path of the cap (pile spacing 8D).

5 Conclusions

A 3-D finite element model was developed to examine the effect of pile/cap geometry on the rigidity of the cap. The modified Drucker-Prager cap plasticity model provides an excellent basis for analyzing the stress-strain of soil continuum constitutive law, and the software "ABAQUS", which is based on pure solid mechanics principles has been used successfully to simulate the problem stated. The following was concluded:

1. The cap rigidity is essential to uniformly distribute the column load on the piles within the group.
2. A thin cap will perform as a flexible structure, which will unevenly distribute the load on the piles within the group. This will cause overloading of some piles, and less to others and significant differential settlement within the cap, which further will offset the principle of pile foundation design.
3. Flexible caps settle more at the center and less at the edges, in the form of a parabolic shape of a bowl or dish, and accordingly, the cap will experience differential settlement.
4. Rigid caps will uniformly distribute the column load on the piles within the group, which will lead to uniform settlement. However, the required thickness to achieve rigidity, i.e., any increase of the cap thickness, will impose additional load on the system due to the cap's own weight and accordingly, additional settlement of the system, which were not considered during the design stage.
5. The larger the cap size, the larger the settlement for all geometrical properties, accordingly, a thicker cap will be required.

References

Adebar, P., Kuchma, D., Collins, M.P.: Strut-and-tie models for the design of pile caps: experimental study. ACI Struct. J. **87**(1), 81–92 (1990)

Ai, Z.Y., Han, J., Yan, Y.: Elastic analysis of single pile-rigid circular raft system in layered soils. Adv. Deep Found. **132**, 1–14 (2005)

Clancy, P., Randolph, M.F.: Simple design tools for piled raft foundations. Geotechnique **46**(2), 313–328 (1996)

Griffiths, D.V., Clancy, P., Randolph, M.F.: Piled raft foundation analysis by finite elements. In: Proceedings of the 7th International Conference on Computational Methods and Advances in Geomechanics, Carins, vol. 2, pp. 1153–1157 (1991)

Jiu, Y., Huang, M.: A simplified nonlinear method for pie group analysis considering pile cap flexibility. New Front. Geotech. Eng. **243**, 38–46 (2014)

Hibbit, H.D., Karlsson B.L, Sorrensen: ABAQUS Theory Manual and all Manuals, Guide, Online support (2007)

Katzenbach, R., Arslan, U., Moormann, C.: Piled Raft Foundation Projects in Germany. Design Applications of Raft Foundations. Thomas Telford, London (2000)

Katzenbach, R., Reul, O.: Design and performance of piled rafts. In: Proceedings of XIV ICSMFE, Hamburg, vol. 4, pp. 2253–2256 (1997)

Kim, K.N., Lee, S.H., Kim, S.K., Chung, C.K., Kim, M.M., Lee, H.S.: Optimal pile arrangement for minimizing differential settlements in piled raft foundations. Comput. Geotech. **28**, 235–253 (2001)

Poulos, H.G.: Piled raft and compensated piled raft foundations for soft soil sites. Adv. Des. Test. Deep Found. **129**, 214–234 (2005)

Poulos, H.G.: Methods of analysis of piled raft foundations, TC18 Report. International Society of Soil Mechanics and Geotechnical Engineering (2001)

Randolph, M.F.: Design of piled raft foundations. In: Proceedings of International Symposium on Recent Developments in Laboratory and Field Tests and Analysis of Geotechnical Problems, Bangkok, pp. 525–537 (1983)

Sabnis, G.M., Gogate, A.B.: Investigation of thick slab (pile cap behavior). J. ACI **8**(1), 35–39 (1984)

Sinha, A., Hanna, A.: 3-D numerical model for piled-raft foundation. Int. J. Geomech. (2016). https://doi.org/10.1061/(ASCE)GM.1943-5622.0000674

Small, J.C., Zhang, H.H.: Behaviour of piled raft foundations under lateral and vertical loading. Int. J. Geomech. **2**(1), 29–45 (2002). https://doi.org/10.1061/(ASCE)1532-3641

Yamashita, K., Yamada, T., Hamada, J.: Investigation of settlement and load sharing on piled rafts by monitoring full-scale structures. Soils Found. **51**(3), 513–532 (2011)

Geotechnical Issues of Testing Piles on Construction Site of Astana

A. S. Tulebekova[1(✉)], A. Zh. Zhussupbekov[1], T. Mussabayev[2], and S. Mussina[2]

[1] Department of Civil Engineering, L.N. Gumilyov Eurasian National University, Astana, Kazakhstan
krasavka5@mail.ru

[2] Eurasian Technological Institute, L.N. Gumilyov Eurasian National University, Astana, Kazakhstan

Abstract. Pile foundations become more essential during construction of megaprojects of new capital of Kazakhstan – Astana. Many advanced pile technologies are appearing today. Unfortunately Kazakhstan Standards have not recommendation for modern pile design, testing and quality control methods. Therefore some aspects of modern pile design concept of Kazakhstan construction is presented in this paper, as well as contemporary approach to the testing methods and monitoring. The papers include the results of static tests of piles by using different standards have some differences. Discussion of using control equipment, technological features, advantages and disadvantages of aforementioned methodic might be important for understanding the difference points. Also given recommendation for the future modernization of Kazakhstan Standards.

1 Introduction

At present time pile foundations are widely used for problematic soil grounds of structures. Such problematic soil grounds in North Kazakhstan, from 1998 were known new capital city Astana. Construction development of the capital city is important for the country which its aim is to be among of developed countries. In this direction of our country foreign investors use their techniques and technology in big project constructions and constructions of high rise buildings and structures.

One of the topical problems of constructions from the point of economy effective designing and laying foundations in the difficult soil conditions. The big attention is given to the increase of labor productivity, quality and reliability of erected structures. With proper planning and design, efficient equipment and experienced personnel, high production rates can be achieved.

Unfortunately, present Standards are confined application of modern technology of pile foundation installation, indicating incomplete usage of advanced technology. The results of research will directed to modernization of the Kazakhstan Standards and oriented to advanced geotechnologies adaptation.

H. El-Naggar et al. (Eds.): GeoMEast 2018, SUCI, pp. 116–124, 2019.
https://doi.org/10.1007/978-3-030-01902-0_11

2 Methodologies of the Field Tests Dynamic Load Test Regarding GOST 5686-94 (Kazakhstan)

In Kazakhstan dynamic test is carry out by using many types of pile driving machine. Before pile driving special marks through 1 m are painted on its surface along the whole length of the pile, but on last meter through each 0.1 m.

In process of the pile driving the numbers of blows is recording for every each meter of pile penetration and last one meter through each 0.1 m. At the same time it is necessary to fix height of the fall of the blowing part of hammer. Test pile driving terminated till designing refusals (cm/blows). For definition of the bearing capacities of piles it is required to use average refusal which are obtained during redriving of the piles after their "rest". The rest time depend on soil condition of site: for clayey soil 6–10 days, for sandy and gravel soil up to 3 days. Redriving of test piles carry out two step-by-step guarantees at three and five blows of the hammer. Before beginning redriving on test piles attached measured tape, with lengths of 10 cm, divided by 1 mm. The control of the test pile penetration conducted through theodolit.

3 Static Load Test Regarding GOST 5686-94 (Kazakhstan)

Static load test should be carry out for driving piles after the "rest" and for bored piles after achievements of the concrete strength more than 80%. For static load tests the following equipment is used:

- hydraulic jack SMJ-158A - 200 ton;
- caving in-measurers of the type 6PAO.

The pressure in the jack was created by the help of manual oil pump station MNSR-400 with power up to 800 kg/cm^2, the moving of steel piles was fixed by caving in-measurers of the type 6-PAO, which were installed on the both sides of unmovable bearings with the benchmark system.

The first record was performed just after putting the loading, then consequently 4 records with an interval of 15 min, 2 records with an interval of 30 min and further for every hour until the conditional stabilization of pile settlement. For the criterion of conditional stabilization of pie settlement was taken when the speed of settlement of piles on the given stage of loading did not exceed 0.1 mm during the last 1–2 h of observations. Reloading (unloading) conducted half stages of the loading.

4 Dynamic and Static Load Test on Construction Site in Astana by American and Kazakhstan Standards

Dynamic static test provided for nine steel testing pile regarding project. Steel pile length is 12 m.

Fig. 1. Dynamic pile test with driving machine Junttan PM-20, hydraulic hammer HHK-5A

Dynamic pile tested by pile driving machine Junttan PM-20 showed in Fig. 1 with hydraulic hammer HHK-5A with the following characteristics:

- weight of hydraulic hammer – 8755 kg;
- mass of the hammer – 5000 kg;
- mass of pile cap – 835 kg.

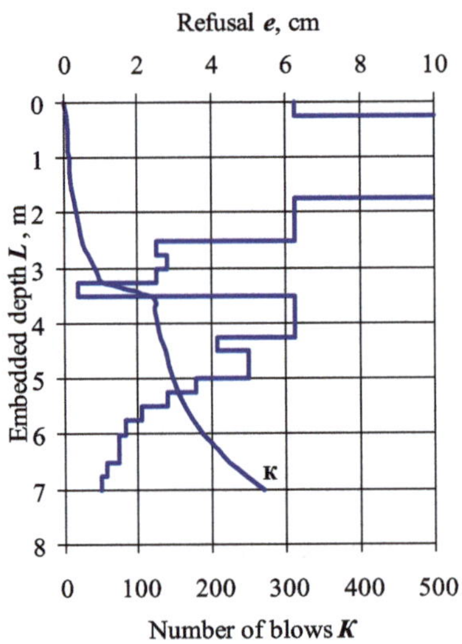

Fig. 2. Graphics before heightening of pile length for testing pile LT-1

Steel piles marked with paint marks every 25 cm on all pile length. Dynamic pile test shown in Fig. 2. Penetration of steel pile as follows:

- for design load of testing pile equal 600 kN, refusal of pile equal 1.25 cm;
- for design load of testing pile equal 400 kN, refusal of pile equal 1.67 cm;

Results of dynamic pile test shown in Table 1 and Figs. 2, 3, 4 and 5.

Static tests of steel piles showed negative results: refusal of pile 1.25 cm for 600 kN and failure of pile 1.67 cm for 400 kN.

So steel piles length heightened as follows:

- Refusal is 0.33 cm for design load for pile equal 600 kN.
- Refusal is 0.83 cm for design load for pile equal 400 kN.

Table 1. Results of driving of steel piles before heightening of pile length

Number of pile	Embedded depth in soil, m	Refusal of pile at driving, cm	Project refusal of pile, cm	Design load, kN
LT-1	7.00	1.00	1.25	600
LT-2	9.00	1.09	1.25	600
LT-3	10.00	1.14	1.25	600
LT-4	9.25	1.25	1.25	600
LT-5	8.25	1.00	1.25	600
LT-6	10.25	1.25	1.25	600
LT-7	8.00	1.56	1.67	400
LT-8	8.25	1.47	1.67	400
LT-9	7.75	1.67	1.67	400

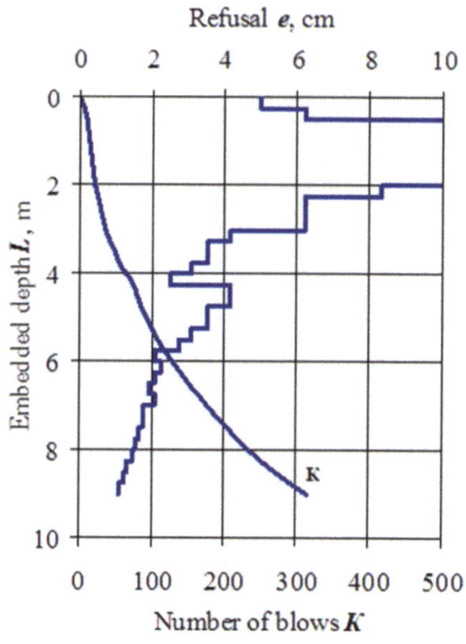

Fig. 3. Graphics before heightening of pile length for testing pile LT-2

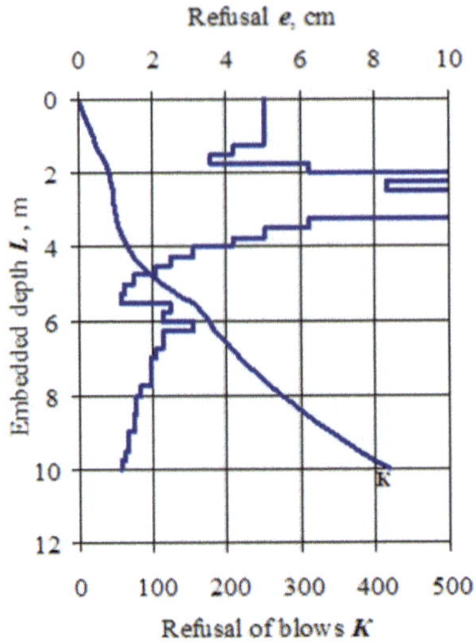

Fig. 4. Graphics before heightening of pile length for testing pile LT-3

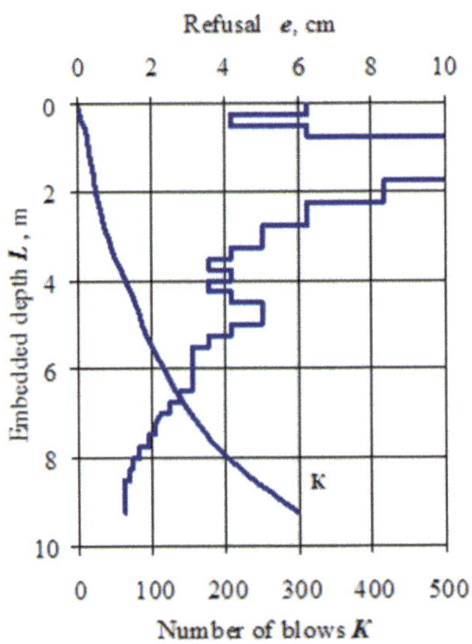

Fig. 5. Graphics before heightening of pile length for testing pile LT-4

Results of static pile tests after heightening of pile length presented in Table 2. A static pile test of steel H-piles was carried out after the “rest” of pile after driving. Results of static test of steel piles before heightening of pile length shown in Table 3 and Fig. 6. Results of static test of steel piles after heightening of pile length shown in Table 4.

Table 2. Results of driving of steel piles after heightening of pile length

Number of pile	Embedded depth before heightening pile length, m	Embedded depth of after heightening of pile length, m	Refusal of pile at driving, cm	Project refusal of pile, cm	Design load, kN
LT-1	7.00	9.75	0.31	0.33	600
LT-2	9.00	15.00	0.26	0.33	600
LT-3	10.00	13.00	0.32	0.33	600
LT-4	9.25	12.75	0.27	0.33	600
LT-5	8.25	11.00	0.27	0.33	600
LT-6	10.25	16.00	0.30	0.33	600
LT-7	8.00	11.25	0.78	0.83	400
LT-8	8.25	11.25	0.74	0.83	400
LT-9	7.75	9.75	0.63	0.83	400

Table 3. Results of static test of steel piles before heightening of pile length

Number of pile	LT-1	LT-4	LT-6	LT-7
Embedded depth, m	7.00	9.25	10.25	8.00
Refusal of pile at driving, cm	1.00	1.25	1.25	1.56
Settlement, mm	43.03	42.55	40.25	40.88
Design load, kN	600	600	600	400
Applied load, kN	900	900	900	600
Max. load, kN	1200	1200	1200	8000

5 Comparative Analyses

Technical specificity of ASTM D1143/D1143M-07 has some differences with national Kazakhstan standard GOST 5686-94. Some devices and control equipment didn't use by requirement of GOST 5686-94 "Soils. Field test". Principal differences between American Standard and Kazakhstan norms showed in Table 5.

In practice requirements by American Standard showed that results of test more safety [1]. Because use many control equipment and devices for determination, measurement reverberated.

Also GOST regulated only two out of six measurements with reducers provided by ASTM. By requirement of both standards loading on pile transferred by jack. But GOST does not consider that each jack need to provide by manometer if we used more than one jack. It's important for control of work of jack and to do correct test [1].

In other case sometimes reducers of axial displacement of pile showed undirected displacement and regarding requirement of GOST isn't instructions. In this case regarding ASTM need to fix reducers of sway.

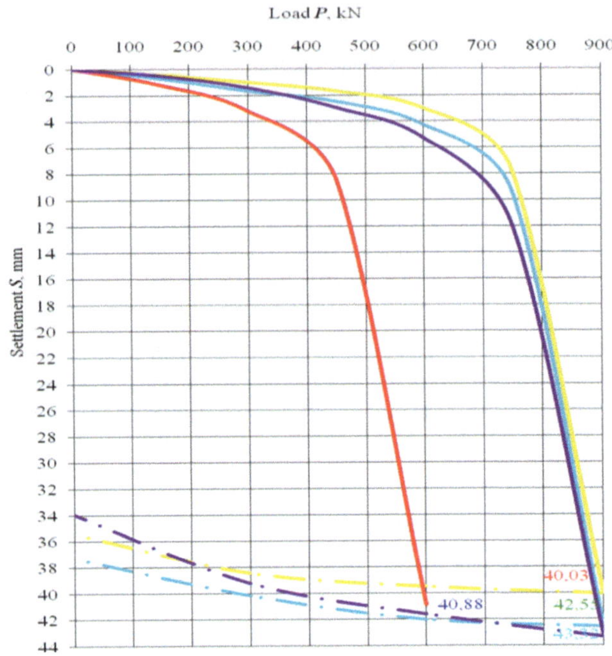

Testing pile LT-1, driving depth L=7,0 m;

Testing pile LT-4, driving depth L=9,25 m;

Testing pile LT-6, driving depth L=10,25 m;

Testing pile LT-7, driving depth L=8,0 m.

Fig. 6. Graphics of dependence of settlement S from load P regarding results of static pile tests before heightening of pile length (by ASTM D/D 1143M-07)

Transducer of measurement of axial displacement serve for measure the movement of the bottom of the pile on its top. This measurement more exactly determined settlement of soil.

For the purposes, ASTM proposes to use a system of measurement of the indicator displacement, installed on top of the pile.

Force transducer to measure the lateral resistance pile used to measure the friction force on his side the entire length of the pile. Thus, we can determine the distribution of the lateral resistance of piles.

Table 4. Results of static test of steel piles after heightening of pile length by GOST 5686-94

Number of pile	Embedded depth, m	Driving depth, m	Refusal of pile at driving, after heightening pile, cm	Settlement, mm	Design load, kN
LT-1	7.00	9.75	0.31	4.80	600
LT-2	9.00	15.00	0.26	7.34	600
LT-3	10.00	13.00	0.32	7.00	600
LT-4	9.25	12.75	0.27	4.96	600
LT-5	8.25	11.00	0.27	4.42	600
LT-6	10.25	16.00	0.30	6.27	600
LT-7	8.00	11.25	0.78	3.38	400
LT-8	8.25	11.25	0.74	2.32	400
LT-9	7.75	9.75	0.63	3.17	400

Table 5. Principal differences between American Standard and Kazakhstan norms

	GOST	ASTM
Parameter of experimental stand for test		
Distance between testing pile till anchoring pile	5d < L_1 > 2.5m	3d < L_1 > 1.5m
Distance between testing pile till	5d < L_2 > 2.5m	L_2 < 2m
Devices and equipment		
For loading	Jack	Jack with Spherical prop
Measurement of load on top pile	Manometer	Manometer
	-	Dynamometer (more 100 tc) fixed for each jack
Measurement of load on all length pile	-	Tensometer
Measurement of axial displacement of top pile	Transducer of axial displacement of cap	Transducer of axial displacement of cap
	-	Visual control
	-	Optical instrumental control
Measurement of sway top pile	-	Transducer of sway top pile
	-	Optical control

6 Conclusion

Nowadays many international projects are realized in Kazakhstan. For example Italian and American companies which are work on western part of Kazakhstan met some complexity in non-correspondences of Kazakhstan codes to international. Kazakhstan engineers also feel some complexity during design modern projects where generally used advanced technologies.

American Standards seems to be more reliable for adaptation for Kazakhstan construction condition. ASTM showed new improvement in pile tests. Method of pile testing regarding Kazakhstan Standard comparison with American Standard it is becomes obvious disadvantage.

Results shown that ASTM methodic gives detailed information about the process testing and makes more reliable results. Technical specificity of ASTM D/D 1143M-07 has some differences with national Kazakhstan standard GOST 5686-94 «Soils. Field test methods by piles».

References

1. GOST 5686-94 Soils. Field test methods by piles
2. ASTM D 4945-00 Standard Test Method for High Strain Dynamic Testing of Piles
3. Smolin, B.S.: Features pile load test. In: ASTM Proceedings of International Symposium, Russia (2010)

An Appraisal of Pile Capacity Interpretation Methods Using Model Piles

Abideen Adekunle Ganiyu[1(✉)], Ahmad Safuan A. Rashid[2], Mohd Hanim Osman[2], and Atef Badr[1]

[1] Department of Civil Engineering and Quantity Surveying, Military Technological College, Muscat, Oman
abideen.ganiyu@mtc.edu.om
[2] Faculty of Civil Engineering, Universiti Teknologi Malaysia, Skudai, Johor, Malaysia

Abstract. An appraisal of pile load test interpretation methods for their suitability in unit- gravity model test was carried out in this research. A replicate of the full scale Constant Rate of Penetration (CRP) pile load capacity test was simulated in the laboratory using model piles and reconstituted Kaolin S300 soil model. Eight model tests were carried out on long and short model piles. The measured capacities, Qm of the piles were obtained from the load-settlement curve using six different interpretation methods and were compared with the predicted capacity, Qp calculated using pile static formula. The results revealed that Hansen 80% and Chin methods produced the best results for the model tests relative to the predicted capacity, Qp of the model piles. In contrast, the 10% of pile diameter displacement method yielded the lowest capacity values for the model piles. This implies that the capacity obtained by this method supposed to be used as a baseline value and not necessarily the actual capacity value for the piles. This study will serve as a yardstick for future researchers on model piles for their choice of suitable model pile load test interpretation method.

1 Introduction

The installation of piles to buttress structures is a foremost example of the art and science of the civil engineer. The art is the search for the most appropriate type for specific soil and load conditions; while the science is the prediction of the pile behaviour after installation. The primary role of piles is to transmit loads from superstructures via less robust soil layers or water onto a more robust soil or rock (Tomlinson and Woodward 2014). The loads may be horizontal, axial, or uplift in monotonic, inclined or cyclic forms. Piles are also used to limit settlements in spread footings, compact loose cohesionless soils, or stiffen soils underlying structures that are subjected to vibrations; the latest use of piles is within ground source heating and cooling systems, as an innovative renewable energy development strategy (Caulk et al. 2016).

The capacity of a pile is influenced by different parameters such as the pile type, installation methods, soil properties, type of load application and also its behaviour during its lifetime (Li et al. 2017). Due to the empirical nature of some of the pile design methods and the influence of variation in soil parameters; the resistance of a

H. El-Naggar et al. (Eds.): GeoMEast 2018, SUCI, pp. 125–140, 2019.
https://doi.org/10.1007/978-3-030-01902-0_12

pile, calculated based on different methods may differ significantly even when the methods follow the same analytical concept. Thus the main challenge associated with pile foundation design is the ability to accurately predict and verify pile capacity (Eslami et al. 2014). Pile capacity is defined as "the load for which rapid movement occurs under sustained or slight increase of the load". It is the total ultimate soil resistance obtained from the measured load-displacement behaviour of the pile (Fellenius 2001).

The capacity of piles is commonly verified through full scale axial compression load test. The test can be for the design load or proof load, and the test type can be static, dynamic or statnamic. The principal goal of pile testing is to validate the design before construction and to confirm compliance with the specifications during construction. However, in some other cases testing is used for design development in order to identify optimum solutions. The constant rate of penetration (CRP) and maintain load test (MLT) are the basic methods for the static load test. The constant rate of penetration, CRP, as a form of static test is basically aimed at determining ultimate load on a pile, and mainly used for preliminary test piles and for research purposes (Brown 2012; Nazir et al. 2015).

However, executing full-scale load test for research purposes is highly demanding and costly. In-situ tests, like standard penetration test (SPT), pressure meter test, dilatometer test and the cone penetration test (CPT) are generally used to determine the bearing capacity of piles. The CPT is prevalent because its apparatus closely simulate a pile, the measurements are continuously taken over depths thereby giving more credibility; further, there is flexibility in the choice of its probe sensors, and it can be carried out both expeditiously and economically (Baziar et al. 2014).

In addition, reduced scale tests that duplicate the actual geotechnical structure in some proportions are increasingly being adopted as alternatives to full-scale tests. This is because Geotechnical engineers depend on reduced-scale physical modelling to assess the performance or to verify the behaviour of geotechnical systems. Moreover, physical modelling can offer an attractive alternative to a wider understanding of the behaviour and mechanisms behind complex soil-structure interaction problems (Chung and Yang 2016). Meanwhile, physical modelling can be done at elevated gravity level in a centrifuge, calibration chamber (CC), frustum confining vessel (FCV); or directly on the laboratory floor i.e. unit-gravity or 1-g testing (Zare and Eslami 2014).

Unit-gravity test can be an efficient and moderate cost alternative to performing geotechnical tests under well-defined conditions. Unit-gravity models clearly depict complex, nonlinear geotechnical mechanisms and phenomena that are otherwise difficult to visualize; they have often served as basis for the formulation or justification of design approaches (Madabhushi et al. 2010). An inherent weakness of small scale 1-g physical models is the lack of similarity of stress profile between the model and prototype. However, this effect is less problematic when modelling with fine grained soil such as clay, since the soil strength is not stress dependent under undrained deformation. The dissimilarity of stress profile within the model then only impacts upon the apparent self-weight of the soil within the model (Ozkahriman and Wartman 2007).

While pile behaviour is considerably established by load tests; construction issues, geotechnical factors, and other unavoidable ambiguities are witnessed during pile load tests. These result into substantial deviations in the behaviour of pile to the applied load. Therefore, load-settlement curves of comparative variations often appear and the maximum resistance of piles is not distinctively established. Hence, "the capacity needs to be defined by an 'interpreted failure load' (Hirany and Kulhawy 2002)". Conventionally, the bearing capacity of a pile is defined as the load corresponding to a settlement equal to 10% of pile diameter (Jabatan Kerja Raya Malaysia 2005; British Standards Institution 2004). However, a number of other methods have been recommended for the interpretation of pile capacity from load test results. The capacity values obtained from these methods can vary appreciably (Chen and Fang 2009).

The Institution of Civil Engineers, ICE specification for piling (Institution of Civil Engineers 2007) specifies 15% of pile diameter settlement among the "failure criteria". Seven methods were presented by each of Tomlinson and Woodward (2014) and Fellenius (2014) for the interpretation of static load test results. Marcos et al. (2013) utilised eight different interpretation methods to analyse a database of 152 field compression load tests. Wang et al. (2011), Abdelrahman et al. (2003) and Lastiasih and Sidi (2014) also presented six different interpretation methods each; while Titi and Elias (2010) employed nine interpretation methods. Ng et al. (2001) gave a lucid description of most of the available methods and employed thirteen methods in their study. It is pertinent to mention that the above mentioned studies utilised different database of full scale pile load tests to obtain the data employed in their analysis. The databases are somewhat distinct with different soil parameter values; they exhibited diverse elucidation of similar tests thereby amplifying the variations in their outputs (Karlsrud 2014).

In this research, unit-gravity model tests were employed to simulate Constant Rate of Penetration (CRP) load capacity test on model piles using reconstituted Kaolin S300 soil model. Eight model tests were carried out on long and short model piles. From the load – displacement data obtained from the tests, the measured capacity values (Q_m) were obtained from six interpretation methods, and were compared with the predicted capacity (Q_p) calculated using static pile design formula. An appraisal of the selected interpretation methods were made with a view to identifying their suitability for evaluating model pile load tests.

2 Methodology

2.1 Materials

Kaolin Clay

The Kaolin S300 that was used in this research is the commercially manufactured powdered kaolin clay supplied by Kaolin (Malaysia) Sdn Bhd. Kaolin, which is chemically known as hydrated Aluminium Silicate is structurally unmodified, platy in structure, white in colour, hydrophilic, readily water dispersible with a pH range between 3.5 and 6.0 and loss on ignition of 2–6% at 1025 °C (Kaolin (Malaysia) Sdn. Bhd. 2014). The choice of a manufactured clay is premised on its ability to lessen variability and enhance repeatability of the results. Table 1 shows the chemical composition of this product obtained by XRF method as supplied by the manufacturer (Kaolin (Malaysia) Sdn. Bhd. 2014).

Table 1. Chemical composition of Kaolin (S300)

Element	Aluminium	Silica	Iron	Potassium	Magnesium
Oxide formula	Al_2O_3	SiO_2	Fe_2O_3	K_2O	MgO
Composition	7–12%	80–90%	<1%	<2%	<0.5%

Model Piles
Eight stainless steel moulds; four each of lengths 100 mm and 150 mm with overall cross sectional area of 452.4 mm^2 were fabricated. A pair each was made in circular, square, hexagonal and octagonal shapes (Fig. 1). The model piles were cast in the moulds with mortar in the laboratory according to BS EN 998-2. Ordinary Portland Cement (OPC) was used to make the mortar, and the fine aggregate was the naturally occurring river sand. A mix ratio of 1:3 for cement: sand; and a water: cement (w/c) ratio of 0.45 was used. The castings were thoroughly vibrated by tamping to prevent any honeycomb, and were thereafter smoothed with a spatula. The castings were demoulded after 24 h and cured in water for 28 days. Table 2 presents the detailed geometrical properties of the model piles. The lengths corresponded to short and long piles based on Brom's classification (l/d < 5 and l/d > 5 accordingly).

Fig. 1. Stainless steel moulds for casting the model piles

Table 2. Geometrical properties of model piles

S/N	Code	Shape	L	D	L/D
1	SP2	Square	150	21.27	7.05
2	HP2	Hexagonal	150	22.86	6.56
3	OP2	Octagonal	150	23.36	6.42
4	CP2	Circular	150	24.00	6.25
5	SP1	Square	100	21.27	4.70
6	HP1	Hexagonal	100	22.86	4.37
7	OP1	Octagonal	100	23.36	4.28
8	CP1	Circular	100	24.00	4.17

Where, 'L' is the vertical length, and 'D' is the diameter of the piles. (All dimensions are in mm).

2.2 Testing Equipment

Testing Chamber
The testing chamber is a rigid box with dimensions (320 × 150 × 430) mm length, width and height respectively. The front face of the chamber is a removable 20 mm thick rigid Perspex panel with dimensions (320 × 300) mm length, and height respectively. The Perspex panel facilitates real time view of the testing while the remaining three sides were made from Aluminium panels. All the sides were fixed tightly using 'M8' bolts to ensure lateral rigidity. Four steel rods 20 mm in diameter were attached to the chamber's base plate and were used to mount the loading plate. The loading plate and the base plate were also made of Aluminium with dimensions 660 mm by 300 mm in area and 20 mm thickness. The driving unit was fixed on the loading plate.

Driving Unit
A driving system that generates a constant rate of penetration of the model pile into the soil was provided for the loading test. The drive shaft was made from a solid stainless steel rod. The rod is 260 mm long with a diameter of 23.75 mm. Four sets of ball bearings were assembled to support the gear that rotated the drive shaft. This gear was housed between two stainless steel disks. A stopper was installed inside the drive shaft to prevent the rod from rotating during the settlement process. At the lower end of the shaft is the shaft connector which connects the shaft with the load cell and the round tipped adaptor. The shaft connector is 50 mm in diameter and 50 mm long, while the round tipped adaptor is 120 mm long with a diameter of 20 mm. A two-way AC motor was used to rotate the central drive shaft via a rubber belt, and a speed controller was connected to the AC motor as the power supply. The controller has variable speed control, and a two-way switch that could change the movement of the shaft; either upwards or downwards. The rate of drive unit penetration was 50 mm/min as obtained in the calibration test. Figure 2 shows a schematic diagram of the testing chamber equipped with the driving unit.

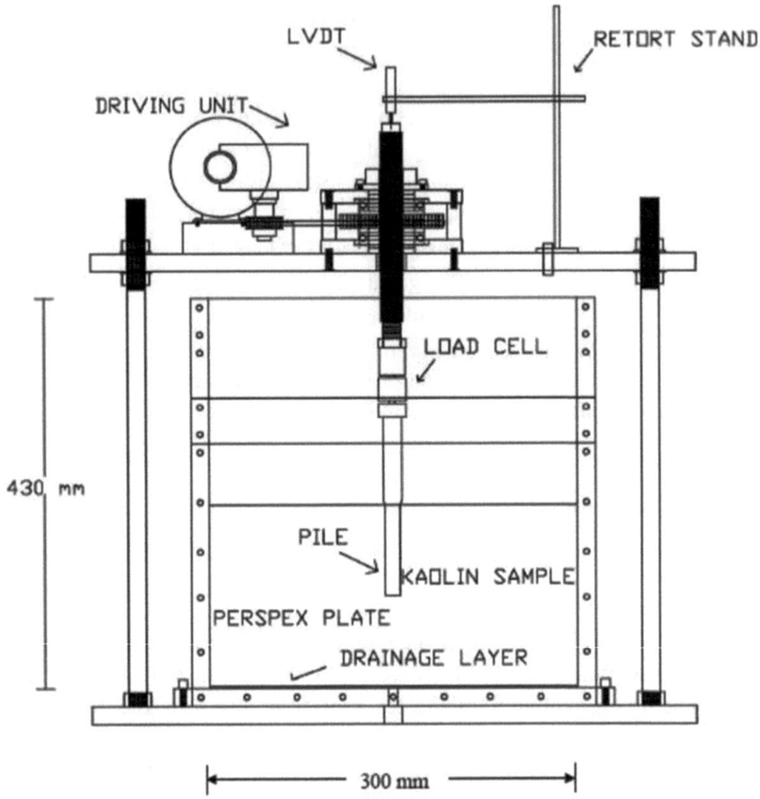

Fig. 2. A schematic diagram of the testing chamber equipped with the driving unit

Transducers and Data Acquisition System
A 25 mm stroke length Linear Variable Displacement Transducer (LVDT) was utilised for measuring the displacement of the model piles during the loading stage. Additionally, a 2.0 kN load cell transducer was employed to measure the force during the loading. Both the LVDT and load cell were connected to the CR800 data logging device and the input signals were recorded on a computer connected to the data logger. For the logging software, 'Loggernet 4.0" was installed on the desktop and used for the data acquisition (Load Cell and LVDT).

2.3 Kaolin S300 Soil Model

Kaolin S300 clay was initially kept in the oven at 105 °C for 24 h to prevent any biological action. For each test, 14 kg of the air cooled Kaolin S300 sample was weighed in the mixing tray, 2.1 kg of water was intermittently added and the sample was thoroughly mixed using hand trowels until a uniform consistency was attained. The moisture contents of the soil model was designed to be 15% and this was confirmed from the results of the moisture content tests carried out on the sample for all the eight test models. A porous plastic (320 × 150 × 5) mm was placed at the bottom of the testing chamber to prevent soil particles from leaking from the testing chamber. After this, the sample was added in 5 layers of comparable mass, each layer was compacted with 25 blows of hammer; the wooden hammer has (100 × 60 × 200) mm head, (40 × 20 × 300) mm handle, and weighs 1.17 kg. The spatula was used to create uneven surface in between the layers. Another porous plastic of the same dimension was placed over the final compacted soil layer to ensure a flat surface. After this, a hole of 24 mm in diameter was made at the centre of this porous plastic and this was marked as the position for the installation of the pile, the top porous plastic was removed afterwards. The model pile was then driven by the wooden hammer through the marked position into the soil to reach the same level with the top of the soil model. Figure 3 shows one of the piles (hexagon) during installation into the soil model. The average undrained shear strength cu of the soil models assessed at the end of each of the tests using hand-held Vane Shear was 60 kPa. The soil model is therefore a stiff clay based on the value of cu obtained.

Fig. 3. Hexagonal pile during installation

2.4 Model Pile Load Test

The driving unit was thereafter mounted on the soil model; the load cell and the round tipped adaptor were placed at the end of the driveshaft, while the LVDT was mounted on top of the driving shaft by attaching it to a retort stand. Then, the drive shaft was carefully lowered until the tip of adaptor came into contact with the top of the pile.

To commence the load test, the driving unit was released to allow its downward movement and subsequent driving of the pile into the soil model. The values of the loads and settlements of the pile during the test were recorded by the logging system. The test was terminated when the penetration of the pile reached approximately 20 mm depth; a point at which the pile must have failed based on all available failure calculation methods. The process was repeated for each of the eight model piles to replicate full scale pile load test using the Constant Rate of Penetration (CRP) method. Figure 4 shows the typical set-up for the load tests, while Fig. 5 shows the details of the driving shaft, shaft connector, load cell, and round tipped adaptor during the loading test of the hexagonal pile. Also, Fig. 6 shows the position of the hexagonal pile, round tipped adaptor, load cell and shaft connector after the loading test.

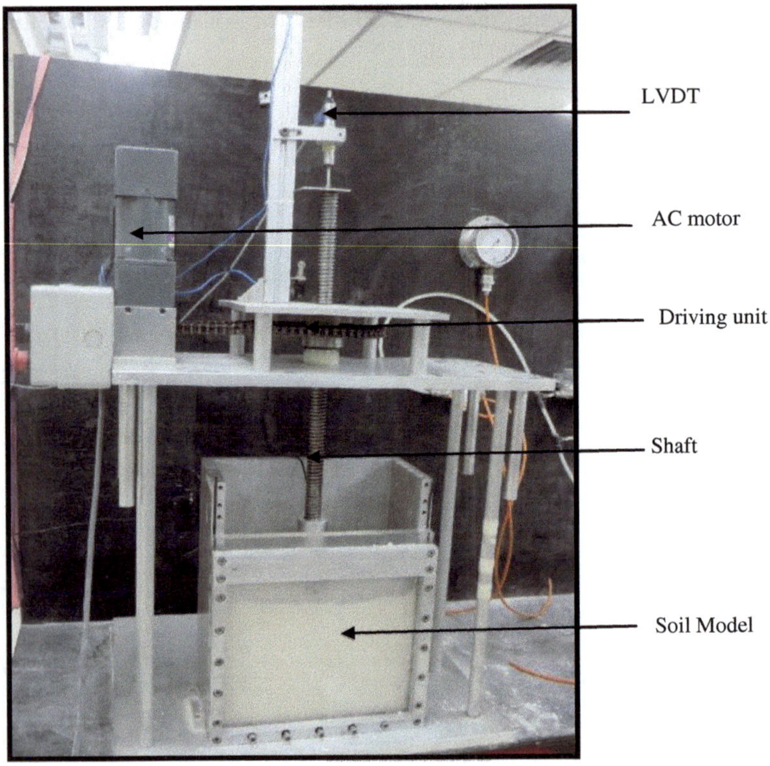

Fig. 4. Typical set-up for the load test

2.5 Failure Load Determination Methods

In order to obtain the measured capacities (Qm) of the piles, six of the pile load test interpretation methods were used. The methods are briefly highlighted below:

(i) 10% of pile diameter settlement: This is the failure load that corresponds to 10% of the pile diameter settlement. In this research the diameter of the circular piles

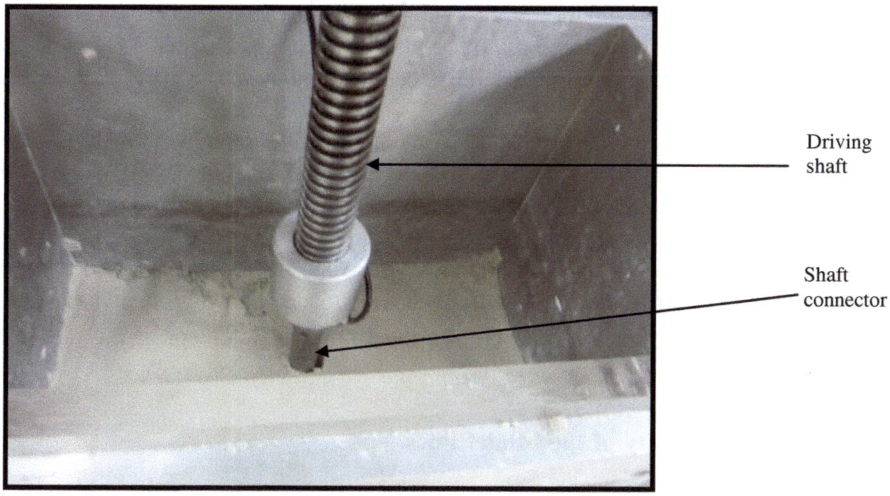

Fig. 5. Pile load test in progress

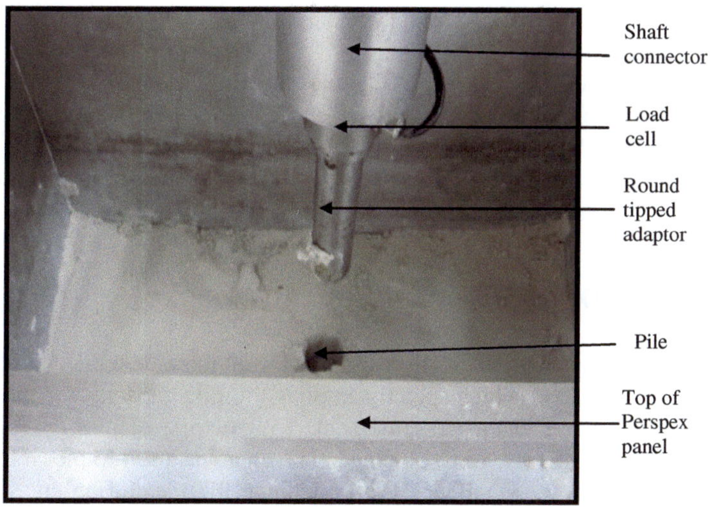

Fig. 6. Pile displacement after load test

is 24 mm, and this value is utilised for other shapes for consistency. Hence, the failure load was measured at a settlement of 2.4 mm (Jabatan Kerja Raya Malaysia 2005; British Standards Institution 2004).

(ii) 15% of pile diameter settlement: This is similar to the above, 15% of pile diameter was determined at a settlement of 3.6 mm (Brown 2012; Institution of Civil Engineers 2007).

(iii) 6 mm settlement: This was measured at a settlement of 6 mm (Tomlinson and Woodward 2014).

(iv) DeBeer (1970): It is the load at the change in slope on a log load - log settlement curve. It was obtained at the intersection of two line approximations connecting the data at before and after the failure load (Fellenius 2001, 2014; Marcos et al. 2013).

(v) Chin (1970): It is the inverse slope (1/m) of a line s/p = ms + c, where p is load and s is total settlement. It was obtained by dividing each settlement by its corresponding load and plotting the resulting value against the settlement (Fellenius 2001, 2014; Marcos et al. 2013).

(vi) Hansen 80% (1963): A plot of the square root of settlement divided by its load value against the settlement that was made. The ultimate load equals $\{1/(2\sqrt{C1C2})\}$ where C1 and C2 are the slope and intercept of the plot (Fellenius 2001, 2014; Ng et al. 2001).

2.6 Predicted Capacity of the Model Piles

The predicted capacity (Q_p) of the piles were computed using the static formula (Eq. 1). The static formula is the basic approach to the design of piles based on soil mechanics principles (Dithinde et al. 2010).

$$Q_{ult} = A_b(c_u N_c) + (\alpha c_u)A_s \tag{1}$$

where, Q_{ult} is ultimate pile capacity; A_b is base area; A_s is shaft area; N_c is bearing capacity factor; c_u is undrained shear strength and α is adhesion factor. In this research, $\alpha = 1 - 0.0011(c_u - 25)$ for $25 \leq c_u \leq 70$ was adopted (Viggiani et al. 2012), N_c was taken as 9 and a factor of safety (FOS) of 2 was adopted, the predicted capacity, $Q_p = Q_{ult}/\text{FOS}$.

3 Results and Discussions

Figure 7 shows the load-settlement curve obtained from the model pile load tests. From the figure, the failure load which directly implies the pile capacity for the first three methods i.e. 10% of pile diameter settlement, 15% of pile diameter settlement and 6 mm settlement respectively, were obtained. The data was used to plot the curves for the last three methods i.e. DeBeer (1970), Chin (1970) and Hansen 80% (1963) methods, respectively. Figures 8, 9 and 10 show the curves for the latter methods, respectively. Table 3 shows the predicted capacity Q_p obtained from the static formula and the measured capacity Q_m obtained for each of the model piles based on the selected methods, as derived from Figs. 7, 8, 9 and 10.

Table 4 depicts the values of the ratio (Q_m/Q_p) in percentage for the model plain piles. Q_m is the measured capacity of the piles based on the selected interpretation methods, while Q_p is the predicted capacity of the piles, based on Eq. 1. Also tabulated are the mean Q_m/Q_p, the standard deviation (SD) and the coefficient of variation (CoV) obtained for the Q_m/Q_p ratio of the piles for each of the selected interpretation methods. The ratio Q_m/Q_p has been utilised in previous researches (Dithinde et al. 2010, Lehane et al. 2013) to compare the capacities of pile based on full scale load tests.

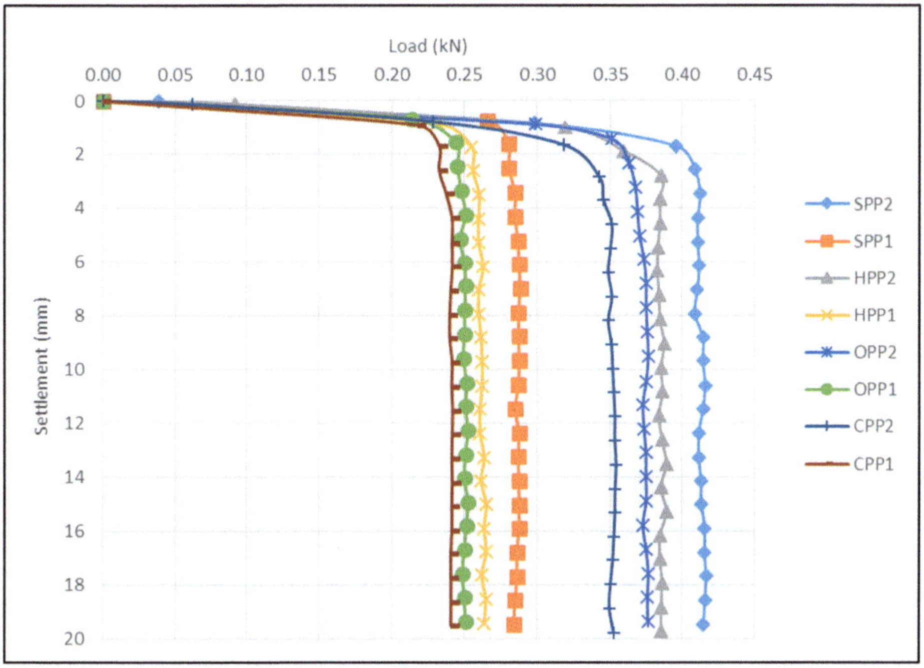

Fig. 7. Load-settlement curve for the model tests

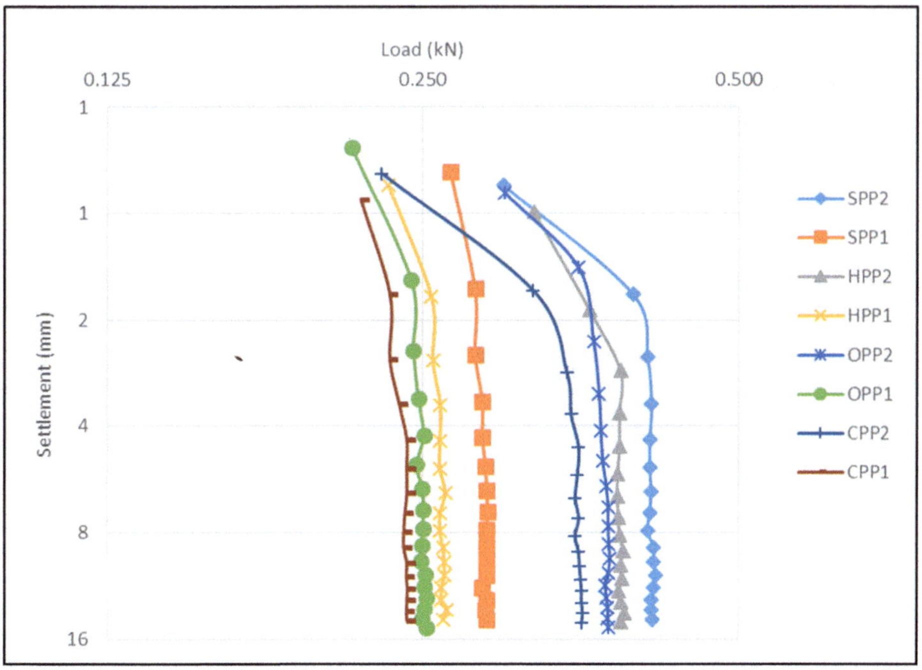

Fig. 8. Load-settlement plot based on De-Beer (1970) method

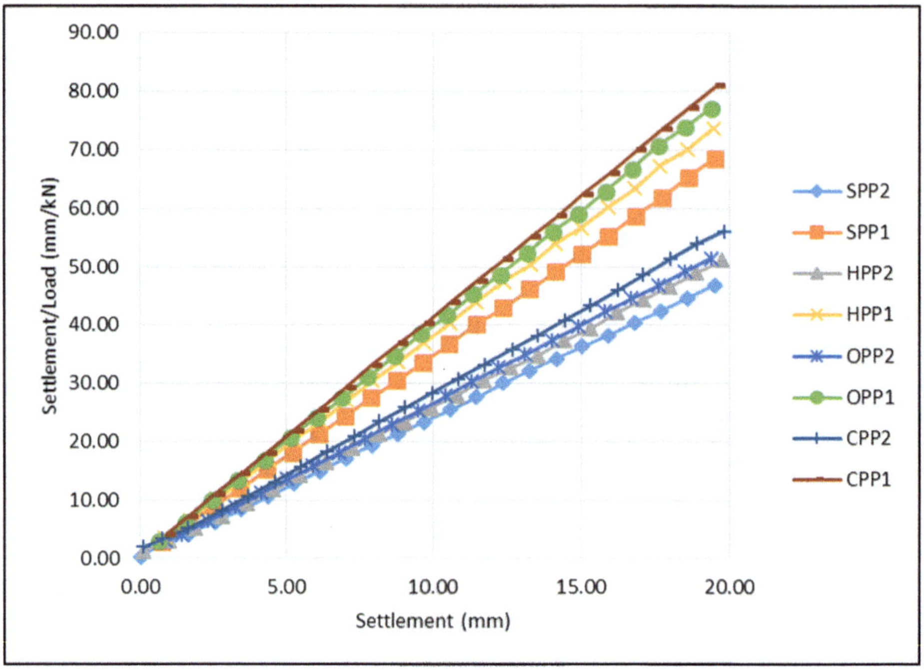

Fig. 9. Load-settlement plot based on Chin (1970) method

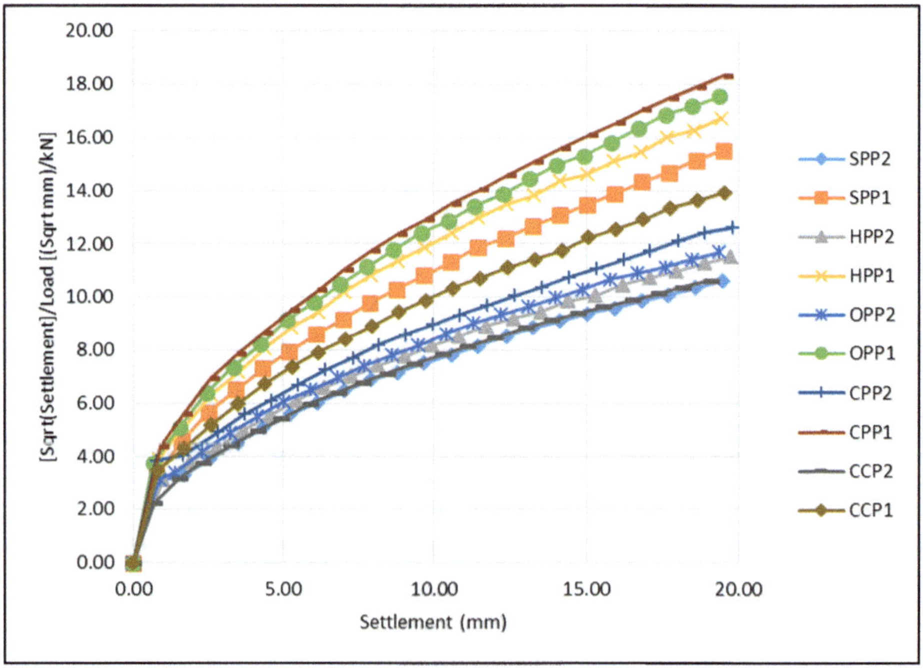

Fig. 10. Load-settlement plot based on Hansen 80% (1963) method

Table 3. Measured capacity of piles (in kN)

	Q_p	10% of diameter	15% of diameter	6 mm settlement	DeBeer (1970)	Chin (1970)	Hansen 80% (1963)	Average Q_m
SP2	0.490	0.404	0.412	0.412	0.417	0.433	0.454	0.422
HP2	0.465	0.372	0.384	0.384	0.390	0.389	0.424	0.391
OP2	0.457	0.364	0.368	0.372	0.364	0.378	0.413	0.377
CP2	0.448	0.332	0.348	0.348	0.347	0.357	0.384	0.353
SP1	0.368	0.280	0.284	0.288	0.290	0.286	0.320	0.291
HP1	0.351	0.256	0.260	0.264	0.258	0.265	0.290	0.266
OP1	0.346	0.245	0.250	0.250	0.246	0.251	0.279	0.254
CP1	0.340	0.232	0.240	0.240	0.232	0.242	0.267	0.242

Table 4. Ratio of measured and predicted pile capacities of model piles from different methods

	Q_m/Q_p in percentage					
	10% of diameter	15% of diameter	6 mm settlement	DeBeer (1970)	Chin (1970)	Hansen 80% (1963)
SPP2	82.28	83.91	83.91	84.93	88.19	92.46
HPP2	80.00	82.58	82.58	83.87	83.66	91.18
OPP2	79.48	80.35	81.22	79.48	82.53	90.17
CPP2	74.11	77.68	77.68	77.46	79.69	85.71
SPP1	76.09	77.17	78.26	78.80	77.72	86.96
HPP1	72.93	74.07	75.21	73.50	75.50	82.62
OPP1	70.81	72.25	72.25	71.10	72.54	80.64
CPP1	68.24	70.59	70.59	68.24	71.18	78.53
Mean (Q_m/Q_p)	75.49	77.33	77.71	77.17	78.87	86.03
SD	4.87	4.81	4.80	5.90	5.80	5.11
CoV	6.44	6.22	6.18	7.64	7.36	5.94

Whereas the lowest percentage ratio of capacity values were for the 10% of pile diameter settlement method with a mean Q_m/Q_p of 75.49%; it is the most popular interpretation method among pile designers, the standard method specified by the Malaysian authorities, and the Eurocode 7 (Jabatan Kerja Raya Malaysia 2005, British Standards Institution 2004). Conversely, it has been ascertained that the method is a misrepresentation of the proponent's – Terzaghi – opinion who stated that "the capacity of a pile should not be considered until a pile has moved at least 10% of the diameter" (Fellenius 2014). This implies that the capacity obtained by this method supposed to be used as a baseline value and not necessarily the actual capacity value for the piles.

The next lower percentage ratio of capacity values is from DeBeer (1970) method with a mean Q_m/Q_p of 77.17%. It is pertinent to mention that in prior researches based on full scale tests such as (Chen and Fang 2009; Marcos et al. 2013); the method gave the lowest capacity for the piles. In fact, Marcos et al. (2013) described the method as a "significant underpredictor" while Burland (2012) described it as "very conservative". Likewise, it has been identified that the method only measures the 'yield limit load' and not 'ultimate limit load' for piles (Fellenius 2014; Wang et al. 2011; Abdelrahman et al. 2003).

Similarly, the 15% of pile diameter settlement produces the next set of low percentage ratio of capacity values with a mean Q_m/Q_p of 77.33%. However, it is observed that its values linearly correlate with the 10% diameter method. Hence, it can also be employed as a baseline value whenever other methods produce higher capacity values. Although the 6 mm settlement method gave a slightly high percentage ratio of capacity values with a mean Q_m/Q_p of 77.71%, its utilisation has to be with caution based on the fact that the value is fixed and will not reflect the scale effect. For example, a 6 mm displacement for a model pile of 150 mm length, will not necessarily translate to a 6 mm displacement for another model test on a pile of 300 mm long.

Chin (1970) method produces good values of Q_m/Q_p with a mean value 78.87%; however the most correlated result is from Hansen (1963) 80% method with a mean Q_m/Q_p of 86.03%. According to Fellenius (2001), the Hansen (1963) 80% method perfectly matches the natural understanding of "plunging failure". This is further substantiated with the result of this study where the method produced the best result, with an individual value of Q_m/Q_p up to 92.46%. The method also yields the lowest obtained scatter of 5.94 and this implies its preciseness. The method is the foremost recommended method for the interpretation of failure of model pile load test with Chin (1970) method as a good alternative.

Additionally, ideal curves can be obtained from equations for both Hansen (1963) 80% and Chin (1970) methods and these equations can facilitate adequate comparison with the load-settlement curves obtained in model tests (Fellenius 2001). Moreover, both methods have been found suitable for the same purpose in full scale pile load tests (Abdelrahman et al. 2003). However, the scatter obtained for all the methods investigated as depicted by the coefficient of variation (CoV) is reasonable (5.94–7.64), and this indicates that none of the methods considered will produce outrageous results if implemented for the interpretation of failure load of model pile load tests.

Although, the capacities of the tested piles varied for the different cross sections in both categories of long and short piles, this variation have no impact on the analysis of the interpretation methods described above. In fact, it translates into multiple scenarios for the measurements of the tested piles. The exposition of the theoretical rationale, and profound analysis of the obtained capacities vis-à-vis the different cross sections of the piles are discussed in a separate manuscript.

4 Conclusions

Since it has been established that the different load test interpretation methods often yield different capacity values for full scale tests on the same pile in the same soil condition (Chen and Fang 2009), it is highly essential to identify the most appropriate method for the specific circumstance of model piles. This paper presented an alternative to the determination of pile capacity through full scale load tests by using 1-g testing of model piles. The measured capacity, Q_m of the model piles were derived using six different 'failure load' interpretation methods based on load-settlement behaviour of the piles, and the methods were assessed vis-a-vis the predicted capacity, Q_p of the piles, calculated using the pile static formula. Hansen (1963) 80% method yielded the most correlated result, followed by Chin (1970) method. It is suggested that the 10% of pile diameter method should be used as a baseline and not necessarily for the actual capacity value for model pile tests. All the methods examined produced results within tolerable limits. The recommendations will serve as a yardstick for future researchers on model pile for their choice of suitable load test interpretation method. The distinctive circumstances of the model piles and soil conditions as described in this study should also be considered relative to the analysis of the interpretation methods.

Acknowledgements. Appreciation goes to the Ministry of Higher Education, Malaysia for the MTCP scholarship awarded to the first author during his PhD program.

References

Abdelrahman, G., Shaarawi, E., Abouzaid, K.: Interpretation of axial pile load test results for continuous flight auger piles. In: 9th Arab Structural Engineering Conference, pp. 791–802 (2003)

Baziar, M.H., Saeedi Azizkandi, A., Kashkooli, A.: Prediction of pile settlement based on cone penetration test results: an ANN approach. KSCE J. Civ. Eng. **19**(1), 98–106 (2014)

British Standards Institution: Eurocode 7: Geotechnical Design—Part 1: General Rules. British Standards Institution, London (2004)

Brown, M.: Pile capacity testing. In: Manual of Geotechnical Engineering, Institution of Civil Engineers, United Kingdom, vol. 2, pp. 1451–1469 (2012)

Burland, J.B.: Behaviour of single piles under vertical loads. In: ICE Manual of Geotechnical Engineering, Institution of Civil Engineers, United Kingdom, pp. 231–246 (2012)

Caulk, R., Ghazanfari, E., Mccartney, J.S.: Parameterization of a calibrated geothermal energy pile model. Geomech. Energy Environ. **5**, 1–15 (2016)

Chen, Y.J., Fang, Y.C.: Critical evaluation of compression interpretation criteria for drilled shafts. J. Geotech. Geoenviron. Eng. **135**(8), 1056–1069 (2009)

Chung, S.-H., Yang, S.-R.: Numerical analysis of small-scale model pile in unsaturated clayey soil. Int. J. Civ. Eng. **15**, 877–886 (2016)

Dithinde, M., Phoon, K., De Wet, M., Retief, J.: Characterization of model uncertainty in the static pile design formula. J. Geotech. Geoenviron. Eng. **137**(1), 70–85 (2010)

Eslami, A., Tajvidi, I., Karimpour-Fard, M.: Efficiency of methods for determining pile axial capacity-applied to 70 cases histories in Persian Gulf northern shore. Int. J. Civ. Eng. **12**(1), 45–54 (2014)

Fellenius, B.H.: What capacity value to choose from the results of a static loading test. In: Deep Foundation Institute, Fulcrum, pp. 19–22 (2001)

Fellenius, B.H.: Basics of Foundation Design. Electronic Edition edn., Canada (2014). www.Fellenius.net

Hirany, A., Kulhawy, F.H.: On the interpretation of drilled foundation load test result. In: Deep Foundations: An International Perspective on Theory, Design, Construction, and Performance, pp. 1018–1028. ASCE, San Diego (2002)

Institution of Civil Engineers: ICE Specification for Piling and Embedded Retaining Walls, 2nd edn. Thomas Telford Publishing, London (2007)

Jabatan Kerja Raya Malaysia: Standard Specification for Building Works, Malaysia (2005)
Kaolin (Malaysia) Sdn. Bhd.: Kaolin clay (S300) technical data sheet (2014)
Karlsrud, K.: Ultimate shaft friction and load-displacement response of axially loaded piles in clay based on instrumented pile tests. J. Geotech. Geoenviron. Eng. **140**(12), 04014074 (2014)
Lastiasih, Y., Sidi, I.D.: Reliability of estimation pile load capacity methods. J. Eng. Technol. Sci. **46**(1), 1–16 (2014)
Lehane, B.M., Li, Y.N., Williams, R.: Shaft capacity of displacement piles in clay using the cone penetration test. J. Geotech. Geoenviron. Eng. **139**(2), 253–266 (2013)
Li, X., Cai, G., Liu, S., Puppala, A. J., Zheng, J., Jiang, T.: Undrained shear strength and pore pressure changes due to prestress concrete pile installation in soft clay. Int. J. Civ. Eng. (2017)
Madabhushi, S.P.G., Cilingir, U., Stringer, M.E.: Development of teaching resources for physical modelling community. In: Springman, S., Laue, J., Seward, L. (eds.) 7th International Conference on Physical Modeling in Geotechnics, pp. 79–84. CRC Press, Zurich (2010)
Marcos, M.C.M., Chen, Y.J., Kulhawy, F.H.: Evaluation of compression load test interpretation criteria for driven precast concrete pile capacity. KSCE J. Civ. Eng. **17**(5), 1008–1022 (2013)
Nazir, R., Tourtiz, A., Moayedi, H., Mosallanezhad, M.: Appraisal of reliable skin friction variation in a bored pile. Proc. ICE Geotech. Eng. **168**(1), 75–86 (2015)
Ng, C.W.W., Yau, T.L.Y., Li, J.H.M., Tang, W.H.: New failure load criterion for large diameter bored piles in weathered geomaterials. J. Geotech. Geoenviron. Eng. **127**(6), 488–498 (2001)
Ozkahriman, F., Wartman, J.: Investigation of 1-g Similitude laws by "modeling-of-models" exercise. In: Advances in Measurement and Modeling of Soil Behavior, pp. 1–9. ASCE (2007)
Titi, H.H., Elias, M.B.: Effect of load test interpretation on the resistance factors of driven piles. In: 2nd International symposium on cone penetration testing, California, USA (2010)
Tomlinson, M., Woodward, J.: Pile Design and Construction Practice, 6th edn. CRC Press, Boca Raton (2014)
Viggiani, C., Mandolini, A., Russo, G.: Piles and Pile Foundation. Spon Press, London (2012)
Wang, D., Zhou, Y., Zhao, J., Yin, Y., Liu, J.: Comparisons of interpreting ultimate capacity of piles based on axial static load testing data between Chinese and Western methods. Electron. J. Geotech. Eng. **16** (2011)
Zare, M., Eslami, A.: Study of deep foundation performances by frustum confining vessel (FCV). Int. J. Civ. Eng. **12**(4), 271–280 (2014)

Bearing Behaviors and Reasonable Rock-Socketed Depth of Bridge Pile Foundation in Karst Areas

Yunxiu Dong[1,2](✉), Zhongju Feng[1], Yumeng Hao[3], Hongsheng Yao[4], and Qi Wan[1]

[1] Highway School, Chang'an University, Xi'an, China
465095940@qq.com
[2] School of Civil Engineering, Longdong University, Qingyang, China
[3] China Communications Construction Company Second Highway Consultant Co., Ltd., Wuhan, China
[4] China Construction Sixth Engineering Division Co., Ltd., Tianjin, China

Abstract. The force of bridge pile foundation in karst areas is complicated. In order to determine the bearing capacity of bridge pile in these areas, the static load test is carried out. Bearing behaviors of the bridge pile in karst areas is analyzed using the displacement meter and strain gauge set respectively at the pile top and bottom. Furthermore, in combination with results of the static loading test for the rock-socketed pile, an optimization method for the socketed depth of the pile is proposed aiming at overcoming the shortcomings of existing pile-designing methods of bridge piles in karst areas. The results show that (1) The Q–s curve has a slowly-varying development under various vertical loads which shows a typical friction pile characteristic though the pile has been embedded in the depth of the limestone. Under a load of 8400 kN, the maximum settlement on the pile top is 3.69 mm, which is much less than 0.03 D or 40 mm, and the pile tip resistance is only 1.6% of the load on pile top. The tip resistance and shaft resistance of the pile are synchronized but with fairly different degrees of response. (2) The hyperbolic curve model is of high precision, and the forecast value is conservative when predicting the ultimate bearing capacity of a single pile, therefore, the hyperbolic model could be better for prediction in similar projects. (3) Based on the wide beam-pile mechanical model and the anti-bending failure mode, an optimal calculation method for rock-socketed depth of bridge piles in karst areas is proposed, which provide a theoretical reference for the design of pile foundation in karst areas.

Keywords: Bridge pile foundation · Karst area · Static loading test
Bearing capacity · Rock-socketed depth · Roof thickness

1 Introduction

Karst is widely distributed in China and is one of the common geological disasters in engineering construction. In recent years, with the rapid expansion of transportation infrastructure in China, many large bridges and tunnels have been constructed in karst

H. El-Naggar et al. (Eds.): GeoMEast 2018, SUCI, pp. 141–161, 2019.
https://doi.org/10.1007/978-3-030-01902-0_13

areas, where the geological condition is complex and changeable. Therefore, the pile foundation is widely used in these areas because of its high bearing capacity and small settlement (Feng 2008). Nevertheless, there are many problems with the application of pile foundation in karst areas, such as oblique hole, leakage slurry and hole-wall collapse, as exposed in the construction process of the Guangzhou-Foshan lines, as well as buried hammer, and ground collapse (Wu et al. 2006). The development of karst tends to make the surface of pile-embedded strata uneven, forming a temporary surface, so that sediment and groundwater would accumulate in the foundation pit and pile hole, which may cause hazards like site collapse (Waltham et al. 2005; Dogan and Yilmaz 2011; Zhou et al. 2012). This can seriously affect the quality and bearing performance of piles, and threaten the safety of superstructure.

At present, research of pile foundation in karst areas has been mainly concentrated on the analysis of mechanical characteristics and stability of the karst cave roof (Perrier et al. 1990; Garlanger 1991; Abdulla and Goodings 1996). In these research papers, different mechanical models were established based on the characteristics of the interaction between the karst roof and the pile, and the formulas to calculate the roof safety thickness were put forward accordingly (Zhao et al. 2004). In addition, some scholars have found that the bearing capacity of bridge piles in karst areas is closely related to factors such as roof thickness of a karst cavity, types of karst features, degrees of karstification, stress characteristics of surrounding rocks and sizes of rock-socketed piles (Huang et al. 2004; Panon et al. 2013). The relationship between karst load, cavern height and the relative parameters of roof safety thickness was analyzed using the finite element method (Andrei et al. 2002; Ahmed 2002; Goodings and Abdulla 2002). Based on a large number of field experiments, Rowe and Armitage (1987) summarized the relationship between the pile-side resistance and the unconfined compressive strength of rocks based on the empirical relationship between the roughness of the hole wall and the lateral friction of the rock-socketed piles. Sowers (1996) drew a conservative conclusion that the safety thickness of the roof of a karst cave should be greater than 70% of the width of the karst cave under the general engineering load in karst areas using numerical methods such as ZD-FLAC. Based on the model tests of rock-socketed piles, Benmokrane et al. (1994) explained that different bearing angles of rock-socketed piles have different effects on the ultimate bearing capacity of rock-socketed piles when there is a weak intercalated layer of rock mass. Considering the influence of the counter-moment on the contact between the pile-bottom and the basement rock, quite a few scholars have deduced the formula to calculate the rock-socketed depth of the pile foundation in karst areas (Pells and Turner 1978). Based on the physical engineering of the bridge in the karst area, the optimization method of the pile foundation in the karst area was proposed by Qubain et al. (2008), which take into account the bearing capacity of the roof and the interaction between the pile and soil.

However, previous studies on the bearing behaviors of pile foundation in karst areas are mainly focused on the high-rises. The calculation of rock-socketed depth of the pile foundation in karst areas is based on the assumption that the cave roof is punctured or sheared, while the other failure mode of the cave roof is seldom considered. In addition, there are few studies on the theoretical basis for bridge foundation, forcing the design of the bridge pile foundation in karst areas to mechanically adhere to current codes and technical regulations. Therefore, it is essential to study the bearing characteristics and designing method of bridge pile foundation in karst areas. This paper carried out a case study on the *Xixie* overpass project in Pingdingshan city, where the ZK37 project pile was chosen for the vertical static loading test, and the bearing behaviors of the pile in the karst area were studied. Furthermore, a method was proposed to calculate the rock-socketed depth of the bridge piles in this area, which can provide theoretical reference for the design and calculation of bridge pile foundations in karst areas.

2 Materials and Methods

2.1 Project Overview and Site Conditions

The *Xixie* overpass in Pingdingshan city is about 470 m long, with the upper part of the bridge composed of continuous box girders, while the lower part of the bridge being the column pier. The foundation of the bridge is the bored pile foundation. There are two 6-pile abutments and twenty-two 4-pile piers in this bridge. The bridge terrain is relatively flat, and geomorphic units are classified as the piedmont impact alluvial plain. The ground elevation is about 89.30 m.

Layers of geotechnical characteristics of the site are as follows: (a) Artificial filling: the layer is generally distributed, composed of crushed stone filling and gravel stone filling with a loose structure. It is either brown yellow or grayish yellow, with an elevation and thickness of, respectively, 88.5 m and 0.8 m. (b) Clay: the layer is mainly composed of slightly wet clay with good plasticity, and is dotted with a small amount of gravel in a brown-grayish yellow color. The elevation and thickness of the layer are 84.4–85.8 m and 2.7–4.1 m respectively. (c) Calcium-containing clay: the layer is distributed with a large number of black iron manganese nodules and 20%–30% of the white calcium nodules, of which the general diameter is of 20–30 mm and the largest particle size is 60 mm. The bottom of the layer is composed of sandstone fragments containing calcium-nucleation of up to 50%–60%, which have been semi-cemented into rocks. The layer is brown red, with an elevation and thickness of, respectively, 64.4–70.3 m and 14.4–20.5 m. (d) Karstized dolomitic limestone: the layer is mainly composed of dolomite and a small amount of mud, containing brecciated debris with either a medium thickness structure or a honeycomb structure. Rock fragmentation is caused by serious dissolution. The pores are awash with clay, and the part of rock mass turns into a cave. The layer is gray, with an elevation and thickness of, respectively, 60.5 m–66.6 m and 1.6 m–6.8 m. (e) Weakly weathering dolomitic limestone: the green-gray layer is mainly composed of calcite and dolomite with either a medium thickness structure or a cryptocrystalline structure. The rock is relatively intact and rigid, where the joint fissure phenomenon is developed. This layer is sandwiched with the karstized dolomitic limestone layer, which displays similar characteristics as described above, but is of brown-yellow color.

According to the previous geological exploration data, a large number of caverns are distributed in the area between the No. 1 and No. 16 piers. The test site was located at No. 6 Pier, where karst developed strongly, and karst features are shown in Table 1.

Table 1. Karst features of pier No. 6

Pier No.	Bedrock roof elevation (m)	Bedrock roof fluctuation degree (m)	Number of caves	Cave height (m)	Minimum cave floor elevation (m)	Cave development degree
6	66.8–68.7	1.9	12	0.3–13.9	33.9	Strong

The karst rock formation in this area is covered by the quaternary soil layer, with a thickness of less than 30 m, which is buried karst. Once a karst cave is formed, it is filled with quaternary sediments, which is a temperate modern karst type. The basic forms of karst growth include gully, dissolved-fissure, water-eroded groove and cave.

2.2 Test Parameters

Taking into account the actual conditions of the study site, the ZK37 pile of pier No. 6 was selected as a subject for the single pile static loading test, of which the parameters are shown in Table 2. Samples of drilling indicate that the bottom of the pile is mainly dolomitic limestone, of which the physical and mechanical properties are shown in Table 3.

2.3 Test Equipment and Test Element Layout

The Heavy platform reaction system used for the static loading test was assembled using the channel, surrounded by a protective stack made of welded steel pipes. The counter-force beam was made of rebar U-shaped hanger and installed under the pile platform. The datum beams were made of strongly magnetic, high-stiffness and consistent steel to ensure the test accuracy, which were fixed at one end with the other end being freely supported. The stacking material was made of steel, and the testing materials were fixed on the pile platform before the static loading test. The platform reaction system is shown in Fig. 1.

The hydraulic system was loaded with four hydraulic jacks with a regulator, which was operated through a four-stage shunt valve. The hydraulic pressure was measured using a high precision oil-pressure gauge connected in parallel with the jack to control the loading. The load was measured via the strain gauge pressure sensor placed on the jack. The lower part of the jack was fixed with steel plate gaskets of sufficient rigidity, and the position of the jack was rigorously set before the test to align the action line of the jack with the center line of the pile.

The displacement gauges and strain gauges were mounted on the pile top and bottom respectively, as shown in Figs. 2 and 3. The distance between the datum beam (the internal datum beam) and the test pile was limited by the conditions of equipment for the loading test as well as the site. In order to eliminate the influence of foundation settlement on the datum beam, two datum beams were arranged on both sides of the test device (the outer datum beam), which could be approximately considered that the outer datum beam have no displacement during the test. A displacement gauge was mounted on each outer datum beam to measure the displacement of the internal datum beam, as shown in Fig. 3.

Table 2. Design parameters

Pile No.	Pile length (m)	Pile diameter (m)	Concrete grade	Design ultimate bearing capacity (kN)	Design bending moment (kN m)	Rock-socketed depth (m)	Cave roof thickness (m)	Cave height (m)	Cave geometry vertical × horizontal (m × m)
ZK37	28	1.5	C25	8120	560.5	6.4	2.8	6.2	8 × 2

Table 3. Physical and mechanical properties of rock

Pile No.	Sample lithology	Sampling depth (m)	The degree of weathering	Gravity density (kN/m^3)	Saturation limit compressive strength standard value (MPa)	Ultimate tensile strength standard value (MPa)
ZK37	Dolomitic limestone	29.0–29.8	Weak	25.7	15.6	1.04

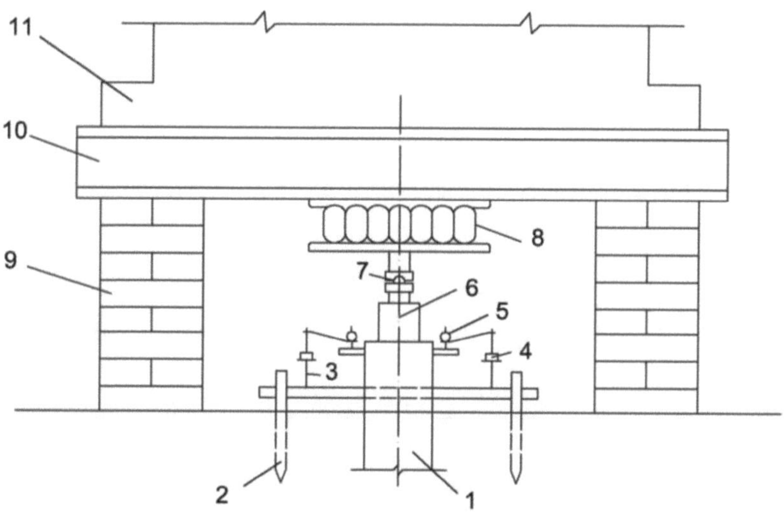

1-Pile; 2-Datum beam; 3- Datum beam; 4- Magnetic table seat; 5- Displacement meter; 6- jack; 7- Ball seat; 8- Joists; 9- Support pier; 10- Beam platform; 11- Ballast

Fig. 1. Pile test platform and reaction device

2.4 Experiment Method

The slow maintenance loading method was selected for the test. On the one hand, the data were recorded for five times within one hour after each stage of loading, with intervals of 5, 10, 15, 15 and 15 min, and then every 30 min. On the other hand, the data were recorded for three times within one hour after each stage of unloading with intervals of 15, 15 and 30 min, and then the next stage of unloading can be started. The data were recorded for 4 h after the unloading was complete.

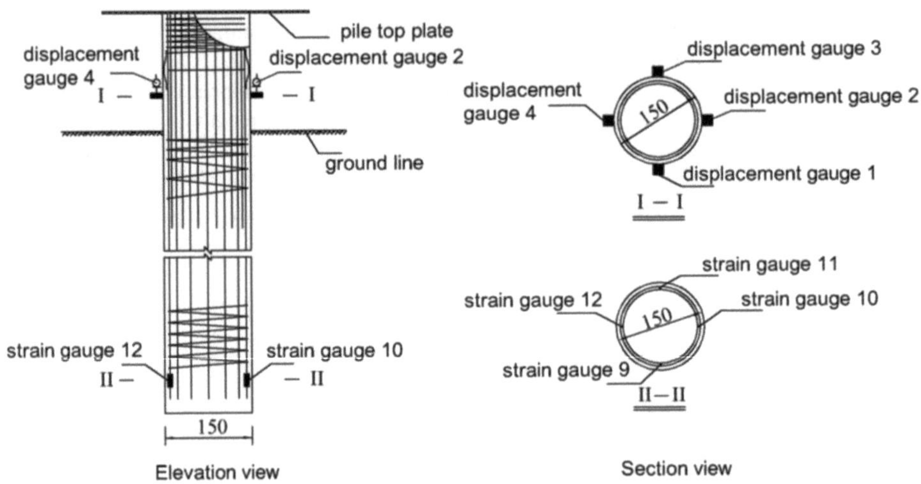

Fig. 2. Displacement gauge and strain gauge layout (unit: cm)

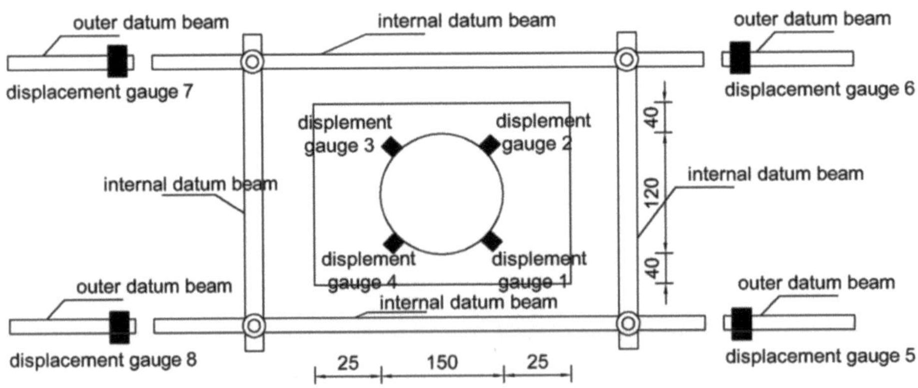

Fig. 3. Displacement gauge and datum beam layout (unit: cm)

If the settlement of the pile is slightly more than 0.1 mm per hour at its maximum, and the average sedimentation rate keeps decreasing within half an hour, the settlement is considered to be substantially stable. The termination procedures of the test could be carried out in accordance with the provisions of the pile foundation engineering manual (The National Standards Compilation Group of People's Republic of China, GB 50007-2011).

3 Test Result Analysis

3.1 Analysis of Load-Settlement Relationship

The relationship curve of the load-settlement (Q–s) and curves of the settlement-time (s–t) of pile ZK37 are shown respectively in Figs. 4 and 5.

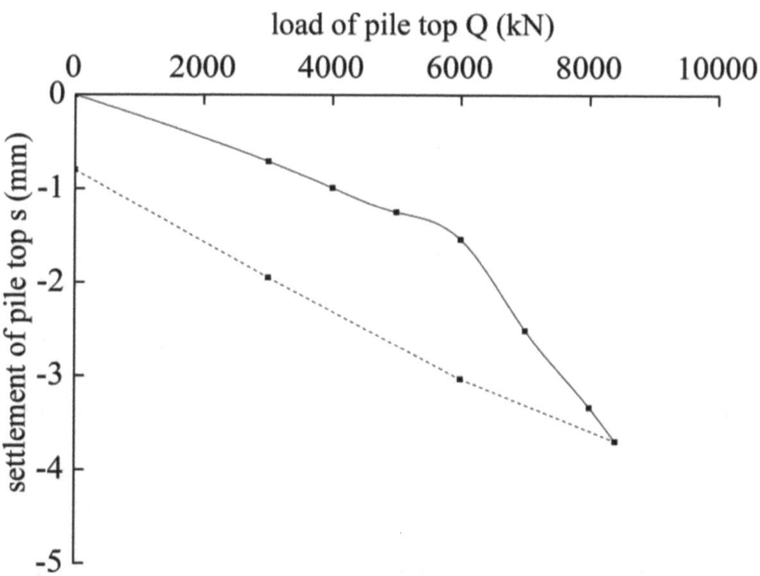

Fig. 4. Q–s curve of pile loading test

Figure 4 shows that the Q–s curve develops slowly without obviously steep descents. When the load is less than 6000 kN, settlement of pile top increases approximately linearly with the increase of load. As the load increases, the curve gradually becomes non-linear. The turning point of the curve appears when the applied load reaches 6000 kN, for the relative settlement (0.97 mm) at the load of 7000 kN is 3.3 times larger than that (0.29 mm) of 6000 kN. However, the termination conditions of the test are still not met, so it cannot be regarded as the termination criterion of the test.

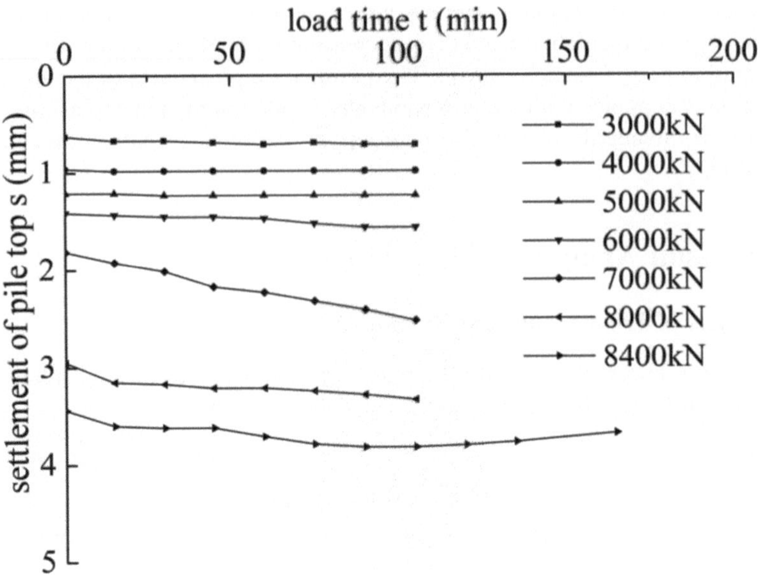

Fig. 5. s–t curves of pile loading test

After the settlement of the applied 7000 kN load is stable, the settlement of the pile top begins to slow down with the further increase in load. It can be seen in Fig. 5 that the slope of the s–t curve increases slightly at the loading stage of 7000 kN, and that the end of the curve does not show an obvious downward trend. At the last stage of the 8400 kN load, the settlement of the pile top shows a slight rebound. Authors believe that it is caused by the incomplete contact between the pile tip and the bearing stratum, and a small amount of sediment at the bottom of the pile. When the loading level is low, the settlement of the pile top is extremely small, which is mainly caused by the deformation of the pile concrete; when the load is increased to 7000 kN, the load at the pile top is transmitted to the bottom, making the latter settle down, which makes the settlement of the pile top increases rapidly until the pile bottom is completely in contact with the bearing layer. Therefore, when the load on pile top continues to increase, the s–t curve gradually becomes smooth.

When the load reaches 8400 kN, the pile still works normally without damage. Additionally, the settlement of the pile top is merely 3.69 mm, which is much lower than the destruction standard (if the Q–s curve is a slow-varying type, the load will generally reach its ultimate when the settlement of the pile top is 40–60 mm), indicating that the pile has good working performance under the load of 8400 kN. The pile could hardly reach the ultimate status, as stated in the specification, caused by the impact of the platform reaction system. Therefore, the ultimate bearing capacity of the pile is far greater than 8400 kN.

In the actual project, it is of great practical significance to predict the ultimate bearing capacity of the pile with a limited amount of measured data. Many domestic and foreign scholars have studied prediction methods and put forward variety of mathematical models. In this paper, the frequently-used prediction models, such as polynomial model, hyperbolic model and exponential curve model, were chosen to predict the ultimate bearing capacity of the pile. Equations of these models are as follows:

Quadratic polynomial model:

$$s = aQ^2 + bQ + c \tag{1}$$

Cubic polynomial model:

$$s = aQ^3 + bQ^2 + cQ + d \tag{2}$$

Hyperbolic model:

$$s = Q/(a + bQ) \tag{3}$$

Exponential curve model:

$$s = ae^{bQ} \tag{4}$$

where a, b, c and d are the fitting parameters of each model. Based on the test data, the regression analysis method was used to calculate the parameter values of each function model via the 1stOpt software.

In this study, the ultimate bearing capacity of the single pile was controlled by deformation. The vertical load corresponding to when the pile top underwent a settlement of 0.03 D or 40 mm are used as its ultimate bearing capacity (The National Standards Compilation Group of People's Republic of China, GB 50007-2011). The test data was substituted into different prediction formulas, and the predicted values of the pile ultimate bearing capacity were calculated as shown in Table 4.

Table 4. Calculation results of different predicting models

Predicting models	Model parameters	Fit accuracy	Predictive value of ultimate bearing capacity (kN)	
			$Q_{0.05\ D}$	$Q_{40\ mm}$
Quadratic polynomial	a = 9.519 × 10^{-8}, b = −5.384 × 10^{-4}, c = 1.5217	0.9933	18729.9	17475.1
Cubic polynomial	a = 9.661 × 10^{-12}, b = −7.150 × 10^{-8}, c = 3.650 × 10^{-4}, d = 0.0031	0.9953	18700.3	18035.2
Hyperbolic	a = 6334, b = −0.4864	0.9901	12452.4	12384.7
Exponential curve	a = 0.2574, b = 3.183 × 10^{-4}	0.9923	16220.3	15850.3

All the four prediction methods performed well in precision, showing that they can effectively describe the relationship between the load and the settlement of the pile top in terms of numerical calculation in this paper. The results represent that the predictive value of the ultimate bearing capacity of the pile is the largest in the polynomial model, followed by that in the exponential curve model, while the value in the hyperbolic model is the smallest.

Due to the complex geological conditions, the high degree of karstification, and the influence of the properties of the rock and soil around the pile, the hyperbolic model was chosen to predict the ultimate bearing capacity of the pile. Specifically, when s = 0.03 D or 40 mm, the ultimate bearing capacity of the pile would be 12452.4 kN or 12384.7 kN (take the smaller value). The safety factor was set as 2, and then the design bearing capacity of the pile was 6226.2 kN or 6192.4 kN (take the smaller value).

3.2 Analysis of Shaft Resistance of Pile

This test was aimed to verify the vertical bearing capacity of the project pile. Therefore, the strain gauges were not placed along the pile body, but near the pile bottom. In order to analyze the characteristics of the pile, pile shaft resistance was obtained using the load on pile top subtracted by the pile tip resistance (calculated from the data of the stain gauges at the pile bottom), which neglected the interaction between shaft resistance and tip resistance of the pile. The results are shown in Figs. 6 and 7.

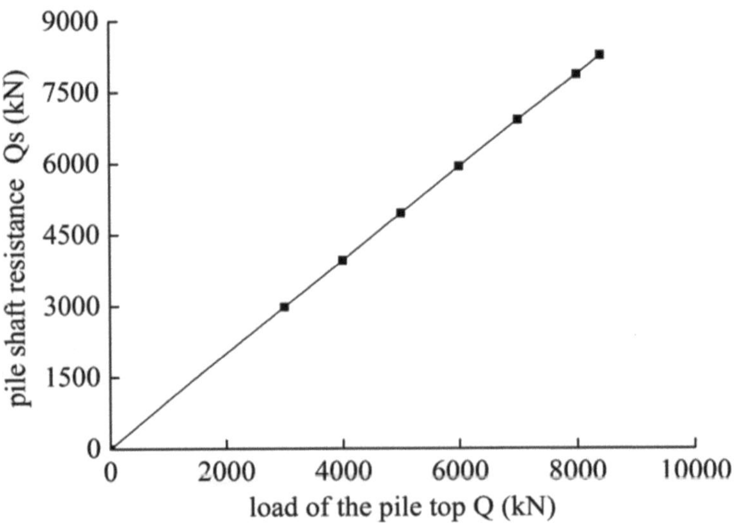

Fig. 6. Curve of pile shaft resistance under various loads

As demonstrated in Figs. 6 and 7, pile shaft resistance increases approximately linearly with the increase of the load on pile top, and that the load is mainly borne by shaft resistance, which accounts for 98.4%–99.3% of the former. This shows that merely a small portion of the load on pile top was transferred to the bottom of the pile, which represents typical properties of the friction pile, in the first stage of the resistance of the rock-socketed pile. Furthermore, with the increase in the load on pile top, pile shaft resistance is gradually decreased. Although the ratios of pile shaft resistance to the load on pile top at the loads of 5000 kN and 8400 kN increase slightly (0.21% and 0.08% respectively), the overall variation of the curve is not affected.

3.3 Analysis of Tip Resistance of Pile

The elastic moduli of concrete and steel are 3.5×10^4 MPa and 2.0×10^5 MPa respectively. The curve of pile tip resistance under various loads is shown in Fig. 8, where the average of the strain data is converted into pile tip resistance.

As demonstrated in Fig. 8, the pile tip resistance increases nonlinearly with increasing load. Specifically, at the beginning of the loading process, the pile tip resistance increases slowly. When the load exceeds 4000 kN, the growth rate gradually increases with the increasing load, but the pile tip resistance is relatively small compared with the load on pile top. When the load on pile top reaches 8400 kN, the tip resistance is 122.9 kN, accounting for merely 1.6% of the load on pile top, due to the large rock-socketed depth of the pile and the under-exerted tip resistance. This can be attributed to the fact that the rock strength of the test pile is too large and that the pile tip resistance is far from being fully exerted, while the test fulfils the termination condition.

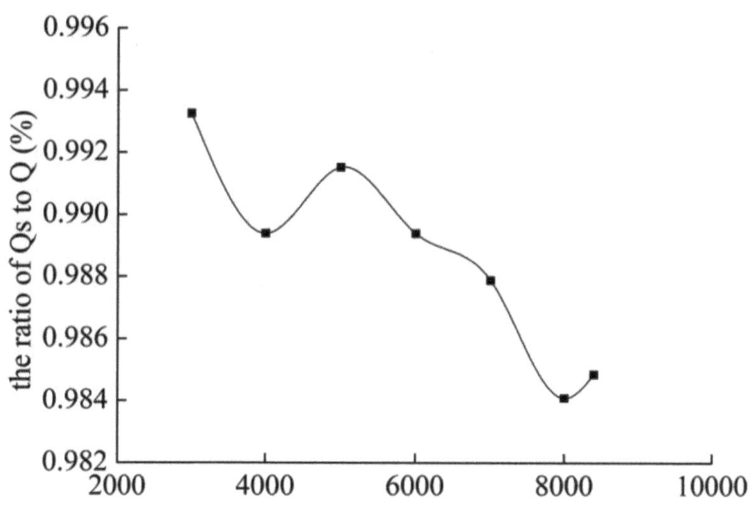

Fig. 7. Curve of the ratio of Qs to Q under various loads

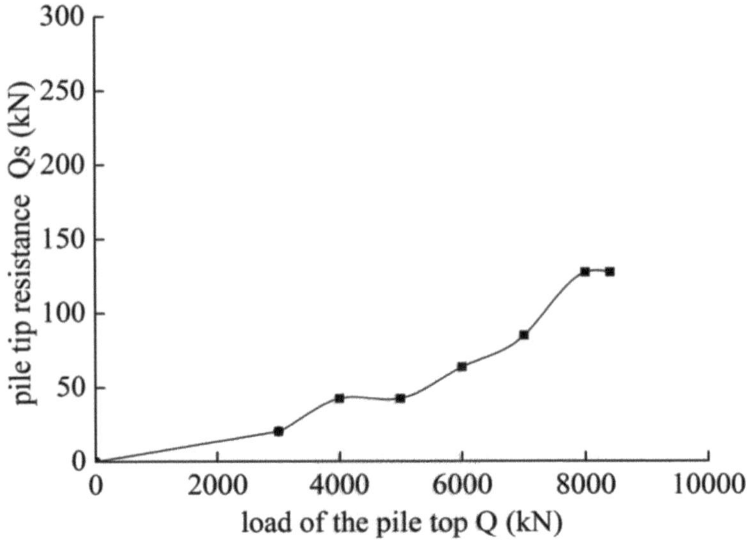

Fig. 8. Curve of pile tip resistance under various loads

4 Analysis of Load Transfer Behavior for the Pile-Rock (Soil) System

The traditional view is that, when the vertical load is gradually applied to the pile top, the compressed pile body is subjected to the upward frictional resistance of the soil around the pile, transmitting the load on pile top through pile shaft resistance to the surrounding soil. Therefore, the axial force and pile compressive deformation of the pile decrease with depth. In addition, the frictional resistance of the middle and lower soil layers increases with the pile-soil relative displacement, until it reaches a certain extent when the vertical displacement occurs at the pile tip and pile tip resistance is generated. The frictional resistance of the upper soil layer around the pile is preceded by that of the lower soil layer, and pile shaft resistance is preceded by the pile tip resistance. For drilling rock-socketed pile with mud wall, when the ratio of pile length to diameter is greater than 15–20 ($L/d > (15–20)$), the load transfer characteristic of the friction pile is presented, whether the pile is embedded in decomposed rock or intact bedrock (Xiu and Li 1988).

Kulhawy and Phoon (1993) raised objections and put forward that shaft resistance and tip resistance of the pile develop synchronously with the increasing load. The load-settlement curves obtained by Kulhawy based on a large number of experiments are shown in Fig. 9.

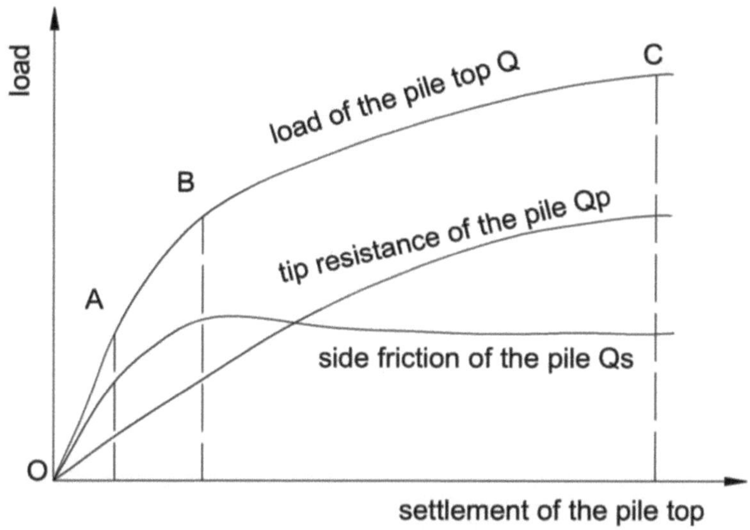

Fig. 9. Curves of load-settlement of bored pile

The load-settlement curves of bored pile shows nonlinear characteristics, which can be divided into three stages.

In the first stage (OA section), small displacement of pile top under load causes pile resistance and pile end resistance to exert. In addition, the load on pile top is jointly borne by shaft resistance and tip resistance of the pile, which is mainly characterized by low settlement of the pile top.

In the second stage (OB section), with the increase of load, pile shaft resistance and pile tip resistance increase correspondingly. However, when the pile shaft resistance reaches the maximum and remains basically stable, the load on pile top is borne by the pile tip resistance. In this stage, a large settlement occurs at the pile top.

In the third stage (BC section), with the further increase of the load on pile top, pile tip resistance reaches the ultimate value, that is, shaft resistance and tip resistance of the pile are fully exerted. The settlement of the pile top increases sharply, and the ultimate bearing capacity of the pile is reached.

The relationship curves of the load on pile top Q, shaft resistance Q_s, tip resistance Q_p and settlement of the pile top s are shown in Fig. 10.

In Fig. 10, under the loading condition, the small settlement of the pile top mobilizes the pile shaft resistance, and the pile tip resistance also slightly responds, which means that the pile tip resistance and the pile shaft resistance synchronously bear the load on pile top. However, due to fact that the pile shaft resistance and the pile tip resistance are obviously affected by the pile-soil (or rock) relative displacement, and that they achieve the ultimate value required for different displacements. With the increase of pile top displacement, the increase of pile side resistance is much greater than that of pile tip resistance, and the load is mainly borne by the pile shaft resistance. The experimental results are inconsistent with those of the traditional view (the pile shaft resistance is preceded by the pile tip resistance), but are consistent with the theory of load-settlement proposed by Kulhawy. The whole test process is in the first stage.

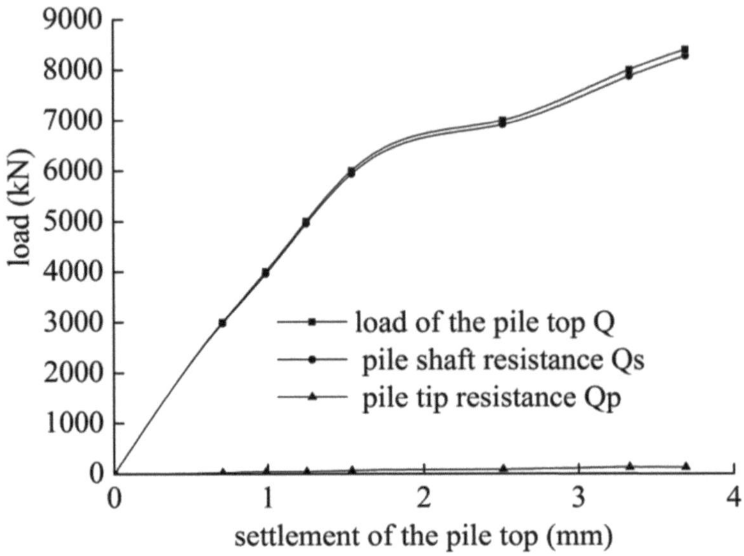

Fig. 10. Curves of load-settlement of pile loading test

5 Optimization of Rock-Socketed Depth of Bridge Pile in the Karst Area

The static load test results show that over 98% of the pile top load is borne by the pile side resistance during the whole test process, and the pile tip resistance is slightly exerting, which could be attributed to the conservative design of bridge pile foundation in the karst area. At present, in the design of bridge pile foundation in karst area, the frictional resistance of the overlying soil is not included in the design. The effect of this part is considered as a safety reserve, which will inevitably lead to the increase of rock-socketed depth. In addition, the thickness of the roof of the cave needs to be checked. Nevertheless, there has not been a consensus on the safety thickness of the roof of the cave. For safety reasons, the method that combines grouting with penetration is generally used in the construction of pile foundation, which not only improves the new requirements for the construction technology, but also greatly increases the cost of the project. Therefore, for both the roof thickness and bearing capacity to meet the requirements, a reasonable reduction in the rock-socketed depth of piles is necessary, which can not only reduce the difficulty of construction, but also greatly reduce the project cost.

Based on this, the rock-socketed depth of the bridge piles in the karst area can be optimized according to the following steps:

(a) Choose a reasonable pile-rock mechanics model to calculate the minimum allowable thickness of cave roof h according to its failure mode;
(b) Calculate the rock-socketed depth h_r by using the rock thickness H minus the minimum allowable thickness of cave roof h;

(c) Check whether the bearing capacity of the pile is satisfied. Provided that the requirements are met, the rock-socketed depth can be measured as described above.

5.1 Selection of Mechanics Model for Pile-Rock

There has been a lot of research on the mechanical characteristics of cave roofs, and a number of mechanical models of the cave roof based on different assumptions have been put forward. Among them, the wide beam model can be applied when h/B (the ratio of the thickness of cave roof to its width) is greater than 0.20 and the width cannot meet the requirements of the beam model (Xie et al. 2007). In this paper, the minimum allowable thickness is calculated by using the mechanical model of simply-supported wide beam. The mechanical model is shown in Fig. 11.

In this model, the effective width e (the value of e depends on the support form, load form and ratio of the width to the span B/L) is used in calculating the strength of the wide beam. The recommended values of the effective width proposed by Holl (1936) are shown in Table 5, taking into account the two types of loads, that is, the uniform load on the full plate and the uniform load on the central area with a radius r. The value of e/L (ratio of the effective width e to the span L) can be obtained from the r/h and B/L values in Table 5.

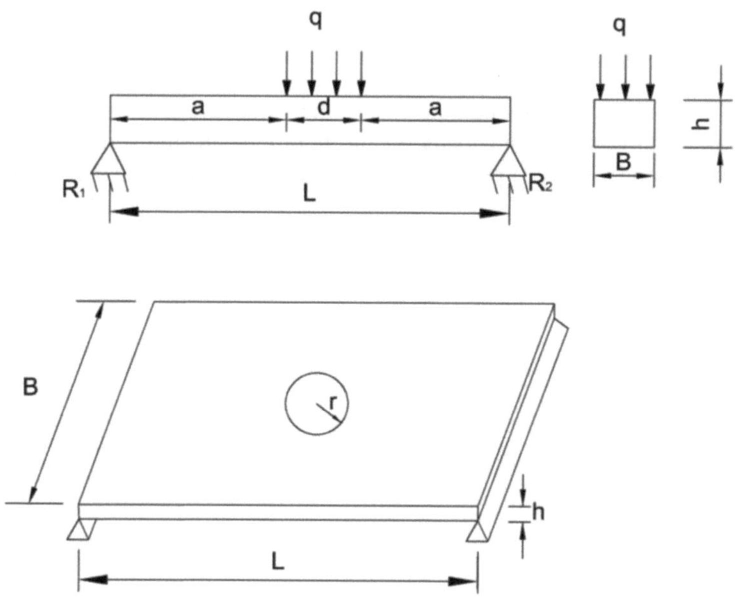

Fig. 11. Mechanical model of simply-supported wide beam

Table 5. Effective width values

Load range	Value of e/L				
	B/L = 1.0	B/L = 1.2	B/L = 1.6	B/L = 2.0	B/L = ∞
Uniform load	0.960	1.145	1.519	1.900	
Center load, r = 0	0.568	0.599	0.633	0.648	0.656
Center load, r = 0.125 h	0.581	0.614	0.649	0.665	0.673
Center load, r = 0.250 h	0.599	0.634	0.672	0.689	0.697
Center load, r = 0.500 h	0.652	0.694	0.740	0.761	0.770

According to the theory of material mechanics, the maximum bending moment of simply supported beam is located in the middle of the beam.

$$\sigma_{\max} = (\sigma)_{z=-h/2} = \frac{M_{\max}}{W} = \frac{3qd(2L-d)}{4Bh^2} \tag{5}$$

where $\sigma_{\max}$ is the maximum normal stress of the simply-supported beam section (kPa). $M_{\max}$ is the maximum bending moment of the simple beam section (kN m) and W is the bending section factor (m^3).

When using the wide beam mechanical model, B is replaced by the effective width e (found in Table 5).

5.2 Selection of Roof Failure Mode

The existing theory holds that when the roof of the karst cave is simplified as the beam-slab structure with concentrated loads on the middle-span, the roof strata may have several failure modes, such as punching failure, shearing failure and bending failure.

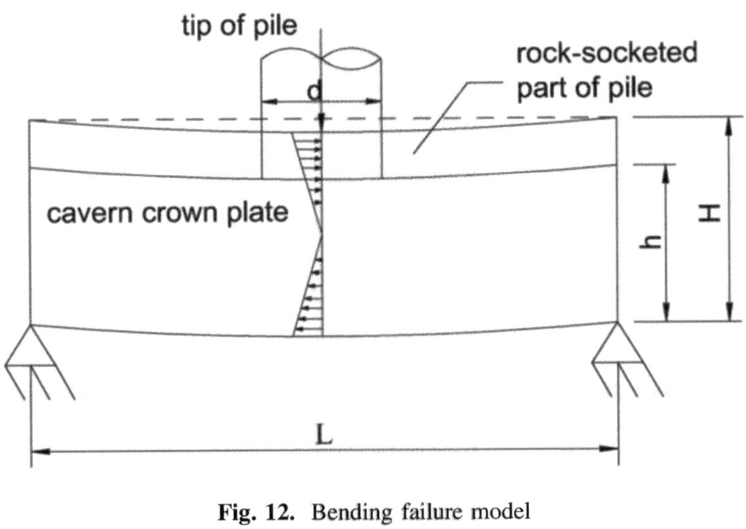

Fig. 12. Bending failure model

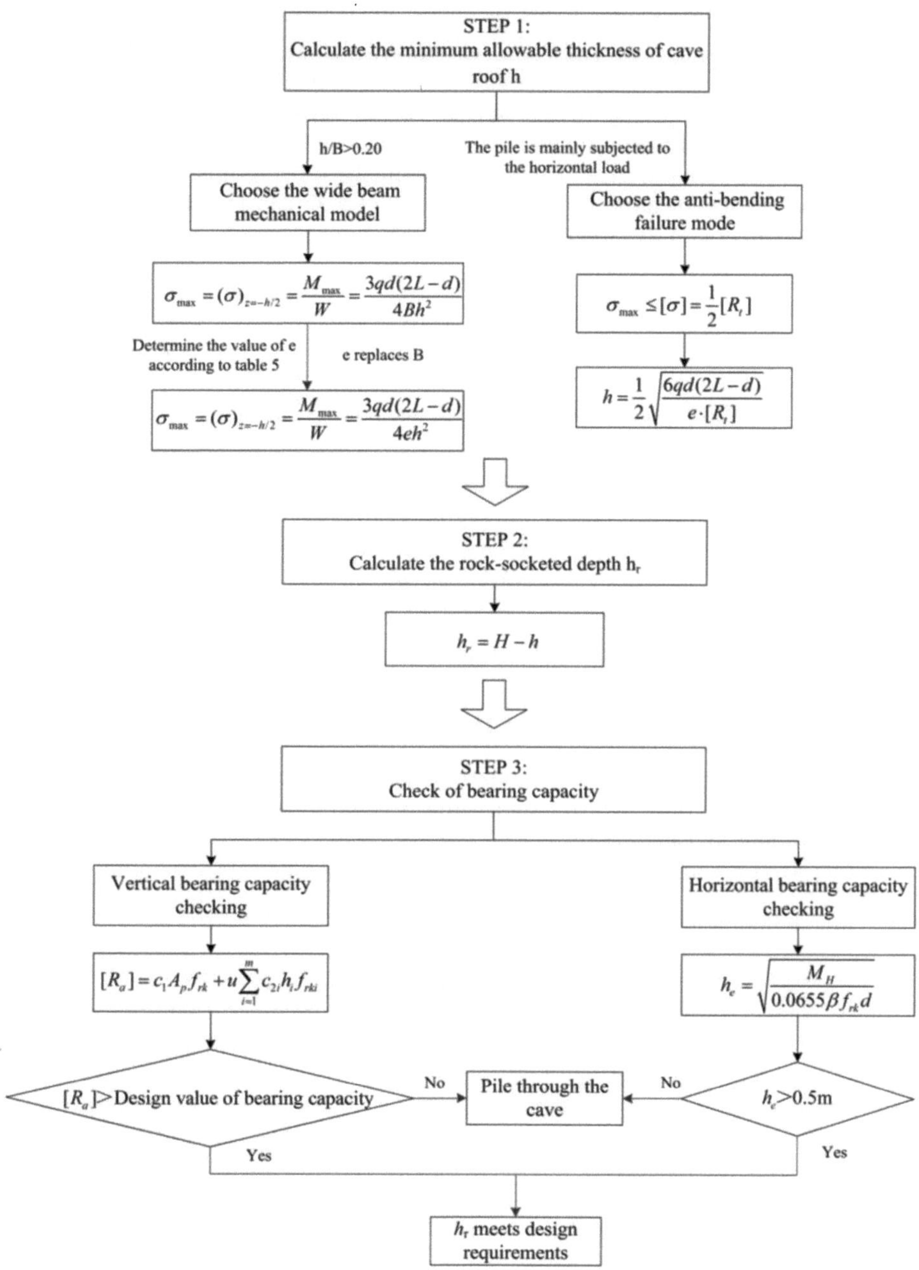

Fig. 13. Flow chart of computing

Generally, the surrounding rocks of caves are mainly dolomite and limestone, which have a higher hardness, with the difference between tensile strength and compressive strength being up to 15 to 20 times. When the pile is mainly subjected to the horizontal load, the roof is more prone to flexural tensile failure. Therefore, in this paper, the anti-bending failure mode is chosen to calculate the allowable thickness of the cave roof under the pile. The bending failure mode of the roof is shown in Fig. 12.

Based on the specification for designs of ground base and foundation of highway bridges and culverts (GB 50007-2011), the checking formula of flexural tensile strength is:

$$\sigma_{\max} \leq [\sigma] = \frac{1}{2}[R_t] \tag{6}$$

where $[\sigma]$ is the allowable bending stress of the rock (kPa) and $[R_t]$ is the ultimate tensile strength (kPa).

The minimum allowable thickness of the roof can be derived by formula (6):

$$h = \frac{1}{2}\sqrt{\frac{6qd(2L-d)}{e \cdot [R_t]}} \tag{7}$$

As mentioned above, the calculation flow chart of this paper is shown in Fig. 13.

5.3 Optimization of Rock-Socketed Depth of Pile

The embedded depth of the pile is optimized by the method proposed in this paper. The cave roof span L is 2.0 m, the cave roof width B is 8.0 m, and the effective width e is 1.38 m according to Table 5.

$q = 8120/(\frac{1}{4}\pi d^2) = 4597.3$ kN, and other parameters in Tables 2 and 3, then:

(1) The minimum allowable thickness of the roof is:

$$h = \frac{1}{2}\sqrt{\frac{6 \times 4597.3 \times 1.5 \times (2 \times 2 - 1.5)}{1.38 \times 1040}} = 4.2 \text{ m}$$

(2) The rock-socketed depth is:

$$h_r = H - h = 9.2 - 4.2 = 5.0 \text{ m}$$

(3) The bearing capacity (excluding soil friction resistance of the pile):
  (a) Checking calculation of the vertical bearing capacity of the pile:

$$[R_a] = c_1 A_p f_{rk} + u \sum_{i=1}^{m} c_{2i} h_i f_{rki} \tag{8}$$

where $[R_a]$ is the allowable value of the axial compressive bearing capacity of a single pile (kN). A_p is the area of the cross section of the pile end (m^2). f_{rk} is

the standard value of axial compressive strength of rock at the pile end (kPa, f_{rki} is the f_{rk} value of the *i-th* layer). c_1 and c_{2i} are exertion coefficient of end resistance at the variable cross-section and the pile end. u is the circumference of the cross section of the steel tube section (m). h_i is the thickness of the pile embedded in each rock (m). m is the layer number of the rock strata.
The allowable value of the axial compressive bearing capacity of a single pile is calculated as:

$$\begin{aligned}[R_a] &= 0.5 \times \frac{1}{4} \times 3.14 \times 1.5^2 \times 15600 + 3.14 \times 1.5 \times 0.04 \times 5 \times 15600 \\ &= 14696.1 \text{ kN}\end{aligned}$$

$[R_a]$ is larger than the design value of the bearing capacity of the pile (4060.0 kN), so the vertical bearing capacity meets the requirements.

(b) Checking calculation of the horizontal bearing capacity of the pile:

$$h_r = \sqrt{\frac{M_H}{0.0655\beta f_{rk} d}} \tag{9}$$

where h_e is the effective depth of pile embedded in bedrock. M_H is the bending moment at the top of bedrock. β is the coefficient which is valued according to the specification. d is the pile diameter.
The effective depth of pile embedded in bedrock is computed as:

$$h_r = \sqrt{\frac{560.5}{0.0655 \times 0.5 \times 15600 \times 1.5}} = 0.86 \text{ m}$$

$[h_r]$ is larger than the standard minimum value (0.5 m), so the horizontal bearing capacity meets the requirements.

Compared with the actual pile, the rock-socketed depth is reduced by 2.4 m, which meets the safety requirements of the roof as well as the bearing capacity and stability of the pile. The method can be used to calculate the rock-socket depth of bridge pile in karst areas with reliable safety and sound economy. Furthermore, it provides a theoretical reference for the future design of bridge pile foundation in karst areas.

6 Conclusions

From what have been discussed so far, the following conclusions can be summarized:

(1) The Q–s curve of the No. 4 pile of Xixie Overpass Bridge in Pingdingshan develops slowly without obvious steep descent under various loads, which shows a typical friction pile characteristic although the pile has been embedded in the depth of the limestone. The pile has good working performance under the load of 8400 kN (larger than the designed ultimate bearing capacity 8120 kN).

Specifically, the maximum settlement on the pile top is 3.69 mm (much lower than the failure criteria of 0.03 D or 40 mm), and the pile tip resistance accounts for only 1.6% of the load on pile top. Therefore, the ultimate bearing capacity of the pile is far greater than 8400 kN.

(2) The pile shaft resistance and the pile tip resistance synchronously bear the load on pile top but their degrees of response are fairly different.

(3) The hyperbolic curve model has a higher accuracy and generates more reliable predictions compared with many other functional models. When the test data of similar engineering are limited, the hyperbolic curve model would be better to predict the bearing capacity of pile foundation.

(4) Based on the wide beam-pile mechanical model and the anti-bending failure mode, an optimal calculation method which is simple and practical with few required parameters for rock-socketed depth of bridge pile in karst areas is proposed, which can serve as a theoretical reference for the design of pile foundation in karst areas.

For practical engineering, the effective pile-rock mechanics model and the reasonable safety thickness of the cave roof should be selected according to the structural characteristics and the physical or mechanical parameters of the surrounding rock. In addition, the feasibility of economy and difficulty in construction should be taken into account.

Acknowledgments. The study is supported by the Key Transportation Science and Technology Program of Henan Province under Grant No. 2014K48. The authors would also like to express their great appreciation to Fugui Xie for help in the development of the experiments and to Zhen Wang for refining and finalising the manuscript.

References

Feng, Z.: Foundation Engineering in Special Areas. China Communications Press, Beijing (2008)

Wu, S., Shi, H., Zhu, Z.: Effects of karst on pile foundations of GuangDong Foshan Lines. Chin. J. Rock Mech. Eng. **25**(Suppl. 2), 3559–3564 (2006)

Waltham, T., Bell, F., Culshaw, M.: Sinkholes and Subsidence: Karst and Cavernous Rocks in Engineering and Construction. Springer/Praxis, Chichester (2005)

Dogan, U., Yilmaz, M.: Natural and induced sinkholes of the Obruk Plateau and Karapmar—Hotamis Plain, Turkey. J. Asian Earth Sci. **40**(2), 496–508 (2011)

Zhou, J., Tang, Y., Yang, P., Zhang, X., Zhou, N., Wang, J.: Inference of creep mechanism in underground soil loss of karst conduits I. Conceptual model. Nat. Hazards **62**(3), 1191–1215 (2012)

Perrier, H., Simon, M., Lacroix, M.: Road reinforcement with geotextile fabric: prevention of Karstic sink holes. In: Proceeding of 4th International Conference, The Hague, pp. 862–869 (1990)

Garlanger, J.E.: Foundation design in floride Karst. Concr. Int. **13**(4), 56–62 (1991)

Abdulla, W.A., Goodings, D.J.: Modeling of sinkholes in weakly cemented sand. J. Geotech. Eng. **122**(2), 998–1005 (1996)

Zhao, M., Cao, W., He, P.: Study on safe thickness of rock mass at end of bridge foundation' s pile in karst and worked-out mine area. Rock Soil Mech. **25**(01), 64–68 (2004)
Huang, S., Mei, S., Gong, W.: Testing study on bearing behavior of piles for NanPan river great bridge in karst areas. Chin. J. Rock Mech. Eng. **23**(5), 809–813 (2004)
Panon, S.V., Kelly, W.R., Angel, J.C., Luman, D.E.: Hydrogeologic and topographic controls on evolution of karst features in Illinois' sinkhole plain. Carbonates Evaporites **28**, 13–21 (2013)
Ahmed, S.: Geophysical response of bedrock fractures, karst conduits, and associated surficial features, Bourbonnais, IL. Northern Illinois University (2002)
Goodings, D.J., Abdulla, W.A.: Stability charts for predicting sinkholes in weakly cemented sand over karst limestone. Eng. Geol. **65**, 179–184 (2002)
Rowe, R.K., Armitage, H.H.: A design method for drilled piers in soft rock. Geotech. J. **24**(1), 126–142 (1987)
Sowers, G.F.: Building on Sinkholes: Design and Construction of Foundations in Karst Terrain. ASCE, New York (1996)
Benmokrane, B., Mouchaorab, K.S., Ballivy, G.: Laboratory investigation of shaft resistance of rock-socketed piers using the constant normal stiffness direct shear test. Can. Geotech. J. **31** (31), 407–419 (1994)
Pells, P.J.N., Turner, R.M.: Elastic solutions for the design and analysis of rock-socketed piles. Can. Geotech. J. **16**(3), 481–487 (1978)
Qubain, B.S., Seksinsky, E.J., Li, J.: Design consideration for bridge foundation in Karst terrain. In: Proceedings and Field Trip Guide of the 49th Highway Geology Symposium, Prescott, Arizona, pp. 339–349 (2008)
Xiu, C., Li, D.: Mathematical description of P–S curves and prediction of ultimate load for vertical static load test of single pile. Chin. J. Geotech. Eng. **10**(06), 64–73 (1988)
GB5007: Code for design of ground base and building foundation. In: The National Standards Compilation Group of People's Republic of China, Beijing, China (2011)
Kulhawy, F.H., Phoon, K.K.: Drilled shaft side resistance in clay soil to rock. In: Design and Performance of Deep Foundations@ Piles and Piers in Soil and Soft Rock. ASCE (1993)
Xie, F.: A study on properties of bridge piles in karst areas. Chang'an University (2007)
Holl, D.L.: Analysis of thin rectangular plates supported on opposite edges. Iowa College (1936)

Analytical Approaches to Predict Pile Settlement in London Clay

Elia Voyagaki[1(✉)], Jamie Crispin[1], Charlotte Gilder[1], Paul Nowak[2], Nick O'Riordan[3], Dinesh Patel[3], and Paul J. Vardanega[1]

[1] University of Bristol, Bristol, UK
e.voyagaki@bristol.ac.uk
[2] Atkins, Epsom, UK
[3] Arup, London, UK

Abstract. Projects involving construction of piled foundations often rely on preliminary full-scale field tests to failure to predict performance under applied load. If these tests are not available, the ensuing uncertainty will naturally lead to conservative design assumptions. Such design assumptions will result in higher construction costs and often in longer construction times. This paper shows how a database of previous pile load tests can be used in conjunction with simple analytical tools to attempt a quantification of performance uncertainty. Data from a series of previously published axial load tests on piles in London Clay is employed to this end. The methodology developed in this paper can arguably be expanded to a wider range of test sites and geological materials.

1 Introduction

Piled foundations 'fail' due to excessive movements. Literature on estimating the magnitude of these movements is extensive; ranging from analytical solutions (e.g., Randolph and Wroth 1978; Mylonakis and Gazetas 1998; Mylonakis 2001; Guo 2012), continuum solutions employing Green's functions (e.g., Poulos and Davis 1968; Butterfield and Banerjee 1971), finite-element studies (e.g., Ottaviani 1975; Baguelin and Frank 1980; Syngros 2004) and semi-empirical approaches (e.g., Seed and Reese 1957; Coyle and Reese 1966; Fleming 1992). Despite these advances, engineers "may never be able to estimate axial pile capacity in many soil types more accurately than about ±30%" (Randolph 2003).

The significance of accessible open source data has been acknowledged by other researchers who have invested in databases of pile performance in various countries e.g., USA (Paikowsky et al. 2004), Ireland (Galbraith et al. 2014) and Egypt (AbdelSalam et al. 2015). Global databases have also been recently compiled for tests in sands by a Chinese-UK-Australian research collaboration (Yang et al. 2015, 2016) and the FindaPile.com (Lemnitzer and Favaretti 2013) database for lateral-load tests by the Deep Foundation Institute and University of California, Irvine in the USA.

In the UK, a large amount of piled foundation construction is in stiff clay deposits. Results of full-scale tests on bored piled foundations are available for London clay (Whitaker and Cooke 1966; Patel 1992), but the raw data is not widely available.

H. El-Naggar et al. (Eds.): GeoMEast 2018, SUCI, pp. 162–180, 2019.
https://doi.org/10.1007/978-3-030-01902-0_14

Design methods used in practice include those reported in Skempton (1959) (see Fig. 1a), Poulos (1989), Fleming et al. (2009) and Tomlinson and Woodward (2014).

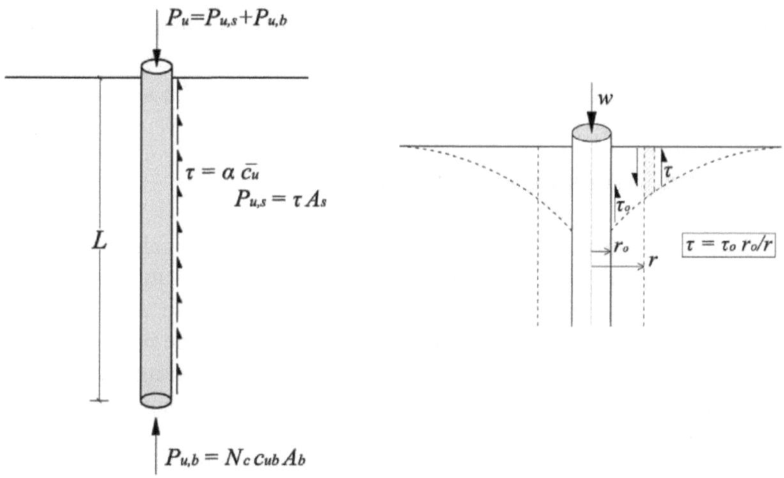

(a) (b)

Fig. 1. (a) Ultimate pile capacity distribution (*based on* Skempton 1959) and (b) displacement of a single pile (*based on* Cooke 1974; Randolph and Wroth 1978).

Settlement prediction is often carried out using numerical software such as Oasys Pile (OASYS 2017), Ensoft TZPile (Reese et al. 2014), Geocentrix Repute (Bond and Basile 2017), Deltares D-Pile group (Deltares 2016) and Fine GEO5 Pile (Fine n.d.). The available software use either three-dimensional boundary element methods based on Mindlin's solution for a point load in the interior of a linear elastic half space, or one-dimensional numerical analysis using theoretical or empirical "*t-z*" curves.

Combining the load test data collected in the stiff clay deposits in the UK in the past is needed to better calibrate geotechnical design models and thus better quantify and hopefully reduce uncertainty. Such data are key for calibrating recent analysis models. This paper presents the preliminary outcomes of an EPSRC funded research project (DINGO: Databases to INterrogate Geotechnical Observations) for the development of an open-source database of pile load tests in UK soil deposits.

2 The Database: Key Features and Structure

At the time of writing the DINGO database was being compiled by a research team based at the University of Bristol. Data from literature includes material published in journals, conference proceedings, textbooks, dissertations and technical reports. Industry data has been sourced from UK-based engineering consultancies and contractors.

The Association of Geotechnical and Geoenvironmental Specialists (AGS) present a data structure, known as the AGS Format (AGS 4.0.4 2017), which is commonly used for transfer of ground investigation data in the UK. This was selected for the DINGO database to allow ease of use and familiarity to the practicing engineer. A similar structure is not available for pile loading tests, therefore, a similar but bespoke format has been used for the pile information. The source metadata is stored in a master spreadsheet with additional spreadsheets describing the site-specific data.

3 Preliminary Analysis and Results

Once the database of tests is compiled, the aim is to use it to calibrate and compare existing settlement prediction approaches available in the literature. In this paper a preliminary investigation focuses into suitable methods to use and the appropriate presentation methods to allow comparison between them. A large number of prediction methods are available both in the literature and in software packages as mentioned in the Introduction. Three families of methods are considered and a representative model is chosen from each.

3.1 Method 1 – Linear Elastic

Linear elastic analyses are commonly used to model piles under axial load. Some software still exclusively uses linear elastic soil models (e.g. PIGLET, Randolph 2006) and most software includes an option to use this model. Simplified Winkler/load transfer models are popular (Randolph and Wroth 1978; Scott 1981; Mylonakis and Gazetas 1998; Guo 2012), and when model parameters are matched to soil properties using more rigorous continuum analyses that typically provide very similar results (Randolph and Wroth 1978; Mylonakis 2001; Syngros 2004; Guo 2012; Anoyatis 2013). Closed-form solutions for piles embedded in inhomogeneous soils are provided, among others, by Scott (1981), Guo (2012) and Crispin et al. (2018). The method by Crispin et al. (2018) has been chosen to represent this 'family' of solutions, in which the Winkler model is employed, and the Winkler modulus is assumed to vary according to the following function of depth, z:

$$k(z) = k_L\left[a + (1-a)\frac{z}{L}\right]^n, a = \left(\frac{k_0}{k_L}\right)^{\frac{1}{n}} \tag{1}$$

where L is the pile length, k_0 and k_L are the Winkler modulus at the pile head and pile base, respectively; a is a dimensionless coefficient accounting for non-zero surface stiffness and n is an inhomogeneity exponent. The pile base is modelled as a spring with stiffness K_b.

The pile head stiffness, K_0, is given by:

$$K_0 = E_p A \lambda_L a^{\frac{n}{2}} \frac{[I_{v-1}(\chi_0) I_{1-v}(\chi_L) - I_{1-v}(\chi_0) I_{v-1}(\chi_L)] + \Omega_L [I_{v-1}(\chi_0) I_{-v}(\chi_L) - I_{1-v}(\chi_0) I_{+v}(\chi_L)]}{[I_{-v}(\chi_0) I_{v-1}(\chi_L) - I_{+v}(\chi_0) I_{1-v}(\chi_L)] + \Omega_L [I_{-v}(\chi_0) I_{+v}(\chi_L) - I_{+v}(\chi_0) I_{-v}(\chi_L)]} \quad (2)$$

where E_p and A denote the pile elastic modulus and cross-sectional area, respectively; λ_L is a load transfer parameter (units of Length^{-1}); Ω_L a dimensionless pile base stiffness constant and $I_v(\chi)$ is the modified Bessel function of the first kind of order v and argument χ. λ_L, Ω_L, v, χ_L and χ_0 are given by:

$$\lambda_L = \sqrt{\frac{k_L}{E_p A}}, \Omega_L = \frac{K_b}{E_p A \lambda_L}, v = \frac{1}{n+2}, \chi_L = \frac{2\lambda_L L}{(1-a)(n+2)}, \chi_0 = \chi_L a^{\frac{n+2}{2}} \quad (3)$$

The Winkler modulus and base spring stiffness must then be related to the soil properties. The simple concentric cylinder model from Cooke (1974) and Randolph and Wroth (1978) (Fig. 1b) is used to relate the Winkler modulus variation, $k(z)$, to the soil shear modulus variation, $G_s(z)$ by defining a "magical" radius, r_m, denoting the radial distance from the pile axis at which displacement due to the load applied on the pile is presumed to reach zero. For a single pile the stiffness of the springs in the shaft can be determined using the following expression (Randolph and Wroth 1978).

$$\frac{k(z)}{G_s(z)} \approx \frac{2\pi}{\ln(2r_m/D_s)}, r_m \approx 2.5L \frac{G_s(L/2)}{G_s(L)}(1 - v_S) \quad (4)$$

where D_s is the pile shaft diameter and v_s is the soil Poisson's ratio. The expression for r_m was fitted to a numerical continuum analysis by Randolph and Wroth (1978).

Pile groups spaced closer than r_m ($\approx$5–10D_s) are interacted and Poulos (1968) superposition method can be applied using the interaction factors of Randolph and Wroth (1979) as modified by Mylonakis and Gazetas (1998), to determine the overall settlement of the group.

The stiffness of the pile base spring, K_b, is approximated as a rigid punch on the surface of a half-space (Randolph and Wroth 1978):

$$K_b \approx \frac{2G_s(L)D_b}{(1 - v_s)} \quad (5)$$

where D_b is the pile base diameter.

Linear elastic models, including the herein presented method, are easy to employ with sufficient accuracy at low strength mobilisations which are often encountered in design. However, these methods do not account for the nonlinear behavior of the soil which can lead to very unconservative predictions at large loads, close to failure.

3.2 Method 2 – Elastoplastic “*t-z*” Analysis Using a Linear Elastic Perfectly-Plastic Soil Model

The simplest non-linear model is linear elastic-perfectly plastic, which is commonly the default for settlement prediction software (e.g. OASYS 2017; Deltares 2016) and allows model parameters to be easily related to soil properties. A simple extension to the linear elastic model has been described by Scott (1981), Guo (2012) and Crispin et al. (2018) that allows this model to be analysed in closed form, without the need for numerical software. The analysis proposed by Crispin et al. (2018) is used in this study.

The model is shown in Fig. 2. Following the assumption that the limiting skin friction is first reached at ground surface, and a plastic length, L_p, propagates down the pile, parametric expressions for the load and settlement at the top of the pile can be developed in terms of L_p. At the interface between this plastic section and the elastic section below, the displacement of the pile is the yield displacement at that depth, $w_y(L_p) = t_u(L_p)/k(L_p)$. The stiffness of the elastic section, K_{el}, can be calculated using any linear elastic method and multiplied by $w_y(L_p)$ to get the axial force in the pile at the interface. The plastic section of pile is then modelled as a column with known base load and distributed shaft load. Once the shaft resistance is exhausted, a displacement at the pile base can be assumed and the same process applied with the elastic section replaced by the base spring. By calculating the head load and settlement for $0 \leq L_p \leq L$ and for the base settlement at failure, a full load-settlement curve can be produced.

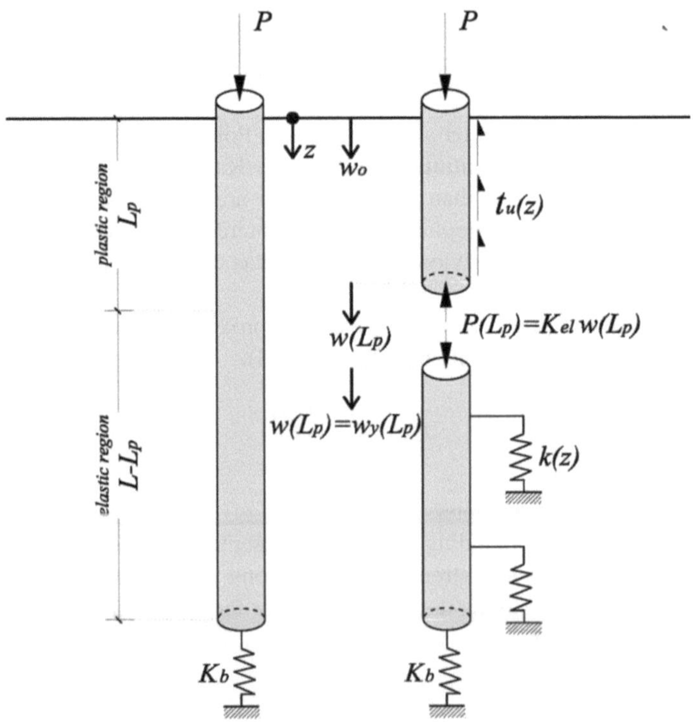

Fig. 2. Proposed linear elastic perfectly – plastic model (method 2) for a pile under axial load.

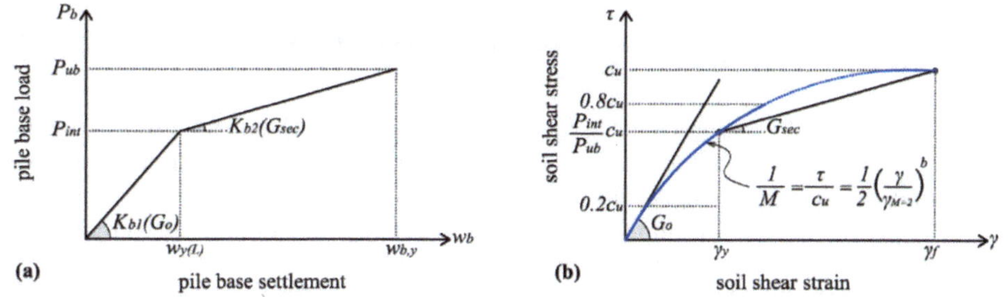

Fig. 3. (a) Bilinear base spring stiffness model; (b) secant shear modulus, G_{sec}, estimation method.

Considered a power-law variation in ultimate shaft resistance per unit length, $t_u(z)$, with exponent, m, and surface value t_{u0}, Crispin et al. (2018) developed Eqs. (6) and (7) to describe the head load and settlement:

$$P = P(L_p) + t_{u0}L_p + \frac{t_u(L_p) - t_{u0}}{m+1}L_p \tag{6}$$

$$w_0 = w(L_p) + \frac{P(L_p)}{E_pA}L_p + \frac{t_{u0}}{2E_pA}L_p^2 + \frac{t_u(L_p) - t_{u0}}{(m+2)E_pA}L_p^2 \tag{7}$$

where $w(L_p)$, $P(L_p)$ and $t_u(L_p)$ are the axial displacement, load and ultimate shaft resistance per unit length at the interface depth, $z = L_p$, respectively.

$t_u(z)$ is calculated using the α-method (Skempton 1959; Patel 1992):

$$t_u(z) = \alpha c_u(z)\pi D_s \tag{8}$$

where α is the pile-soil adhesion factor and $c_u(z)$ is the undrained shear strength profile with depth.

This extension of the linear elastic method gives more reasonable predictions at large loads yet is simpler to employ than full numerical analysis, but careful consideration is therefore required to select a representative stiffness and strength (e.g. Poulos 1999).

3.3 Method 3 – Elastoplastic "*t-z*" Analysis Using a Power–Law Nonlinear Soil Model

A linear elastic perfectly-plastic soil model is a relatively crude approximation of the constitutive behavior of soils, neglecting any nonlinear elastic behavior and plastic behavior such as strain softening. Modelling soil nonlinearity rigorously requires numerical analysis, however, Vardanega et al. (2012a, 2018) proposed a simple analytical method based on mobilised stress design principles that is suitable for hand calculation and can be applied here. The constitutive relationship in Eq. (9) was derived from a database of soil tests in clays (Vardanega and Bolton 2011a).

$$\frac{1}{M} = \frac{\tau}{c_u} = \frac{1}{2}\left(\frac{\gamma}{\gamma_{M=2}}\right), \quad 1.25 \leq M \leq 5 \tag{9}$$

where M is the mobilisation factor, a factor of safety on soil shear strength c_u; τ is the mobilised shear stress at a corresponding shear strain γ; $\gamma_{M=2}$ is the shear strain at 50% of the undrained shear strength and b is a nonlinearity exponent.

By applying this relationship to the concentric cylinder model introduced by Cooke (1974) and Randolph and Wroth (1978) (Fig. 1b) and accounting for pile compression, Vardanega et al. (2012a) derived the following relationship between the average mobilisation of soil strength at the pile circumference and the displacement at the pile head:

$$\frac{w_h}{D_s} = \frac{b \cdot \gamma_{M=2}}{2(1-b)}\left(\frac{2}{M}\right)^{1/b} + \frac{2}{M}\frac{\overline{c_u}}{E_p}\left(\frac{L}{D_s}\right)^2 \tag{10}$$

where $\overline{c_u}$ is the average undrained shear strength over the pile length.

In order to apply Eq. (10), a relationship between the applied load at the pile head and the mobilisation of the soil at the pile circumference is required. Using the adhesion factor α, on soil shear strength (Skempton 1959; Patel 1992) to estimate the shear stress at the pile-soil interface when slip occurs, the mobilisation factor can be related to the factor of safety on the pile shaft resistance, F_{shaft}:

$$F_{shaft} = \frac{P_{u,s}}{P_s} = \alpha M \tag{11}$$

where P_s and $P_{u,s}$ are the applied and ultimate load on the shaft, respectively and M is the average mobilisation factor over the pile length. This indicates that the minimum value of M is $1/\alpha$. Following the conservative simplifying assumption that no base load is mobilised until the shaft resistance is exhausted, the total factor of safety, F_{total}, is given by:

$$F_{total} \cong \frac{P_{u,s} + P_{u,b}}{P_s} = \frac{P_{u,s} + P_{u,b}}{P_{u,s}} F_{shaft} = \frac{P_{u,s} + P_{u,b}}{P_{u,s}} \alpha M \tag{12}$$

where $P_{u,b}$ is the ultimate load resisted by the pile base. Substituting the following expressions for $P_{u,s}$ and $P_{u,b}$ (Skempton 1959) into Eq. (12) yields Eq. (14).

$$P_{u,s} = \alpha \overline{c_u} \pi D_s L, \; P_{u,b} = N_c c_{ub} \pi D_b^2/4 \tag{13}$$

$$F_{total} = \frac{P_u}{P}\left[\alpha + \frac{N_c}{4}\frac{c_{ub}}{\overline{c_u}}\frac{D_s}{L}\left(\frac{D_b}{D_s}\right)^2\right] M \tag{14}$$

where P and P_u are the total applied and ultimate loads, respectively; N_c is the bearing capacity factor, taken as 9, and c_{ub} is the undrained shear strength at the pile base.

The power-law constitutive model is an improvement on a linear elastic assumption and the method employed is still simple to apply. However, once the shaft resistance is exhausted, the concentric cylinder model no longer applies and Eq. (11) cannot be used, therefore this method is limited to predicting settlement before the shaft resistance is exhausted.

3.4 Analysis

The methods described in the previous sections have been used to analyse 51 maintained load tests from 15 sites in London reported by Patel (1989). A summary of the tests is shown in Tables 1 and 2. All piles are embedded in London Clay with a short, commonly sleeved, length through any overlying material.

Table 1. Patel (1989), summary of pile tests presented in this study

Stats of Patel (1989) pile tests			
Site locations	15 sites in London	Pile diameters (m)	0.343–0.914
Number of test piles	51	Pile lengths (m)	6.1–24.2
Pile types	Bored, straight shafted and under reamed	Pile lengths in clay (m)	4.3–22.7
Test types	Maintained load	Max head displacements (mm)	19.4
Geology	London Clay	Max applied load (kN)	3540

Patel (1989) interpreted mean $c_u(z)$ lines for each site using triaxial test results, as is now recommended by the London District Surveyors Association (LDSA 2009). Most of these lines were determined from undrained triaxial shear tests on 100 *mm* samples, therefore, following the recommendations of Patel (1992) and LDSA (2009), $\alpha = 0.5$ has been used directly from back analysis of reported maintained load pile test results. However, for sites A5, A7, A8 and A14 the design lines were based on c_u tests on 38 mm samples. Patel (1992) found that 38 mm samples overestimate the fissured strength of London Clay (found using plate load tests or using 100 mm samples) by a factor of about 1.3, therefore, the design lines for these sites have been reduced by this factor. This is in agreement with experimental results by Whitaker and Cooke (1966) who used 38 mm samples and recommended reducing the base value by a similar factor. The equivalent α value is 0.38, which is similar to the value found by Skempton (1959) using the approach after Patel (1992) to account for the slower rate at which these tests were completed. Skempton's work (1959) was largely based on very fast tests to failure at a time when rate and pore pressure effects in stiff clays were unknown. The design lines are shown in Fig. 4.

Table 2. Pile lengths in London Clay, L, and pile base diameters, D_b, normalised with pile shaft diameters, D_s, for the tests recorded in Patel (1989).

Site	Pile	L/D_s	D_b/D_s	Site	Pile	L/D_s	D_b/D_s
A1	TP1	17.1	1.0	A9	TP3	22.2	1.0
	TP2	25.4	1.0		TP4	28.9	1.0
A2	TP1	13.3	1.0		TP5	26.7	1.0
A3	TP1	11.7	1.0		TP6	20.0	1.0
A4	TP1	26.5	1.0	A10	P1	37.2	1.0
A5	TP1	14.0	1.0		P2	33.3	1.0
	TP2	16.8	2.8		P4	27.6	1.0
A6	24/7 10	22.0	1.0		P5	34.9	1.0
	29/2	16.0	1.0		P6	32.3	1.0
	95 4/25	14.9	1.0	A11	TP1	30.6	1.0
	B4	20.0	1.0		TP2	32.6	1.0
	B5	15.5	1.0	A12	TP1	13.5	2.3
	B6	15.0	1.0		TP2	15.0	1.0
	B7	11.5	1.0	A13	TP1	23.4	1.0
	B8	22.5	1.0		TP2	12.5	1.0
	B9	18.5	1.0	A14	4138	28.0	1.0
	B10	15.0	1.0		A	13.4	1.0
	B11i, ii	14.4	1.0		B	13.4	1.0
	B12	17.6	1.0		C	13.4	1.0
A7	TP1	16.9	1.0		TP1	30.0	1.0
	TP2	16.5	2.0		TP2	30.1	1.0
A8	TP1	30.3	1.0		TP3	26.7	1.0
	TP2	27.5	1.0	A15	TP1	20.0	1.0
A9	TP1	21.4	1.0		TP2	22.0	1.0
	TP2	17.8	1.0		TP3	15.0	1.0

Vardanega and Bolton (2011b) analysed 15 high quality triaxial tests on London Clay samples to determine the deformation behaviour of the material. The tests had an average b value of 0.58 and an average $\gamma_{M=2}$ of 7×10^{-3}, which are used in Method 3. Additionally, the maximum shear modulus, G_0, was found to be approximately $320c_u$. G_0 is used in Methods 1 and 2 until the shaft resistance is exhausted. A secant shear modulus is employed for the remaining settlement in Method 2, resulting in a bilinear spring to represent the base (shown in Fig. 3a). This is estimated using Eq. (9) and the predicted mobilisation of the base resistance once the shaft resistance was exhausted and at failure (shown in Fig. 3b).

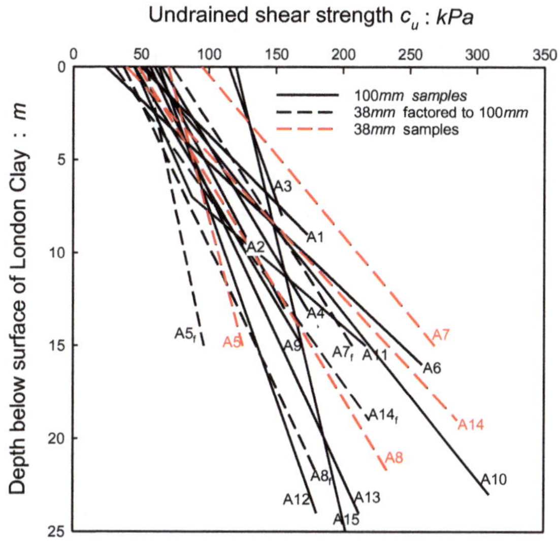

Fig. 4. Undrained shear strength variation with depth for the reported sites in London Clay as determined from triaxial tests on 100 and 38 mm samples (*modified after* Patel 1992). Subscript *f* denotes factored profiles.

3.5 Results

Load-settlement curves for each test are presented in Figs. 5, 6 and 7 comparing the three predicted curves with the measured data. The head displacement is normalised with the pile base diameter and the applied load is normalised with the predicted ultimate load using Eq. (13). This has been equated to the inverse of the design Factor of Safety (F_{total}). The predicted ultimate load for each pile is shown in Table 3 with the maximum load tested and the valid prediction range of each method. All three methods predict the settlement well at low loads, indicating the deformation parameters chosen are realistic. Method 1 under-predicts settlement significantly at high loads when nonlinear soil behaviour dominates. Method 2 accounts for this relatively well, although the linear elastic perfectly-plastic model provides a sharp transition to yielding. Method 3 shows good agreement over the range for which it is valid but should not be used beyond the point where the shaft resistance is exhausted.

The predicted and measured settlements at specific F_{total} values have been interpolated. Predicted versus measured plots are needed to show the bandwidth of error when using a model to predict observations (cf. Koutsoftas et al. 2017; Kootahi and Mayne 2017). Figure 8 shows the predicted vs. measured settlement for each model at F_{total} = 2.5. Vardanega et al. (2012b) shows that various codified design approaches for a pile in London Clay imply F_{total} values around this level while, LDSA (2009) recommends a value of 2.6 for piles in London Clay where no load testing is completed. 50% error bounds and the percentage of points within this range for each method are shown. Higher points indicate unconservative predictions. Figure 9 illustrates similar plots for F_{total} values of 5, 3, 2, and 1.5. These allow a designer to observe

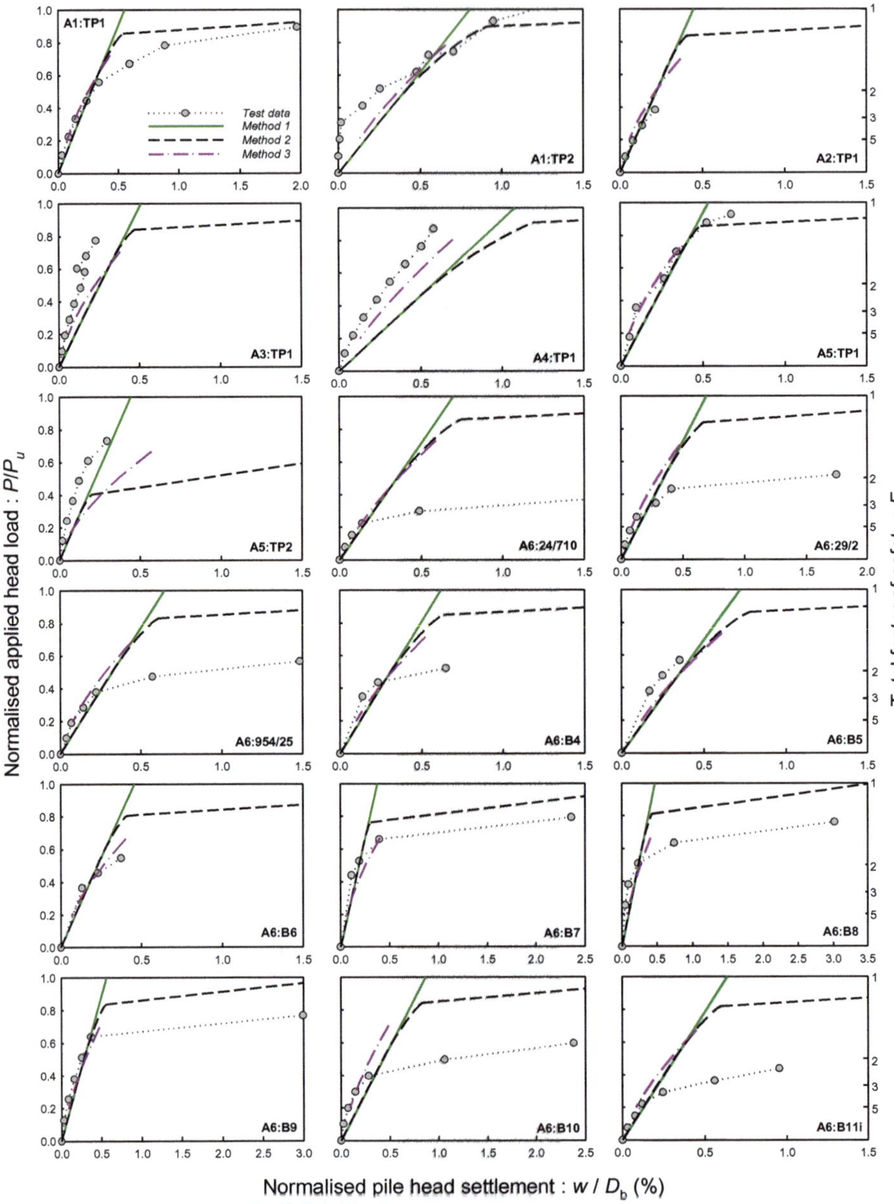

Fig. 5. Comparison of the predicted load–settlement curves using the three methods against the experimental results reported in Patel 1992. Method 1 (Crispin et al. 2018), Method 2 (Crispin et al. 2018), Method 3 (Vardanega et al. 2012a, 2018). (Test piles shown: A1/TP1 to A6/B11i).

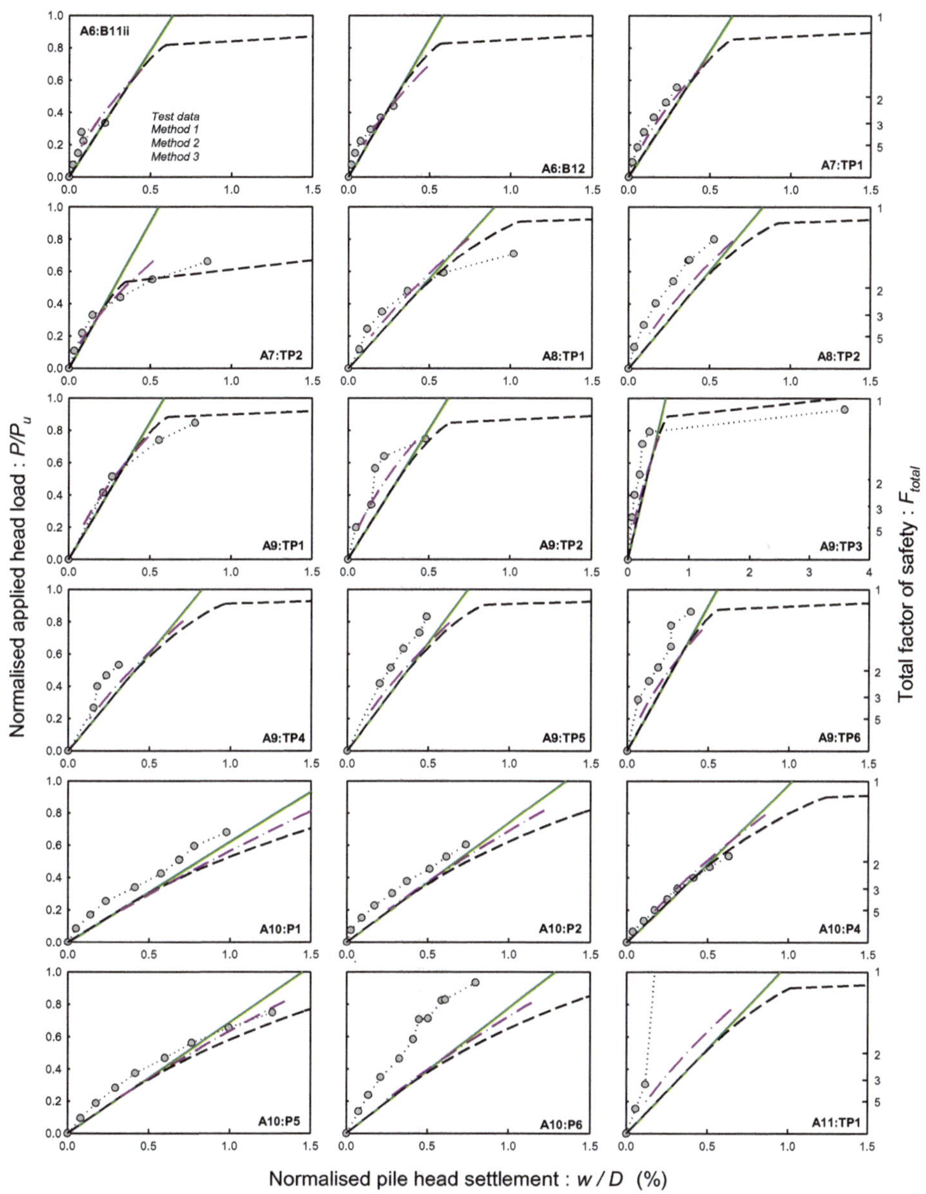

Fig. 6. Comparison of the predicted load–settlement curves using the three methods against the experimental results reported in Patel 1992. Method 1 (Crispin et al. 2018), Method 2 (Crispin et al. 2018), Method 3 (Vardanega et al. 2012a, 2018). (Test piles shown: A6/B11ii to A11/TP1).

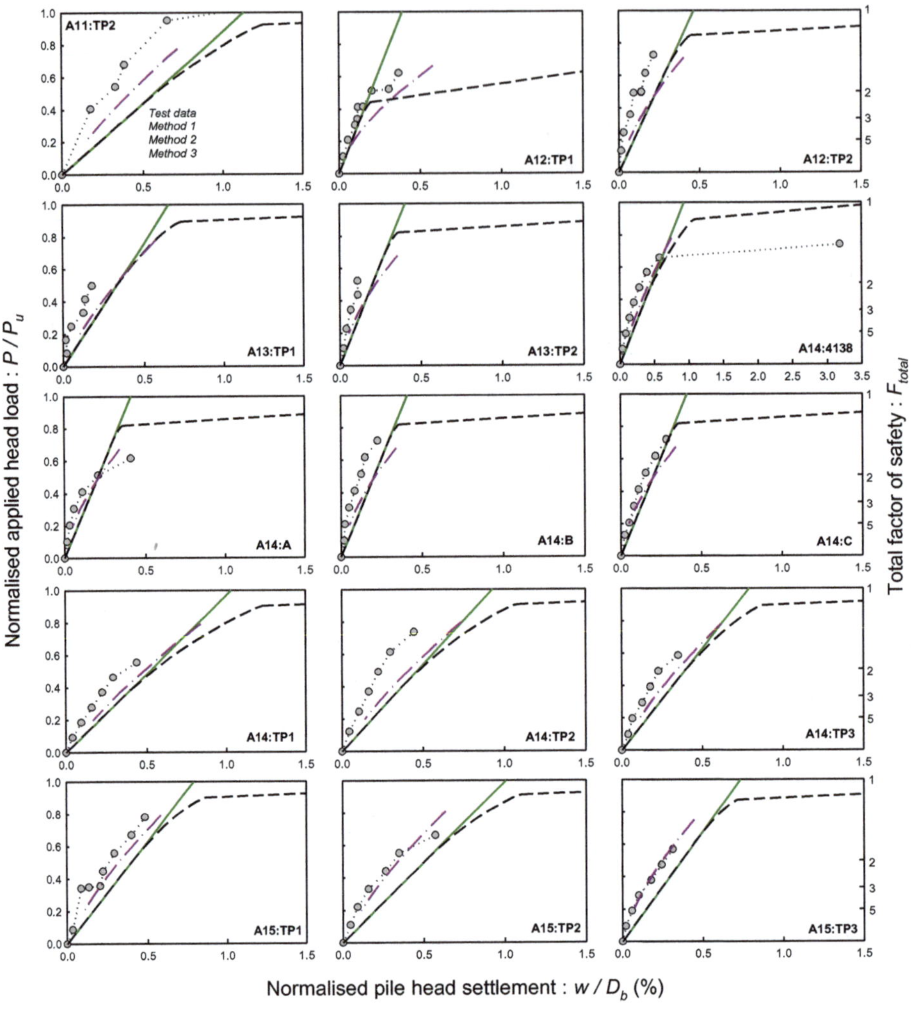

Fig. 7. Comparison of the predicted load–settlement curves using the three methods against the experimental results reported in Patel 1992. Method 1 (Crispin et al. 2018), Method 2 (Crispin et al. 2018), Method 3 (Vardanega et al. 2012a, 2018). (Test piles shown: A11/TP2 to A15/TP3).

the possible variation in pile settlement predictions in London Clay at different factors of safety. The maintained load tests in the database studied here were not carried out to failure so the number of tests on the plot reduces as the applied load increases. All three models seem to initially give conservative predictions, however, as the F_{total} gets lower it is apparent that some results were significantly under-predicted.

Table 3. Applied and predicted loads

Site	Pile	Predicted ultimate load, P_u (kN)	Normalised maximum applied test load, P_{ML}/P_u	Normalised load prediction range, P/P_u		
				Method 1	Method 2	Method 3
A1	TP1	445	0.90	0	0–1	0–0.73
	TP2	724	1.03	0	0–1	0–0.78
A2	TP1	517	0.67	0	0–1	0–0.70
A3	TP1	463	0.78	0	0–1	0–0.71
A4	TP1	1141	0.87	0	0–1	0–0.81
A5	TP1	783	0.93	0	0–1	0–0.73
	TP2	3999	0.73	0	0–1	0–0.67
A6	24/7 10	2156	0.67	0	0–1	0–0.73
	29/2	1871	0.67	0	0–1	0–0.70
	95 4/25	1716	0.67	0	0–1	0–0.69
	B4	1864	0.67	0	0–1	0–0.71
	B5	2235	0.67	0	0–1	0–0.73
	B6	1215	0.67	0	0–1	0–0.67
	B7	841	0.79	0	0–1	0–0.67
	B8	1277	0.76	0	0–1	0–0.67
	B9	1655	0.77	0	0–1	0–0.70
	B10	2134	0.67	0	0–1	0–0.71
	B11i	2907	0.67	0	0–1	0–0.68
	B11ii	2907	0.67	0	0–1	0–0.68
	B12	2875	0.67	0	0–1	0–0.68
A7	TP1	2789	0.67	0	0–1	0–0.72
	TP2	4961	0.67	0	0–1	0–0.67
A8	TP1	2147	0.71	0	0–1	0–0.81
	TP2	1865	0.80	0	0–1	0–0.79
A9	TP1	756	0.85	0	0–1	0–0.76
	TP2	601	0.75	0	0–1	0–0.74
	TP3	794	0.93	0	0–1	0–0.77
	TP4	1127	0.67	0	0–1	0–0.81
	TP5	1010	0.83	0	0–1	0–0.80
	TP6	695	0.86	0	0–1	0–0.75
A10	P1	4711	0.68	0	0–1	0–0.83
	P2	3976	0.67	0	0–1	0–0.81
	P4	3003	0.67	0	0–1	0–0.79
	P5	4269	0.75	0	0–1	0–0.82
	P6	3791	0.93	0	0–1	0–0.81
A11	TP1	980	1.22	0	0–1	0–0.77
	TP2	1103	1.09	0	0–1	0–0.78

(*continued*)

Table 3. (*continued*)

Site	Pile	Predicted ultimate load, P_u (kN)	Normalised maximum applied test load, P_{ML}/P_u	Normalised load prediction range, P/P_u		
				Method 1	Method 2	Method 3
A12	TP1	5428	0.67	0	0–1	0–0.67
	TP2	2673	0.73	0	0–1	0–0.71
A13	TP1	1797	0.67	0	0–1	0–0.78
	TP2	870	0.67	0	0–1	0–0.68
A14	4138	2419	0.74	0	0–1	0–0.78
	A	194	0.67	0	0–1	0–0.67
	B	194	0.72	0	0–1	0–0.67
	C	194	0.72	0	0–1	0–0.67
	TP1	2693	0.67	0	0–1	0–0.79
	TP2	1221	0.74	0	0–1	0–0.80
	TP3	1024	0.67	0	0–1	0–0.78
A15	TP1	3347	0.78	0	0–1	0–0.79
	TP2	2258	0.67	0	0–1	0–0.81
	TP3	1578	0.67	0	0–1	0–0.75

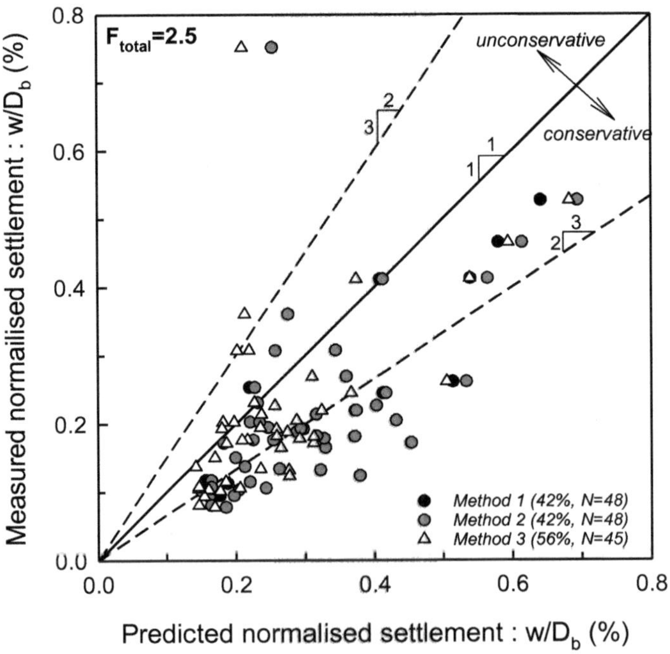

Fig. 8. Predicted vs. measured pile settlement w, normalised with pile base diameter D_b. Measured values for Patel (1992) test data. Predicted values generated by the three presented analytical approaches. Dashed lines represent ±50% prediction intervals. N denotes number of tests, loaded to that level, where methods apply. Percentages within the ±50% bounds are shown in parentheses in the legend. (Factor of safety F_{total} = 2.5).

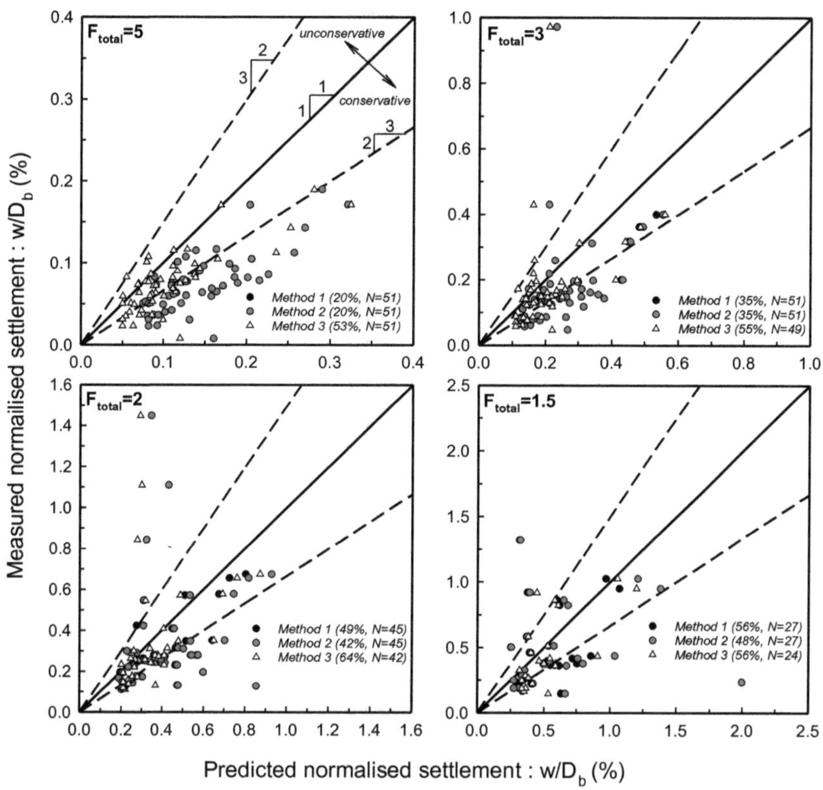

Fig. 9. Predicted vs. measured pile settlement w, normalised with pile base diameter D_b. Measured values for Patel (1992) test data. Predicted values by the three presented analytical approaches. Dashed lines represent ±50% prediction intervals. N denotes number of tests, loaded to that level, where methods apply. Percentages within the ±50% bounds are shown in parentheses in the legends.

4 Conclusions

An opensource, user-friendly database for UK pile tests is being compiled at the University of Bristol. Preliminary analyses were presented concerning the data from Patel's (1989) database of field tests. Experimental measurements were compared to three selected theoretical approaches from three groups: (i) linear elastic, (ii) elasto-plastic "*t-z*" analysis using a linear elastic perfectly-plastic soil model, and (iii) elastoplastic "*t-z*" analysis using a power-law non-linear soil model. All three selected methods, although idealised, show good agreement with the test data. An alternative representation comparing predicted vs. measured normalised pile head settlement proves desirable, as it allows for a better understanding of the accuracy of the theoretical models in a statistical sense.

5 Data Availability Statement

The work reported in this paper has not generated new experimental data.

Acknowledgments. The authors acknowledge the support of University of Bristol and "Databases to INterrogate Geotechnical Observations (DINGO)" EP/P020933/1 EPSRC. The second and third authors were supported by the Engineering and Physical Sciences Research Council [Grant Numbers: EP/N509619/1 and EP/R51245X/1, respectively]. The authors would like to thank Professor George Mylonakis for his valuable support in the preparation of this manuscript and Dr Melvin England for his helpful comments and suggestions.

References

AbdelSalam, S.S., Baligh, F.A., El-Naggar, H.M.: A database to ensure reliability of bored pile design in Egypt. Proc. Inst. Civ. Eng. Geotech. Eng. **168**(2), 131–143 (2015). https://doi.org/10.1680/geng.14.00051

Anoyatis, G.M.: Contribution to kinematic and inertial analysis of piles by analytical and experimental methods. Ph.D. thesis. University of Patras, Patras (2013)

Association of Geotechnical and Geoenvironmental Specalists (AGS): Electronic Transfer of Geotechnical and Geoenvironmental Data, Edition 4.0.4, Bromley, Kent (2017)

Baguelin, F., Frank, R.: Theoretical studies of piles using the finite element method. In: Numerical Methods in Offshore Piling, Institution of Civil Engineers (ICE), London, pp. 83–91 (1980)

Bond, A., Basile, F.: Geocentrix Repute 2.5 Reference Manual – Onshore Pile Design and Analysis. Geocentrix Ltd (2017). http://www.geocentrix.co.uk/repute/downloads/Repute%202.5%20Reference%20Manual.pdf. Accessed 29 June 2018

Butterfield, R., Banerjee, P.K.: The elastic analysis of compressible piles and pile groups. Géotechnique **21**(1), 43–60 (1971). https://doi.org/10.1680/geot.1971.21.1.43

Cooke, R.W.: The settlement of friction pile foundations. In: Proceedings of the Conference Tall Buildings, Kuala Lumpur, no. 3, pp. 7–19 (1974). Also available as building research station current paper CP 12/75

Coyle, H.M., Reese, L.C.: Load transfer for axially loaded piles in clay. J. Soil Mech. Found. Div. (ASCE) **92**(2), 1–26 (1966)

Crispin J.J., Leahy, C.P., Mylonakis, G.: Winkler Model for Axially-Loaded Piles in Inhomogeneous Soil (2018). (under review)

Deltares: D-Pile Group User Manual – 3D modelling of simple pile and pile groups. Deltares (2016). https://content.oss.deltares.nl/delft3d/manuals/DPileGroup-Manual.pdf. Accessed 12 April 2018

Fine (n.d.) Engineering manuals for GEO5 programs – part 2. Fine Ltd. http://geo5-software.de/de/support_files/projektierungshandbucher/geo5-engineering manuals_em2.pdf. Accessed 12 April 2018

Fleming, W.G.K.: A new method for single pile settlement prediction and analysis. Géotechnique **42**(3), 411–425 (1992). https://doi.org/10.1680/geot.1992.42.3.411

Fleming, W.G.K., Weltman, A.J., Randolph, M.F., Elson, W.K.: Piling Engineering, 3rd edn. Wiley, New York (2009)

Galbraith, A.P., Farrell, E.R., Byrne, J.J.: Uncertainty in pile resistance from static load tests database. Proc. Inst. Civ. Eng. Geotech. Eng. **167**(5), 431–446 (2014). https://doi.org/10.1680/geng.12.00132

Guo, W.D.: Theory and Practice of Pile Foundations. CRC Press, Boca Raton (2012)
Koutsoftas, D.C., Karina, K., Hashash, Y.M.A.: Discussion of "index test method for estimating the effective preconsolidation stress in clay deposits" by Karim Kootahi and Paul W. Mayne. J. Geotech. Geoenviron. Eng. American Society of Civil Engineers **143**(10), article no. 07017030 (2017). https://doi.org/10.1061/(ASCE)GT.1943-5606.0001764
Kootahi, K., Mayne, P.W.: Closure to "index test method for estimating the effective preconsolidation stress in clay deposits" by Karim Kootahi and Paul W. Mayne. J. Geotech. Geoenviron. Eng. (ASCE) **143**(10), article no. 07017031 (2017). https://doi.org/10.1061/(ASCE)GT.1943-5606.0001765
LDSA: London District Surveyors Association: Foundations No. 1: Guidance Notes for the Design of Straight Shafted Bored Piles in London Clay, Bromley, UK (2009)
Lemnitzer, A., Favaretti, C.: Find a Pile.com (2013). http://www.findapile.com/. Accessed 29 June 2018
Mylonakis, G., Gazetas, G.: Settlement and additional internal forces of grouped piles in layered soil. Géotechnique **48**(1), 55–72 (1998). https://doi.org/10.1680/geot.1998.48.1.55
Mylonakis, G.: Winkler modulus for axially loaded piles. Géotechnique **51**(5), 455–461 (2001)
OASYS: Pile Oasys Geo Suite for Windows (2017). https://www.oasys-software.com/wp-content/uploads/2017/11/pile19.6_manual.pdf. Accessed 29 June 2018
Ottaviani, M.: Three-dimensional finite element analysis of vertically loaded pile groups. Géotechnique **25**(2), 159–174 (1975). https://doi.org/10.1680/geot.1975.25.2.159
Paikowsky S.G., Brigisson, B., McVay, M., Nguyen, T., Kuo, C., Baecher, G., Ayyub, B., Stenersen, K., O'Malley, K., Chernauskas, L., O'Neill, M.: Load and Resistance Factor Design (LRFD) for Deep Foundations. NCHRP Report 507, National Cooperative Highway Research Program, Transportation Research Board (2004)
Patel, D.C.: A case study of the shaft friction of bored piles in London Clay in terms of total and effective stresses. MSc thesis, Imperial College London, UK (1989)
Patel, D.C.: Interpretation of Results of Pile Tests in London Clay. Piling Europe, Thomas Telford, London (1992)
Poulos, H.G.: Analysis of the settlement of pile groups. Géotechnique **18**, 449–471 (1968)
Poulos, H.G.: Pile behaviour – theory and application. Géotechnique **39**(3), 365–415 (1989). https://doi.org/10.1680/geot.1989.39.3.365
Poulos, H.G.: Common procedures for foundation settlement analysis – are they adequate? Aust. Geomech. J. **34**(1), 13–38 (1999)
Poulos, H.G., Davis, E.H.: The settlement behaviour of singly axially loaded incompressible piles and piers. Géotechnique **18**(3), 351–371 (1968). https://doi.org/10.1680/geot.1968.18.3.351
Randolph, M.F., Wroth, C.P.: Analysis of deformation of vertically loaded piles. J. Geotech. Eng. Div. American Society of Civil Engineers **104**(12), 1465–1488 (1978)
Randolph, M.F., Wroth, C.P.: Analysis of vertical deformation of pile groups. Géotechnique **29**(4), 423–439 (1979)
Randolph, M.F.: Science and empiricism in pile foundation design. Géotechnique **53**(10), 847–875 (2003). https://doi.org/10.1680/geot.2003.53.10.847
Randolph, M.F.: PIGLET – Analysis and Design of Pile Groups. University of Western Australia, Crawley (2006)
Reese, L.C., Wang, S.T., Arrellaga, J.A.: TZPILE 2014 User's Manual - Analysis of Load Versus Settlement for an Axially-Loaded Deep Foundation. ENSOFT INC., Austin (2014). https://www.ensoftinc.com/products/tzpile/doc/tzpile-users-manual.pdf. Accessed 29 June 2018
Scott, R.F.: Foundation Analysis. Prentice Hall, Englewood Cliffs (1981)

Seed, H.B., Reese, L.C.: The action of soft clay along friction piles. Trans. Am. Soc. Civ. Eng. **122**, 731–764 (1957)

Skempton, A.W.: Cast-in-situ bored piles in London clay. Géotechnique **9**(4), 153–173 (1959). https://doi.org/10.1680/geot.1959.9.4.153

Syngros, K.: Seismic Response of piles and pile-supported bridge piers evaluated through case histories. PhD thesis, The City College of the City University of New York, NY, USA (2004)

Tomlinson, M., Woodward, J.: Pile Design and Construction in Practice, 6th edn. CRC Press, Boca Raton (2014)

Vardanega, P.J., Bolton, M.D.: Strength mobilization in clays and silts. Can. Geotech. J. **48**(10), 1485–1503 (2011a) http://doi.org/10.1139/T11-052 and Corrigendum, **49**(5), 631, http://doi.org/10.1139/t2012-023

Vardanega, P.J., Bolton, M.D.: Predicting shear strength mobilisation of London clay. In: Anagnostopoulos, A., et al. (ed.) Proceedings 15th European Conference on Soil Mechanics and Geotechnical Engineering: Geotechnics of Hard Soils – Weak Rocks, vol. 1, pp. 487–492. IOS Press, Amsterdam (2011b). http://doi.org/10.3233/978-1-60750-801-4-487

Vardanega, P.J., Williamson, M.G., Bolton, M.D.: Bored pile design in stiff clay II: mechanisms and uncertainty. Proc. Inst. Civ. Eng. Geotech. Eng. **165**(4), 233–246 (2012a). https://doi.org/10.1680/geng.11.00063; Corrigendum, **166**(5), 518; https://doi.org/10.1680/geng.2013.166.5.518

Vardanega, P.J., Kolody, E., Pennington, S.H., Morrison, P.R.J., Simpson, B.: Bored pile design in stiff clay I: codes of practice. Proc. Inst. Civ. Eng. Geotech. Eng. **165**(4), 213–232 (2012b). https://doi.org/10.1680/geng.11.00062

Vardanega, P.J., Crispin, J.J., Gilder, C.E.L., Voyagaki, E., Shepheard, C.J., Holcombe, E.A.: Geodatabases to improve geotechnical design and modelling. Ce/Papers **2**(2–3), 401–406 (2018). https://doi.org/10.1002/cepa.704

Whitaker, T., Cooke, R.W.: An investigation of the shaft and base resistance of large bored piles in London Clay. In: Large Bored Piles: Proceedings of the Symposium by the Institution of Civil Engineers and the Reinforced Concrete Association; Institution of Civil Engineers, London, UK, pp. 7–49 (1966)

Yang, Z., Jardine, R., Guo, W., Chow, F.: A new and openly accessible database of tests on piles driven in sands. Géotech. Lett. **5**(1), 12–20 (2015). https://doi.org/10.1680/geolett.14.00075

Yang, Z., Jardine, R., Guo, W., Chow, F.: A comprehensive database of tests on axially loaded piles driven in sand. Zhejiang University Press Co. Ltd. Published by Elsevier Inc., New York (2016)

Energy Formulation for Axial Pile Head Stiffness in Inhomogeneous Soil

Jamie Crispin[1(✉)], Huize Su[1], Katherine Round[1], and George Mylonakis[2]

[1] University of Bristol, Bristol, UK
j.crispin@bristol.ac.uk
[2] Chair in Geotechnics and Soil Structure Interaction, University of Bristol, Bristol, UK

Abstract. Approximate closed-form solutions are derived for the settlement of axially loaded piles embedded in inhomogeneous soil by employing an energy method based on the Winkler model of soil reaction. Simple, linear shape functions for the attenuation of pile displacement with depth result in easy-to-use formulas. Two alternative methods are discussed for selecting the slope of the linear function. The proposed methodology is compared with rigorous (yet complicated) closed form solutions and an alternative approximation using a representative homogeneous soil.

1 Introduction

The Winkler model for axially loaded piles is a popular and cost-effective approach as it allows the development of simple, closed form solutions for predicting pile settlement without resorting to numerical analysis (Randolph and Wroth 1978; Poulos and Davis 1980; Scott 1981; Mylonakis and Gazetas 1998; Guo 2012; Crispin et al. 2018). Comparison of these solutions to more rigorous continuum analyses allows the Winkler spring stiffness to be determined by matching the settlement at the pile head between the two models (Mylonakis 2001; Syngros 2004; Guo 2012; Anoyatis 2013). In addition, the model can be extended to inhomogeneous soils in a straightforward manner by employing depth-dependent springs. For soil stiffness varying according to a power-law function of depth, closed-form solutions for pile head stiffness are available in the literature (Scott 1981; Guo 2012; Crispin et al. 2018). However, these solutions are complex, expressed in terms of special functions of mathematical physics, and are limited to a handful of inhomogeneous soil profiles.

This paper investigates an energy method based on a corresponding solution for laterally loaded piles by Karatzia and Mylonakis (2012, 2016). The proposed method applies the theory of virtual work to the Winkler model via an approximate shape function describing the attenuation of pile displacement with depth, to provide an easier to apply yet still realistic solution. By relating the Winkler modulus to soil properties using either the concentric cylinder model (Randolph and Wroth 1978) or by matching the results to the aforementioned continuum analyses, this solution can be applied to design problems.

H. El-Naggar et al. (Eds.): GeoMEast 2018, SUCI, pp. 181–190, 2019.
https://doi.org/10.1007/978-3-030-01902-0_15

2 Method

Equation (1) shows the familiar governing equation for an axially loaded pile attached to a bed of Winkler springs with depth dependent modulus, $k(z)$ (Scott 1981).

$$E_pAw''(z) - k(z)w(z) = 0 \tag{1}$$

where E_p and A denote the pile Young's modulus and cross-sectional area, respectively, and $w(z)$ is the pile settlement at depth z. By considering the work done by a small virtual displacement profile, $w^*(z)$, over the pile length, L, Eq. (1) can be expressed in an integral form as:

$$\int_0^L E_pAw''(z)w^*(z)dz - \int_0^L k(z)w(z)w^*(z)dz = 0 \tag{2}$$

Following the related solution for laterally loaded piles by Karatzia and Mylonakis (2012, 2016), a dimensionless shape function, $\phi(z)$, is used to describe both the real and virtual displacements by normalising with the corresponding value at the pile head. The pile head stiffness, K_0, can then be expressed in terms of this shape function in the following virtual work equation:

$$K_0 = E_pA\int_0^L [\phi'(z)]^2dz + \int_0^L k(z)[\phi(z)]^2dz + \beta^2K_b \tag{3}$$

where $\beta = \phi(L)$ is the fraction of the pile head settlement propagating to the pile base, and K_b is the stiffness of the lumped spring used to model the pile base reaction. Note that the first term expresses the energy stored in the pile body, the second term the energy stored in the distributed Winkler springs that represent the stiffness (mainly through shearing) of the surrounding soil, and the final term the energy stored in the base spring that represents the soil below the pile. The sum of the three energies equals the work done by the force at the pile head under a unit virtual displacement.

Unlike in the lateral case where most piles of practical dimensions may be treated as infinitely long beams, the pile base conditions have a significant effect on the axial response. Therefore, the shape functions derived from the homogeneous soil case are relatively complex and do not lend themselves to the simple formulations desired. As an alternative, this paper investigates the benefits from using a simple linear shape function given in Eq. (4) for predicting pile head stiffness.

$$\phi(z) = \frac{w(z)}{w(0)} = \frac{w^*(z)}{w^*(0)} = 1 - (1-\beta)\frac{z}{L} \tag{4}$$

2.1 Homogeneous Soil

In the case of homogeneous soil, $k(z) = k$, and the normalised pile head stiffness approximated using Eqs. (3) and (4) is given by:

$$\frac{K_0}{E_p A \lambda} = \frac{1}{\lambda L}(1-\beta)^2 + \frac{1}{3}\lambda L(\beta^2+\beta+1) + \beta^2\Omega \tag{5}$$

where λ is a load transfer parameter (units of Length^{-1}) controlling the attenuation of pile displacement with depth and Ω is a dimensionless base stiffness coefficient:

$$\lambda = \sqrt{\frac{k}{E_p A}}, \quad \Omega = \frac{K_b}{E_p A \lambda} \tag{6a, b}$$

The coefficient Ω can take any value between 0, representing a fully-floating pile, and ∞, representing an end-bearing pile. Using Eq. (6a) and $k \approx 0.6E_s$ (Roesset and Angelides 1980), λL can be approximated by (Mylonakis 1995):

$$\lambda L \approx 0.85\left(\frac{E_p}{E_s}\right)^{-\frac{1}{2}}\left(\frac{L}{d}\right) \tag{7}$$

where E_s is the Young's modulus of the soil and d is the pile diameter.

A simple method to estimate β, referred to here as *exact* β, is to use the exact solution for homogeneous soil (Crispin et al. 2018):

$$\beta = \frac{w(L)}{w(0)} = [\Omega\sinh(\lambda L) + \cosh(\lambda L)]^{-1} \tag{8}$$

However, in light of the approximate shape function in Eq. (4), this choice would result in equilibrium of external forces on the pile not being satisfied. Considering equilibrium of vertical forces on the pile yields the following expression for pile head stiffness:

$$K_0 = \int_0^L k(z)\varphi(z)dz + \beta K_b \tag{9}$$

For the homogeneous soil case, using the shape function in Eq. (4), Eq. (9) evaluates to:

$$\frac{K_0}{E_p A \lambda} = \frac{1}{2}(1+\beta)\lambda L + \beta\Omega \tag{10}$$

This expression can be used with the *exact* β in Eq. (8). Alternatively, an improved equation for β that obeys the equilibrium condition, referred to here as *matched* β, can be obtained by equating Eq. (10) with Eq. (5) and solving for β:

$$\beta = \frac{6 - (\lambda L)^2}{6[1 + \Omega\lambda L] + 2(\lambda L)^2} \tag{11}$$

Figure 1a compares the proposed linear shape functions using both the *matched* β and *exact* β against the exact shape function for an example pile in homogeneous soil. It is evident that the *matched* β results in a slightly closer fit to the exact shape function. Figure 1b shows the variation of β with dimensionless length, λL, when $\Omega = 0$. Both estimates are very similar for $\lambda L < 1.5$, which covers a wide range of pile dimensions. However, the *matched* β behaves unexpectedly at larger values, predicting upward displacement of the pile base, which is clearly inadmissible.

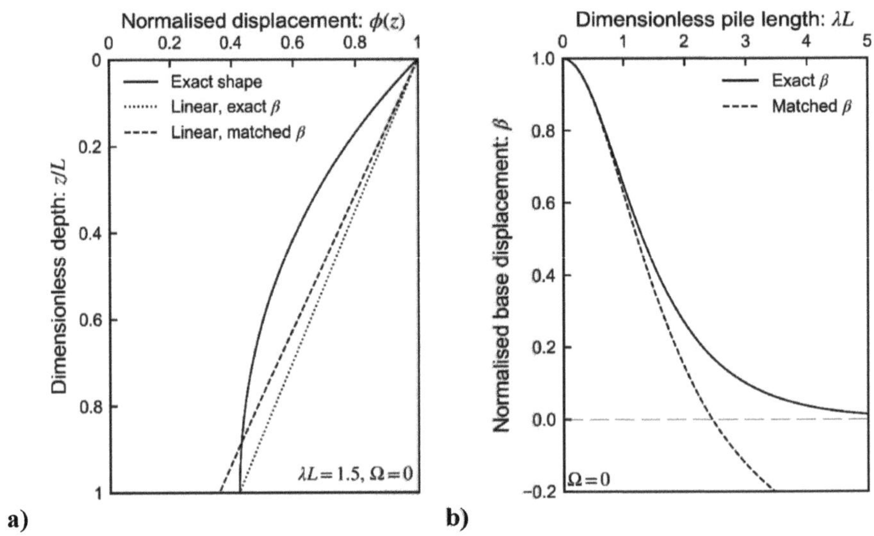

Fig. 1. (a) Exact and approximate shape functions for example pile. (b) Variation of *exact* β and *matched* β with pile length, $\Omega = 0$.

In Fig. 2, Eqs. (5) and (10) with both the *exact* β and *matched* β are compared to the closed form solution to Eq. (1) for homogeneous soil (Mylonakis and Gazetas 1998):

$$\frac{K_0}{E_p A \lambda} = \frac{\Omega + \tanh(\lambda L)}{1 + \Omega \tanh(\lambda L)} \tag{12}$$

Both approximate solutions match Eq. (12) well when $\lambda L < 1$ (i.e. $L/d < 40$ for $E_p/E_s = 10^3$), with the *matched* β providing the closest results.

2.2 Inhomogeneous Soil

For homogeneous soils, the utility of the proposed approximate solution is limited by the simplicity of the exact solution. However, in inhomogeneous soils exact solutions

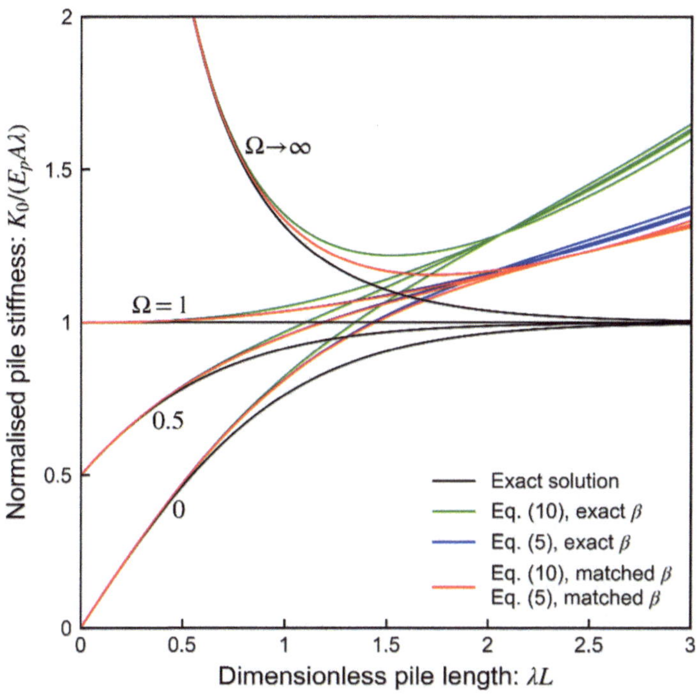

Fig. 2. Variation of exact and approximate pile head stiffness in homogeneous soil with pile length.

have only been found for a limited number of soil stiffness variations with depth. The resulting formulae use special functions, such as those of the Airy or the Bessel type, that are unfamiliar to many practicing engineers. A simpler solution to derive and implement is desired. Extension of the proposed energy solution to inhomogeneous soil profiles only requires integration of the second term in Eq. (3) using a depth varying spring stiffness, $k(z)$, rather than solving the full governing differential equation. This paper investigates Winkler modulus varying according to the power-law function of depth given in Eq. (13). Closed form solutions to similar stiffness variations have been provided by Scott (1981), Guo (2012) and Crispin et al. (2018).

$$k(z) = k_L\left[a + (1-a)\frac{z}{L}\right]^n, \quad a = \left(\frac{k_0}{k_L}\right)^{\frac{1}{n}} \tag{13a, b}$$

where k_0 and k_L are the Winkler modulus at the ground surface and pile base, respectively, a is an inhomogeneity parameter accounting for non-zero surface stiffness and n is an inhomogeneity exponent. Substituting this variation in Winkler modulus into Eq. (3) and performing the integration yields:

$$\frac{K_0}{E_pA\lambda_L} = \frac{(1-\beta)^2}{\lambda_L L} + \frac{\lambda_L L}{(1-a)^3}\left[\frac{(1-a^{n+1})(1-a\beta)^2}{(n+1)} - \frac{2(1-a^{n+2})(1-a\beta)(1-\beta)}{(n+2)} + \frac{(1-a^{n+3})(1-\beta)^2}{(n+3)}\right] + \beta^2\Omega_L \tag{14}$$

where λ_L and Ω_L are the values of λ and Ω considering the soil properties at the pile base:

$$\lambda_L = \sqrt{\frac{k_L}{E_pA}}, \quad \Omega_L = \frac{K_b}{E_pA\lambda_L} \tag{15a, b}$$

An alternative approximate approach to account for soil inhomogeneity is to use the exact solution for homogeneous soil in Eq. (12) with the average Winkler modulus over the pile length, k_{av}. In this case:

$$\frac{k_{av}}{k_L} = \frac{1-a^n}{(n+1)(1-a)}, \quad \lambda = \lambda_L\sqrt{\frac{k_{av}}{k_L}}, \quad \Omega = \Omega_L\sqrt{\frac{k_L}{k_{av}}} \tag{16a, b, c}$$

This is a generalisation of the approximation for linear stiffness variations proposed by Randolph and Wroth (1978).

The error introduced by both approximate solutions, Eq. (14) with the *matched* β in Eqs. (11), (12) and (16), is compared in Fig. 3 for $\lambda L = 1$ (i.e. L/d $\approx$ 40 for $E_p/E_s = 10^3$) and $\Omega = 0$ (when the error in both approximations peaks). The exact solution to Eq. (1) when $k(z)$ is described by Eq. (13) is given by (Crispin et al. 2018):

$$\frac{K_0}{E_pA\lambda_L} = a^{\frac{n}{2}}\frac{[I_{v-1}(\chi_0)I_{1-v}(\chi_L) - I_{1-v}(\chi_0)I_{v-1}(\chi_L)] + \Omega_L[I_{v-1}(\chi_0)I_{-v}(\chi_L) - I_{1-v}(\chi_0)I_{+v}(\chi_L)]}{[I_{-v}(\chi_0)I_{v-1}(\chi_L) - I_{+v}(\chi_0)I_{1-v}(\chi_L)] + \Omega_L[I_{-v}(\chi_0)I_{+v}(\chi_L) - I_{+v}(\chi_0)I_{-v}(\chi_L)]} \tag{17}$$

where $I_v(\chi)$ is the modified Bessel functions of the first kind, of order v and argument χ:

$$v = \frac{1}{n+2}, \ \chi_L = \frac{2\lambda_L L}{(1-a)(n+2)}, \ \chi_0 = \chi_L a^{\frac{n+2}{2}} \tag{18a, b, c}$$

The performance of this solution has been evaluated against pile field test results in London Clay by Voyagaki et al. (2018) and Crispin et al. (2018) by employing the concentric cylinder model and magical radius proposed in Randolph and Wroth (1978).

The average homogeneous solution performs better the closer the stiffness variation is to a uniform distribution ($a = 1$ or $n = 0$), while the accuracy of the proposed solution is almost entirely independent of the inhomogeneity function. For $a > 0.3$, the average homogeneous solution provides a better approximation of the pile head stiffness. However, for $a = 0$ and $n = 1$, the error in the average homogeneous solution is approximately double that for the proposed solution. Similar results have been observed for different pile lengths and base conditions with only the magnitude of the errors varying. Consequently, Eqs. (14) and (9) have been used to derive a new expression for pile head stiffness when $a = 0$ incorporating an inhomogeneous *matched*

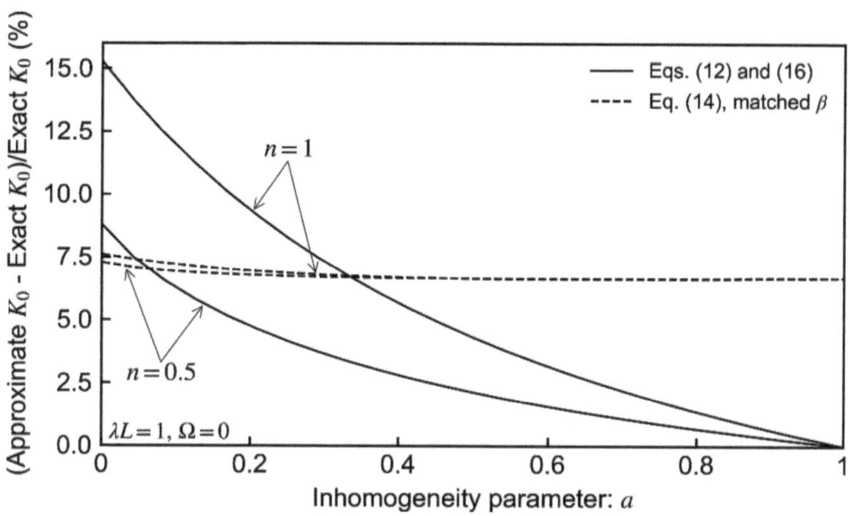

Fig. 3. Error in pile head stiffness estimation for different soil stiffness profiles.

β. The general expression is given in Eq. (19) and simplified expressions for different soil stiffness variations and base conditions are provided in Table 1.

$$\frac{K_0}{E_pA\lambda_L} = \frac{(n+2)\lambda_L L[(n+2)(n+3)+2\Omega_L\lambda_L L]+(\lambda_L L)^3+(n+1)(n+2)^2(n+3)\Omega_L}{(n+1)(n+2)^2(n+3)[1+\Omega_L\lambda_L L]+(n+1)(n+2)^2(\lambda_L L)^2} \tag{19}$$

Table 1. Summary of proposed pile head stiffness expressions (Eq. 19) for different soil stiffness variations and base conditions

$\frac{K_0}{E_pA\lambda_L}$	General case $0<L<\infty$, $0\le\Omega_L<\infty$	Perfectly floating pile, $\Omega_L=0$	Perfectly end-bearing pile, $\Omega_L\to\infty$
$n=0$	$\frac{4\lambda L(3+\Omega_L\lambda_L L)+(\lambda_L L)^3+12\Omega_L}{12(1+\Omega_L\lambda_L L)+4(\lambda_L L)^2}$	$\frac{12\lambda_L L+(\lambda_L L)^3}{12+4(\lambda_L L)^2}$	$\frac{(\lambda_L L)^2+3}{3\lambda_L L}$
$n=1/2$	$\frac{10\lambda_L L(35+8\Omega_L\lambda_L L)+16(\lambda_L L)^3+525\Omega_L}{525(1+\Omega_L\lambda_L L)+150(\lambda_L L)^2}$	$\frac{350\lambda_L L+16(\lambda_L L)^3}{525+150(\lambda_L L)^2}$	$\frac{16(\lambda_L L)^2+105}{105\lambda_L L}$
$n=1$	$\frac{6\lambda_L L(6+\Omega_L\lambda_L L)+(\lambda_L L)^3+72\Omega_L}{72(1+\Omega_L\lambda_L L)+18(\lambda_L L)^2}$	$\frac{36\lambda_L L+(\lambda_L L)^3}{72+18(\lambda_L L)^2}$	$\frac{(\lambda_L L)^2+12}{12\lambda_L L}$

Figure 4 shows the variation of predicted pile head stiffness with length for the proposed expression, the equivalent homogeneous soil approximation using Eqs. (12) and (16) and the exact solution in Eq. (17) when $\Omega=0$. The variation in error with pile length of Eq. (19) is compared to the average homogeneous solution when $\Omega=0$ in

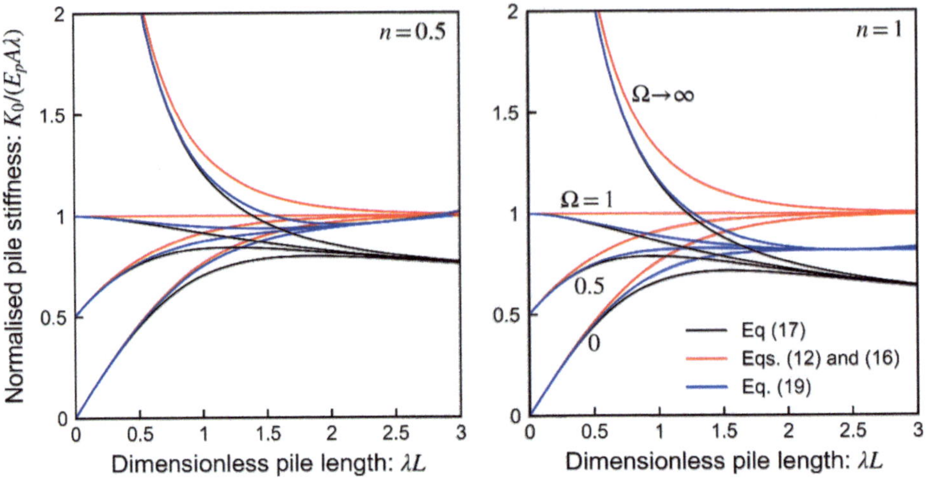

Fig. 4. Variation of exact and approximate pile head stiffness in inhomogeneous soil (described by Eq. 13) with pile length, $a = 0$.

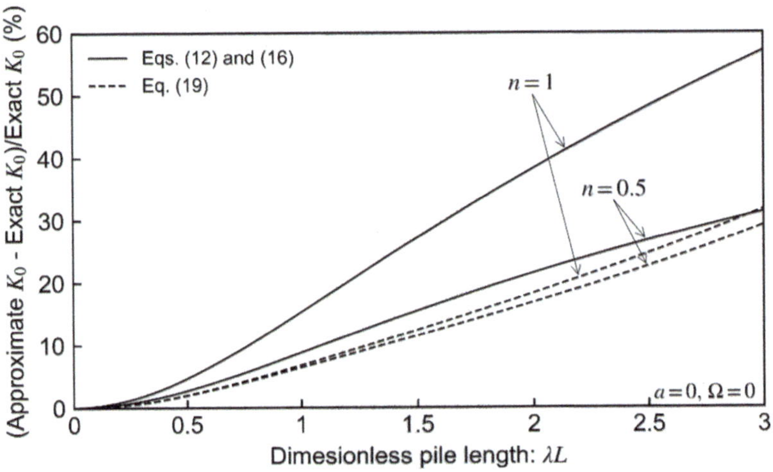

Fig. 5. Error in head stiffness estimation with pile length, $a = 0$.

Fig. 5. The proposed expression has approximately half the error for $n = 1$ and it is less than 12% for $\lambda L < 1.5$, covering most likely pile dimensions (i.e. $L/d < 60$ for $E_p/E_s = 10^3$).

3 Conclusions

An approximate solution has been developed using the Winkler model and virtual work to predict the response of axially loaded piles in inhomogeneous soils. The performance of this solution has been evaluated against an exact closed-form solution and an approximation considering an equivalent homogeneous soil and the same Winkler model. The main findings of this study are:

- The proposed energy solution, in conjunction with an elementary shape function, approximates pile head stiffness well for piles of most practical dimensions. The best results are obtained when equilibrium of the pile is considered to establish the slope of the shape function (*matched* β).
- The homogeneous approximation performs better the closer the soil stiffness profile is to homogeneous. The accuracy of the energy solution is relatively independent of soil inhomogeneity.
- A simple expression for pile head stiffness, derived using the energy method, is provided in Eq. (19) for inhomogeneous soil with vanishing surface stiffness. This expression has half the error when compared to the homogeneous approximation for soil stiffness varying linearly with depth.

Further investigation into this method using more complex shape functions could yield accurate expressions for pile head stiffness using well known elementary functions.

Acknowledgments. The authors would like to thank Dr. Elia Voyagaki for her assistance in preparing the figures. The first author would also like to thank the Engineering and Physical Sciences Research Council for their support (Grant No. EP/N509619/1).

References

Anoyatis, G.: Contribution to kinematic and inertial analysis of piles by analytical and experimental methods. Ph.D. thesis, University of Patras, Patras, Greece (2013)

Crispin, J.J., Leahy, C.P., Mylonakis, G.: Winkler model for axially-loaded piles in inhomogeneous soil (under review) (2018)

Guo, W.D.: Theory and Practice of Pile Foundations. CRC Press, Boca Raton (2012)

Karatzia, X., Mylonakis, G.: Horizontal response of piles in inhomogeneous soil: simple analysis. In: Maugeri, M. (ed.) Proceedings 2nd International Conference on Performance-Based Design in Earthquake Geotechnical Engineering, Taormina, Italy, pp. 1312–1327 (2012)

Karatzia, X., Mylonakis, G.: Horizontal stiffness and damping of piles in inhomogeneous soil. J. Geotech. Geoenviron. Eng. ASCE **143**(4), 04016113 (2016). https://doi.org/10.1061/(ASCE)GT.1943-5606.0001621

Mylonakis, G.: Contributions to static and seismic analysis of piles and pile-supported bridge piers. Ph.D. thesis, State University of New York at Buffalo, New York, NY, USA (1995)

Mylonakis, G.: Winkler modulus for axially loaded piles. Géotechnique **51**(5), 455–461 (2001)

Mylonakis, G., Gazetas, G.: Settlement and additional internal forces of grouped piles in layered soil. Géotechnique **48**(1), 55–72 (1998)

Poulos, H.G., Davis, E.H.: Pile Foundation Analysis and Design. Wiley, New York (1980)

Randolph, M.F., Wroth, C.P.: Analysis of deformation of vertically loaded piles. J. Geotech. Eng. Div. ASCE **104**(12), 1465–1488 (1978)

Roesset, J.M., Angelides, D.: Dynamic Stiffness of Piles. Numerical Methods in Offshore Piling, pp. 75–82. Institution of Civil Engineers, London (1980)

Scott, R.F.: Foundation Analysis. Prentice Hall, New York (1981)

Syngros, K.: Seismic response of piles and pile-supported bridge piers evaluated through case histories. Ph.D. thesis, The City College of the City University of New York, New York, NY, USA (2004)

Voyagaki, E., Crispin, J.J., Gilder, C.E.L., Nowak, P., O'Riordan, N., Patel, D., Vardanega, P.J.: Analytical approaches to predict pile settlement in London clay. These proceedings (2018)

Indirect In-situ Tests During the Execution of Deep Foundations: Relating the Excavation Energies to the Young's Moduli of the Surrounding Soils

Luan C. S. M. Ozelim[1,2(✉)], Darym J. F. Campos[1,2], José Camapum de Carvalho[1], and Andre L. B. Cavalcante[1]

[1] Department of Civil and Environmental Engineering, University of Brasília, Brasília, DF, Brazil
luanoz@gmail.com
[2] Consultant of PiezoGreen Solutions, Brasília, DF, Brazil

Abstract. The quality and reliability of a foundation design strongly relies on how well the designer understands and models the soil mass and its properties. Even though experimental test results are available prior to the design process, the quantity and quality of such tests may still not be adequate. Quantity is an issue because the greater the knowledge about the construction site, the better. Quality, on the other hand, figures as a determinant issue as poorly-executed or poorly-located tests significantly affect the designed solution. Instead of executing more tests, engineers should understand how to take advantage of common procedures to obtain parameters of interest. In the present paper, a study of the execution energy of continuous flight auger piles can be related to the Young's modulus of the surrounding soil mass is conducted. The building process (excavation and concreting) of this type of pile can be fully monitored by collecting data from sensors in the drilling machine. Thus, based on experimental models presented in the literature, the collected data was analyzed and it was possible to estimate the Young's moduli of the excavated soil layers. In order to verify the consistency of the values estimated, soil surveys were carried out to identify the soil strata in the site. The indirect measurements considerably agree with typical Young's moduli of the materials considered. Besides, the sample disturbance during indirect tests is considerably less significant while compared to other test types, which enhances the suitability of the procedure proposed. The estimated parameters can be used to update projects almost instantly, avoiding common spatial interpolation errors. Therefore, allying technology, theory and engineering practice is a good and cheap source of information, which ultimately enhances the quality of the foundation designs.

1 Introduction

In general, in order to correctly design a given foundation, geotechnical engineers must have access to parameters that characterize the underground medium. Such parameters are inserted in physical (constitutive) models, which estimate some behavior of interest. In general, foundation design must account for strength and deformability parameters.

H. El-Naggar et al. (Eds.): GeoMEast 2018, SUCI, pp. 191–205, 2019.
https://doi.org/10.1007/978-3-030-01902-0_16

The most straightforward method to obtain these parameters is by performing laboratory or *in-situ* tests.

Laboratory tests are interesting because controlling environmental variables (temperature, confining pressure etc.) may lead to results that are more reliable. Sometimes these tests are not suitable, as laboratory tests may require undisturbed samples, which may be hard to get.

In-situ tests, on the other hand, tend to be cheaper than laboratory tests. This drop in price comes together with less control over environmental variables as well as experimental procedures, possibly leading to poorly executed or poorly located tests.

In general, contractors are worried about the cost/benefit relationship of tests and since *in-situ* tests have been extensively validated in the literature, this type of test is commonly chosen. Instead of executing tests to solely obtain parameters, engineers should understand how to take advantage of common unavoidable procedures to estimate such parameters.

Estimating strength parameters based on drilling data of geomaterials is a promising *in-situ* method that has been studied by many researchers (Li and Itakura 2012). Initial studies of this topic were not driven by academic interests, but by economic ones. As indicated by Behboud et al. (2017), geomaterials such as rock and soils, understood as the formation to be drilled, play a very essential role in the drilling speed, depreciation of drilling bit, machines, and overall drilling costs. Therefore, understanding the drilling environment and the characteristics of the in-situ rock mass contributes a lot to the selection of the machines and prediction of execution schedules (Behboud et al. 2017).

In short, there are two approaches for estimating strength parameters from drilling data, namely: theoretical and experimental correlations. Warren (1981) presented the development of a penetration rate (P_r) model for soft-formation bits under conditions where cuttings removal does not impede this rate. This model relates P_r to weight on bit, rotary speed, rock strength, and bit size. Later on, the same author presented a P_r model that includes the effect of both the initial chip formation and cuttings-removal processes (Warren 1987).

Li and Itakura (2012) proposed an analytical model to describe rock drilling processes using drag bits and rotary drills, and to deduce the relations among rock properties, bit shapes, and drilling parameters (rotary speed, thrust, torque and stroke). By doing so, these authors could estimate the unconfined compressive strength of rocks from drilling data.

Kahraman et al. (2003), on the other hand, studied percussive blast hole drills and experimentally related the net penetration rates of the drills to strength parameters. In their work, rock samples were collected from the drilling locations and the physical and mechanical properties of the rocks were determined both in the field and in the laboratory. The correlation between penetration rates and uniaxial compressive strength, the Brazilian tensile strength, the point load strength, the Schmidt hammer value, impact strength, Young's modulus (E), natural density and P-wave velocity were studied.

The literature reveals that, besides the penetration rate, the energy required for drilling a given volume of material can also be correlated to strength and deformability parameters. The concepts behind this approach are further explored in the next subsection.

1.1 Energy as a Useful Measured Parameter for Rocks and Soils

Rocks

In his work, Teale (1965) indicated that a useful parameter which can be defined and studied during drilling processes is the specific energy (S_e), defined as the work done per unit volume excavated. Teale (1965) indicates that S_e can be taken as an index of the mechanical efficiency of a rock-working process. Besides, that author states that its minimum value appears to be very roughly correlated with the crushing strength of the medium drilled in, for rotary, percussive-rotary and roller-bit drilling. Therefore, understanding the drilling problem from an energetic point of view is a valid approach.

In rotary non-percussive drilling, work is done both by the thrust, F $[MLT^{-2}]$, and the torque, T $[ML^2T^{-2}]$. If the rotation speed is N $[T^{-1}]$, the area of the hole or excavation A $[L^2]$ and the penetration rate P_r $[LT^{-1}]$, the total work done is $(FP_r + 2\pi NT)$. The volume of rock excavated is (AP_r). Teale (1965) mathematically described the Specific Energy S_e $[ML^{-1}T^{-2}]$ as:

$$S_e = \frac{F}{A} + \frac{2\pi NT}{A\,P_r} \tag{1}$$

Soils

Energetic measurements for soils are also present in the literature. Perko (2001) proposed a theoretical model to relate the energy exerted during installation of helical piles to the energy required to displace the foundation or anchor once in place. Through the equivalence of energy, the model relates bearing and pullout capacity directly to installation torque. Perko (2001) considers the influence of downward force during installation, helical blade geometry, multiple helices, blade pitch per revolution, and hub radius to build his formula.

Silva et al. (2013), on the other hand, discuss the application of the SCCAP methodology, which was developed to control the execution of Continuous Flight Auger (CFA) piles foundations. The SCCAP methodology proposes formulations, routines and criteria for pile acceptance based on the comparison of the necessary energy to excavate a particular pile to statistical characteristics of the energetic population. Silva et al. (2013) claim that this approach enhances the reliability and mitigates involved risks to the geotechnical job as well as indicating that the SCCAP methodology is able to relate the necessary energy to excavate a particular pile and its bearing capacity.

As previously discussed, during a drilling process, the literature reveals that the penetration rate of the drill may be related to the specific energy of the material being drilled. Both parameters, on the other hand, can also be theoretically and experimentally related to deformability parameters as Young's Modulus. In order to estimate the Young's Moduli for different strata being drilled, in the present paper, experimental correlation is considered. In the literature there are some datasets which correlate S_e and E as well as P_r and E. The data and methodology considered in the current paper is presented in the next subsections.

2 Methodology

As indicated, an experimental correlation shall be used to predict Young's moduli data based on the rate of penetration of the drill and on the specific energy calculated during the drilling process. The datasets and methods considered are presented in the next subsections.

2.1 Reddish and Yasar (1996) Approach

In the work of Reddish and Yasar (1996), an experimental setup was created using a modified industrial rotary hammer drill. Samples of different rocks were drilled and their specific energy and penetration rate were recorded. Besides, the rock samples were tested in order to obtain their Young's modulus.

Reddish and Yasar (1996) indicate that two features became apparent: stall out penetration rate reflected the rock strength and the gradient of the line through a particular rock's results also reflected strength. This gradient is obtained for a straight line fitted for the relation $P_r \times S_e$. Their data is presented in Table 1, regarding the Specific Energy and Young's Modulus of different rock types.

Table 1. Reddish and Yasar (1996) data

Rock type	Gradient – Gr ((MJ/m^3)/(mm/s))	Specific energy – SE (MJ/m^3)	E (GPa)
State	2094	2928	15.5
Sandstone	1139	1745	10.49
Siltstone	653	1604	9.04
Mudstone	30	318	6.6
Limestone 2	1106	1617	7.31
Gypsum	339	470	4.29

In order to follow the Reddish and Yasar (1996) methodology, two approaches are considered: relating the gradient to Young's modulus and directly relate specific energy to Young's modulus.

2.2 Moronkeji et al. (2017) Approach

Moronkeji et al. (2017) performed coring operations and monitored the drilling process. They also laboratory tested the samples obtained to obtain their Young's modulus. These authors propose a relation between the penetration rate and Young's modulus. The data obtained by Moronkeji et al. (2017) is presented in Table 2.

Following Moronkeji et al. (2017) methodology is quite straightforward: just model the data in Table 2 to extrapolate and predict the Young's modulus based on the measured penetration rate.

Table 2. Moronkeji et al. (2017) data

Rock type	ROP (cm/min)	E (MPa)
Cement	65.53	7170.55
Shale	33.02	6550.02
Sandstone	29.46	28682.2
Granite	0.254	47091.2

2.3 Checking the Accuracy of Predicted Young's Moduli

Both Reddish and Yasar (1996) and Moronkeji et al. (2017) provide data to estimate Young's moduli of the materials drilled. In order to check for consistency of the values obtained, one shall compare those to the values presented in the literature for the materials drilled. Therefore, simple soil surveys were carried out to identify the type of soil presented. By knowing which materials were drilled, the estimated Young's moduli can be compared to literature results. Obrzud and Truty (2012) compiled data from Kezdi (1974) and Prat et al. (1995), obtaining the usual Young's moduli presented in Table 3.

Table 3. Usual values for Young's moduli of soils.

USCS	Description	Very soft to soft (MPa)	Medium (MPa)	Stiff to very stiff (MPa)	Hard (MPa)
ML	Silts with slight plasticity	2.5–8	10–15	15–40	40–80
ML, CL	Silts with low plasticity	1.5–6	6–10	10–30	30–60
CL	Clays with low-medium plasticity	0.5–5	5–8	8–30	30–70
CH	Clays with high plasticity	0.35–4	4–7	7–20	20–32
OL	Organic silts	-	0.5–5	-	-
OH	Organic clays	-	0.5–4	-	-

Thus, direct comparison of the predicted values and those in Table 3 enable one to check the consistency of the estimates hereby presented.

3 Results and Discussions

In the current section, the results and discussions shall be presented.

3.1 Pile Selection Based on Distance from Soil Surveys

Four surveys were carried out to identify the soil profile for each drilled pile. The results are presented in Fig. 1.

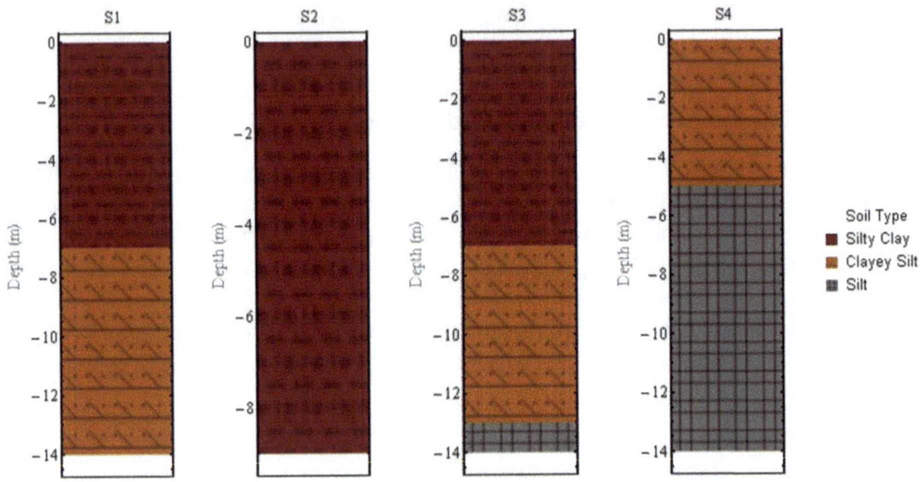

Fig. 1. The soil strata based on survey results

As the piles were executed in the whole terrain, we chose the closest pile to each of the surveys in order to characterize the pile's installation soil profile. Figure 2 presents the surveys locations and the correspondent piles.

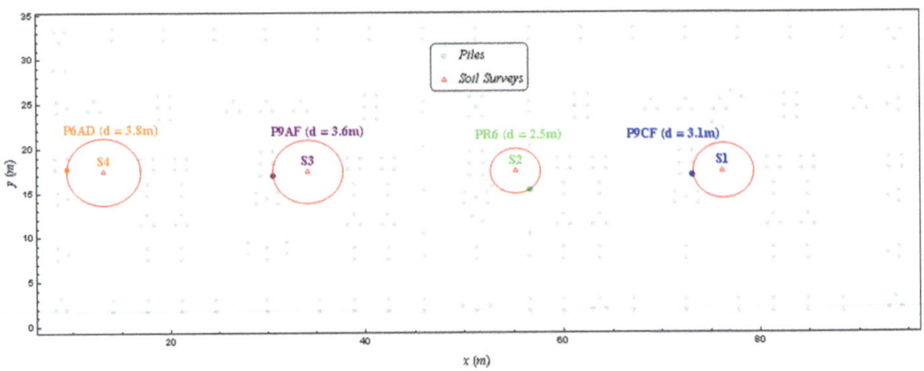

Fig. 2. Closest piles to each soil survey

It can be seen in Fig. 2 that the piles P9CF, PR6, P9AF and P6AD are the closest piles to the soil surveys S1, S2, S3 and S4, respectively. Now that the piles have been chosen, its necessary to estimate the Young's moduli of the soil layers drilled during their executions. The next subsection indicates how to perform the estimation based on Reddish and Yasar (1996) data.

3.2 Modelling Data from Reddish and Yasar (1996)

Reddish and Yasar (1996), presented two approaches to estimate Young's moduli of excavated material. The first one is the indirect approach, which firstly performs a linear fitting to the data relating the penetration rate to the specific energy and obtains the gradient of such line. This gradient is then related to the Young's modulus of the material. In the next subsection the modelling of the data presented in that paper is shown.

***Indirect Approach Based on Data from Reddish and Yasar* (1996)**
As indicated, the first step is to fit a line to the $P_r \times S_e$ data. Figures 3, 4, 5 and 6 present the fitted lines for each 1 m layer of excavated soil, for each pile executed.

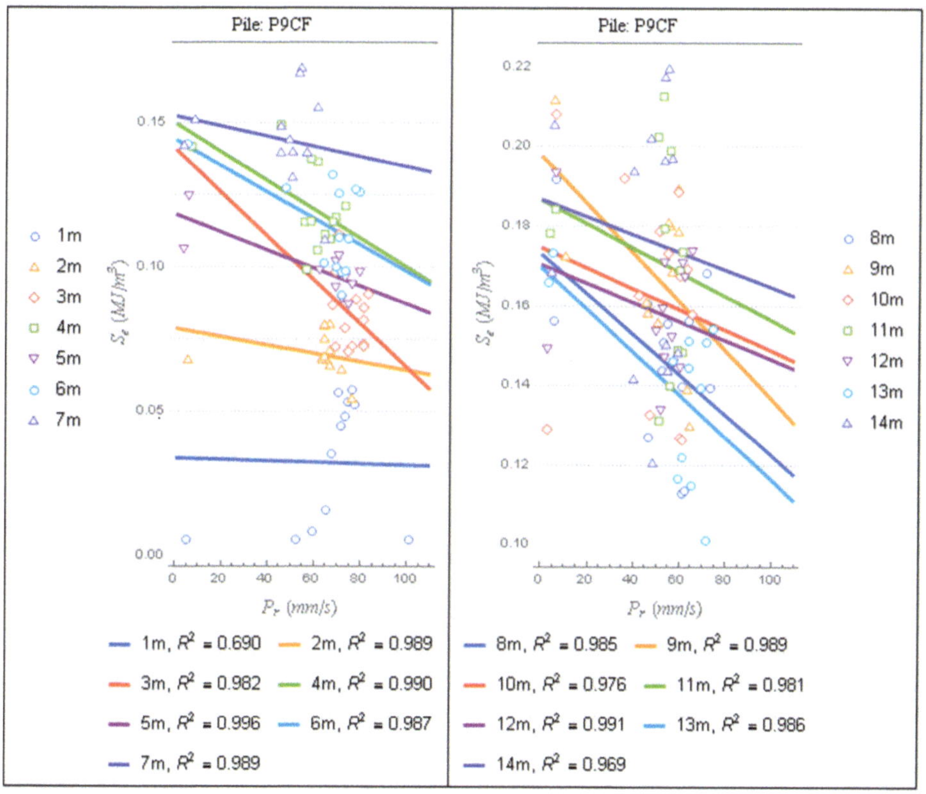

Fig. 3. Gradient estimation for pile P9CF

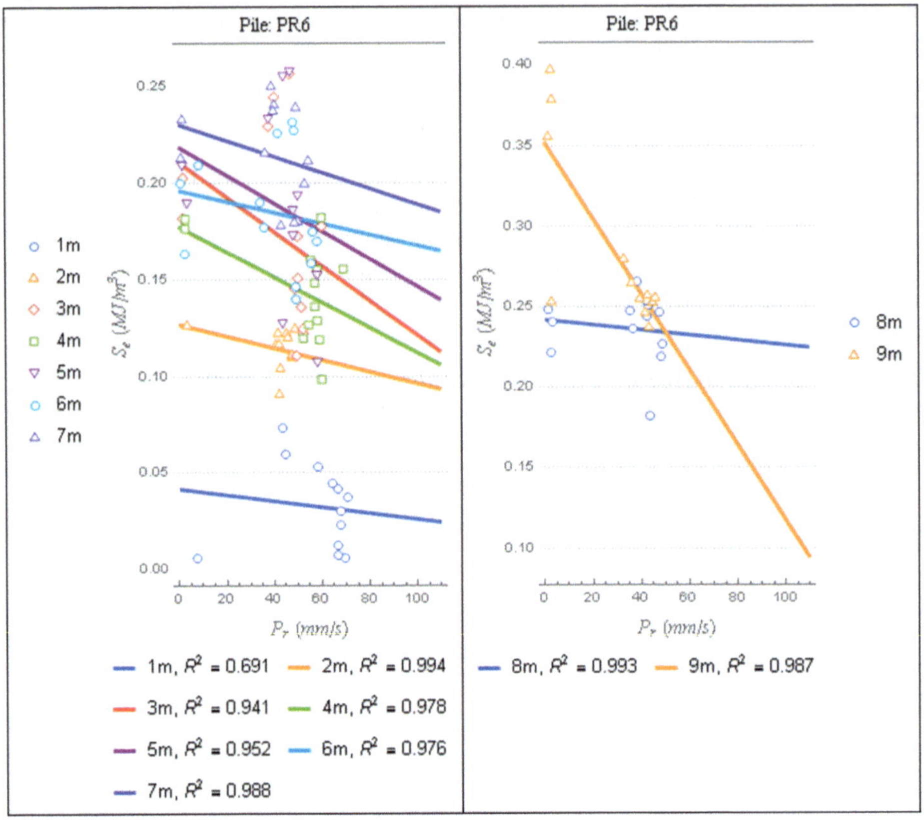

Fig. 4. Gradient estimation for pile PR6

As the data (torque, rotary speed and penetration rate) were recorded for each 8 cm of executed depth, for each meter 12 data points are available. These points were fitted to a straight line.

Based on the intercepts calculated on Figs. 3, 4, 5 and 6, one can estimate the Young's Modulus. This can be achieve by using the modelled relation between Gr and E, based on the data of Reddish and Yasar (1996). Figure 7 presents their data and an exponential model proposed in the present paper.

An exponential model is advised as a linear relation would overestimate Young's Modulus for low gradients. Low gradients are present for soils, as deformation parameters for those are significantly lower than that of rocks.

Instead of the gradients to estimate Young's moduli, one could directly relate the specific energy to E by using the data presented by Reddish and Yasar (1996). This procedure is presented in the next subsubsection.

***Direct Approach Based on Data from Reddish and Yasar* (1996)**

Based on the data presented by Reddish and Yasar (1996), Fig. 8 presents the exponential relation between S_e and E.

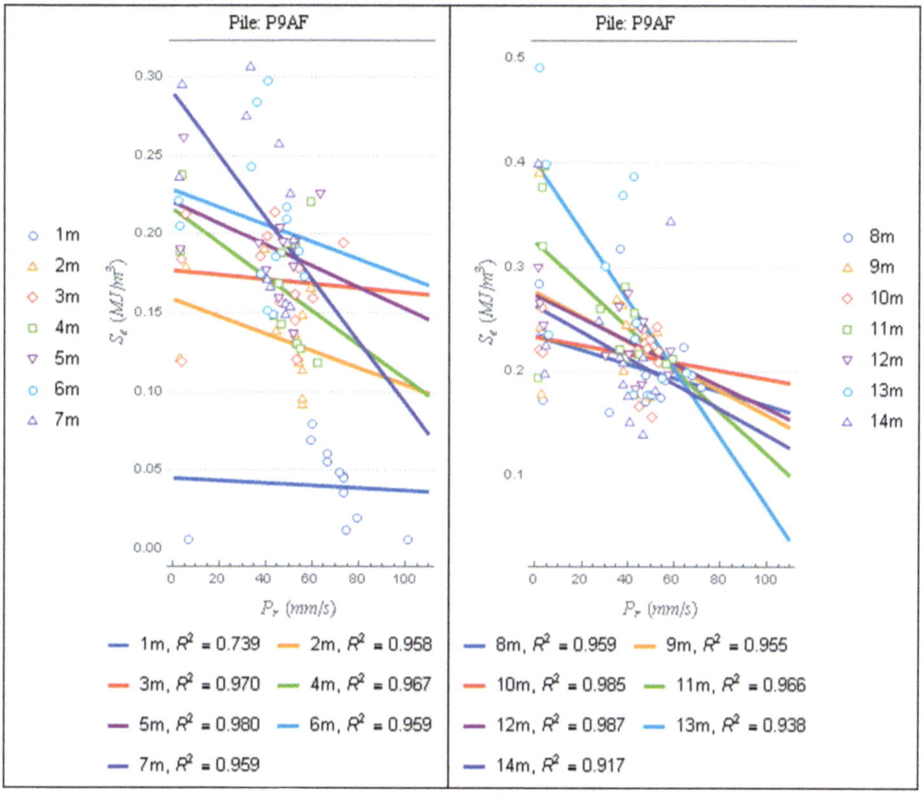

Fig. 5. Gradient estimation for pile P9AF

An exponential model is used for the same reasons indicated above.

3.3 Modelling Data from Moronkeji et al. (2017)

***Direct Approach Based on Data from Moronkeji et al.* (2017)**

Instead of considering the relation between S_e and E, Moronkeji et al. (2017) presented a relation directly between P_r and E. Figure 9 presents the nonlinear model fitted to experimental data.

Even though the relation proposed by Moronkeji et al. (2017) is between P_r and E, it could be converted to a relation between S_e and E by the linear relation between P_r and S_e, as indicated by Reddish and Yasar (1996) experiments.

3.4 Young's Moduli Estimation

Based on the nonlinear models presented in Figs. 7, 8 and 9, the values of the Young's moduli of the excavated 1 m soil layers are presented for each pile considered. Figures 10, 11, 12 and 13 present these results.

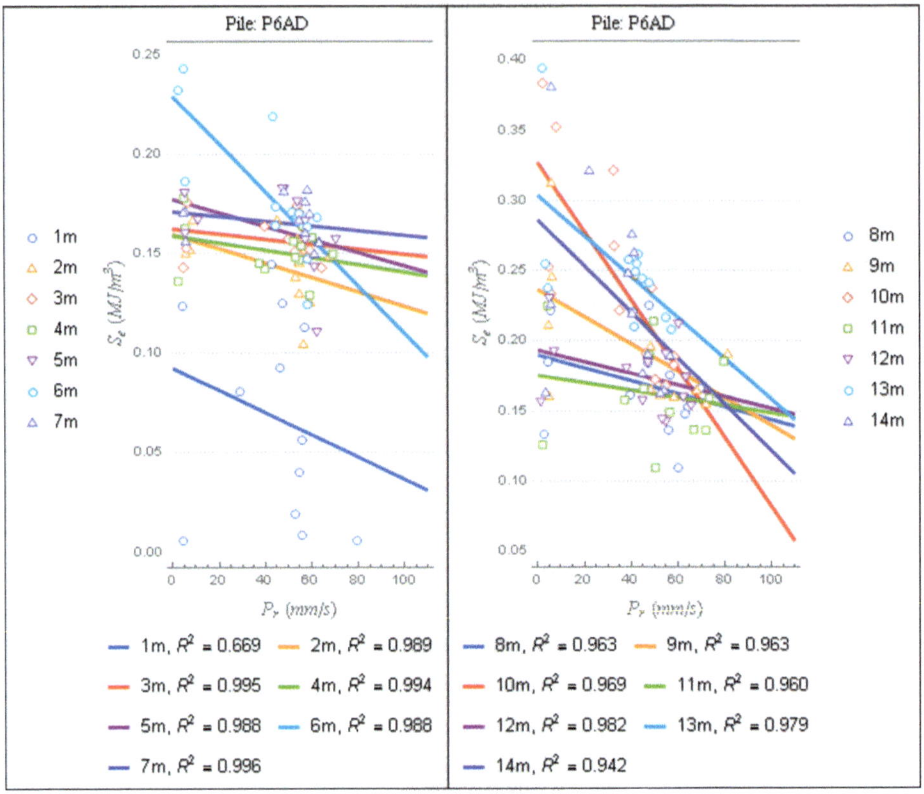

Fig. 6. Gradient estimation for pile P6AD

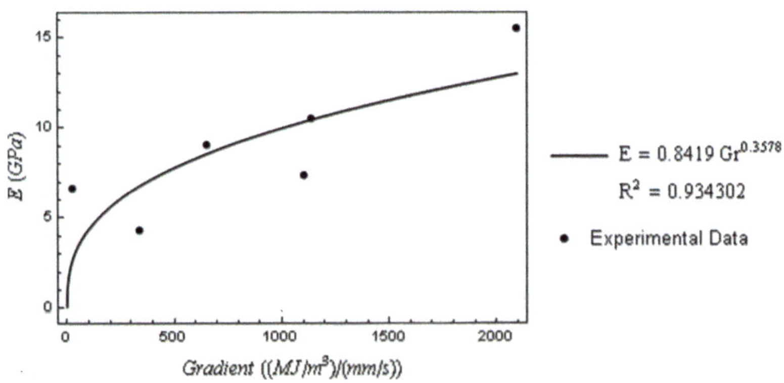

Fig. 7. Nonlinear model proposed for the relation between E and G_r.

Figures 10, 11, 12 and 13 reveal that the estimated Young's moduli for the soil layers obtained from the excavation data are in accordance to the ones predicted in the literature, as Table 3 indicates.

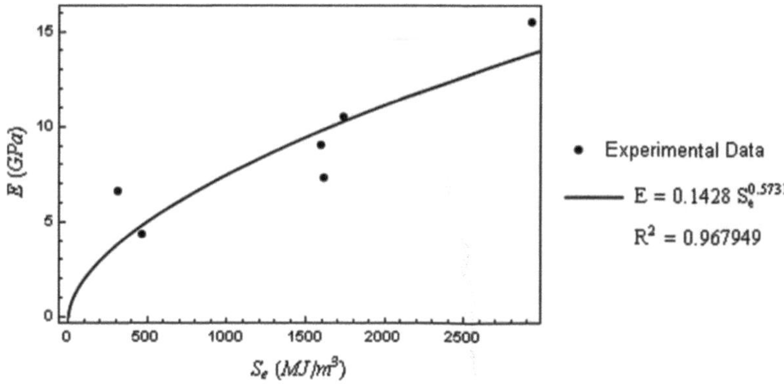

Fig. 8. Nonlinear model proposed for the relation between E and S_e.

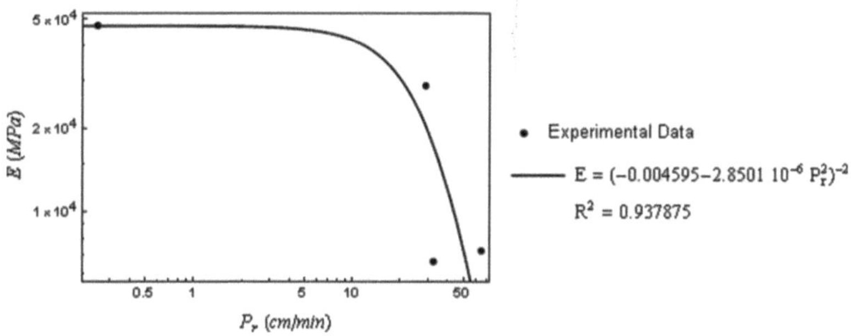

Fig. 9. Nonlinear model proposed for the relation between E and P_r.

For all the values estimated, it is possible to see that the ones based on Moronkeji et al. (2017) data were the lowest. This may be related to the nonlinear behavior of the model proposed to analyze this dataset as the model proposed and used to extrapolate Reddish and Yasar (1996) data increases faster than the one proposed to model Moronkeji et al. (2017) data for small penetration rates/specific energies.

It is worth noticing that the approach hereby presented is a first approximation to the process of estimation of deformability parameters from drilling data of soils. The nonlinear prediction models hereby proposed could benefit from considering more data, specially in the region of low gradients and low specific energies. As indicated, the data used to create the nonlinear models come from rock tests, which may impact the results hereby presented.

As indicated by Teale (1965), it is axiomatic that, to excavate a given volume of rock, a certain theoretically attainable minimum quantity of energy will be required. Its amount will depend entirely on the nature of the drilled material. Real mechanical processes might or might not approach this theoretical minimum: the difference between actual and theoretical requirements would be a measure of work dissipated in, for example, breaking the excavated material into smaller fragments than necessary, in friction between tools and material (which amounts perhaps to the same thing on a

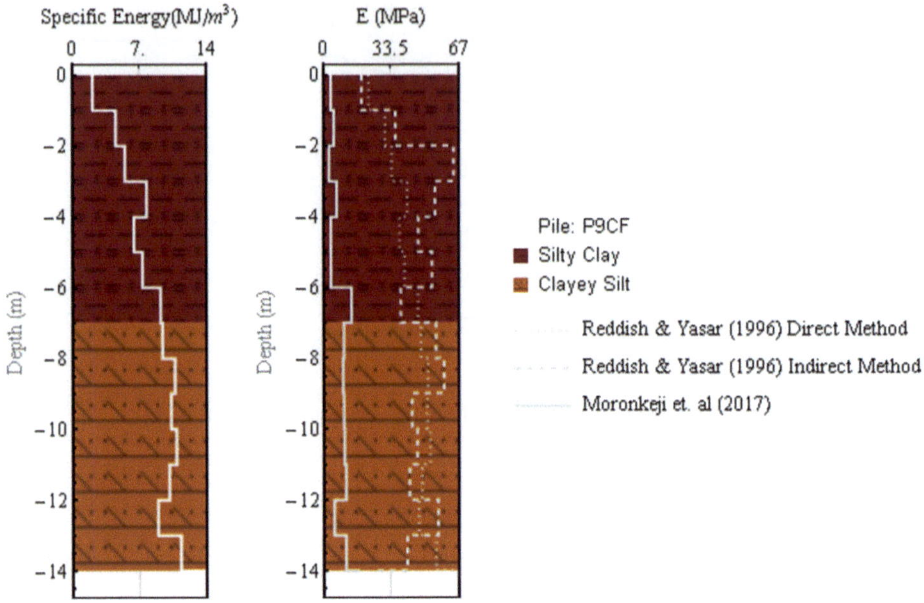

Fig. 10. Young's moduli estimation for each layer of excavated soil for pile P9CF.

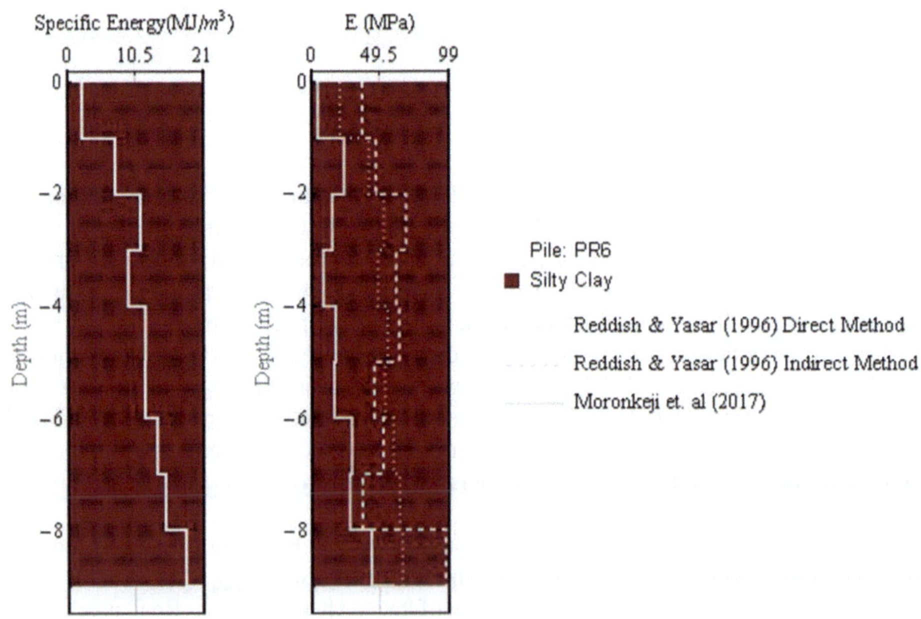

Fig. 11. Young's moduli estimation for each layer of excavated soil for pile PR6.

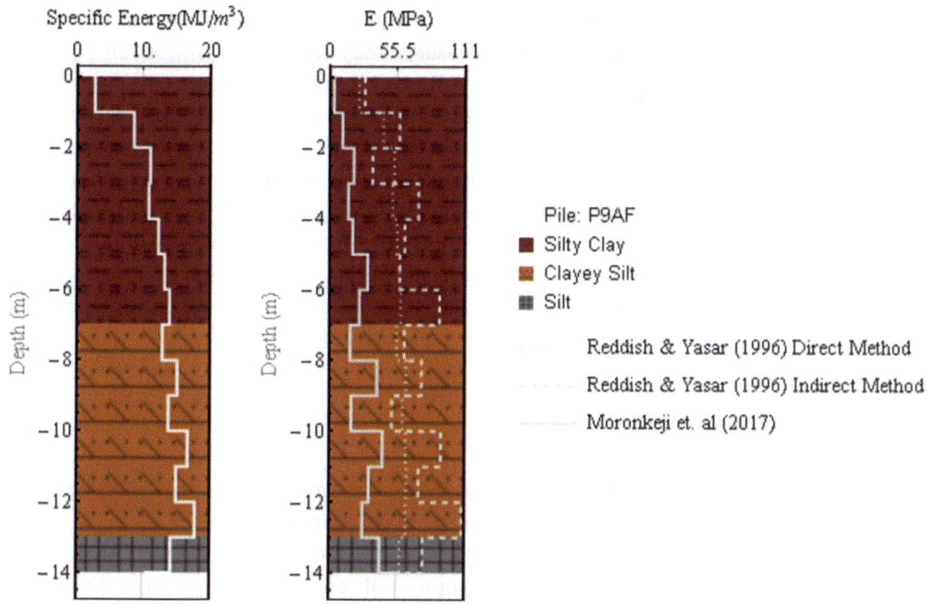

Fig. 12. Young's moduli estimation for each layer of excavated soil for pile P9AF.

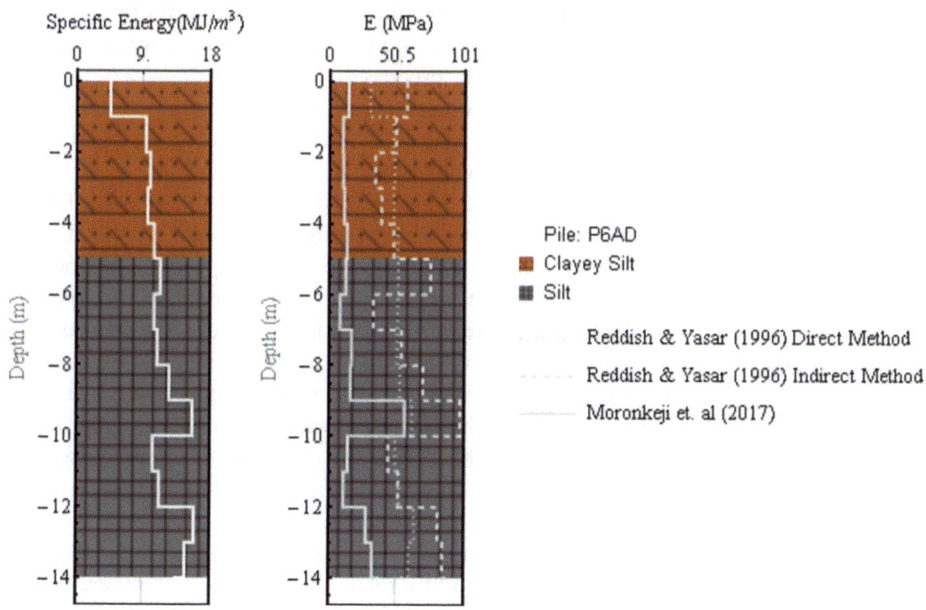

Fig. 13. Young's moduli estimation for each layer of excavated soil for pile P6AD.

microscopic scale); or in mechanical losses quite outside the rock/soil system. Breaking the debris into 'smaller fragments than necessary' may have a disproportionate effect on the energy needed to excavate the given volume. Not only do more particles have to be broken needlessly, but the specific energy itself increases considerably as the particle size is reduced (Teale 1965).

Therefore, it is believed that excavation of soils is much less efficient than rock excavation, implying that the measured specific energies are far greater than the minimum amount required for soil drilling. A smaller specific energy would imply in even lower Young's moduli values than the ones hereby predicted.

Further studies must be carried out to investigate the nonlinearity between energetic measurements and strength and deformability parameters.

4 Conclusions

The quality and reliability of a foundation design strongly relies on how well the designer understands and models the soil mass and its properties.

In order to reduce costs and increase the reliability of experimental results, engineers must understand how to take advantage of common procedures to obtain parameters of interest. In the present paper, it is studied how the execution energy of continuous flight auger piles can be related to the Young's modulus of the surrounding soil mass. Based on two experimental models presented in the literature, the collected data was analyzed and it was possible to estimate the Young's moduli of the excavated soil layers. Those models rely on studying and modeling the penetration rate and the specific energy calculated during the drilling process.

By comparing the estimated and common reported values for the Young's moduli of the layers, it is possible to see that the values estimated considerably agree with typical Young's moduli of the materials considered.

The estimated parameters can be used to update the projects almost instantly, avoiding common spatial interpolation errors. Even though the results obtained were physically consistent, the nonlinear prediction models hereby proposed could benefit from considering more data, specially in the region of low gradients and low specific energies. As indicated, the data used to create the nonlinear models come from rock tests, which may impact the results hereby presented as a extrapolation was considered.

It could be seen that allying technology, theory and engineering practice is a good and cheap source of information, which ultimately enhances the quality of the foundation designs. In order to further validate the results and discussions hereby presented, independent laboratory and in-situ test data must be collected and compared to those considered in the current paper.

It is worth noticing that the results presented are fundamentally empirical. No direct physical relation has been formulated to explicitly relate the parameters compared. This is actually a concern which eventually has to be taken into account in future works. The authors believe that the mechanisms involved during the excavation of the piles are mostly related ultimate strength parameters. Therefore, the key aspect, which must be addressed to properly model the excavation process, is how these strength parameters can be physically correlated to deformation parameters.

Acknowledgments. L. C. de S. M. Ozelim thanks the Brazilian Research Council (CNPq) for funding his postdoctoral fellowship at University of Brasilia (grant number 151778/2018-3). The authors also thank FAP-DF for providing grants to present this paper at the conference.

References

Behboud, M.M., Ramezanzadeh, A., Tokhmechi, B.: Studying empirical correlation between drilling specific energy and geo-mechanical parameters in an oil field in SW Iran. J. Min. Environ. **8**(3), 393–401 (2017). https://doi.org/10.22044/jme.2016.689

Kahraman, S., Bilgin, N., Feridunoglu, C.: Dominant rock properties affecting the penetration rate of percussive drills. Int. J. Rock Mech. Min. Sci. **40**, 711–723 (2003)

Kezdi, A.: Handbook of Soil Mechanics. Elsevier, Amsterdam (1974)

Li, Z., Itakura, K.: An analytical drilling model of drag bits for evaluation of rock strength. Soils Found. **52**(2), 216–227 (2012)

Moronkeji, D., Villegas, R., Shouse, R., Prasad, U.: Rock strength prediction during coring operation. Paper Presented at International Symposium of the Society of Core Analysts Held in Vienna, Austria, 27 August–1 September 2017. SCA2017-048, 9 p (2017)

Obrzud, R., Truty, A.: The Hardening Soil Model - A Practical Guidebook Z Soil. PC 100701 report (2012)

Perko, H.A.: Energy method for predicting the installation torque of helical foundations and anchors. In: Mullins, G., et al. (eds.) New Technological and Design Developments in Deep Foundations, pp. 342–352. American Society of Civil Engineers, Reston (2001)

Prat, M., Bisch, E., Millard, A., Mestat, P., Cabot, G.: La Modelisation Des Ouvrages. Hermes, Paris (1995)

Reddish, D.J., Yasar, E.: A new portable rock strength index test based on specific energy of drilling. Int. J. Rock Mech. Min. Sci. **33**(5), 543–548 (1996)

Silva, C.M., Camapum de Carvalho, J., Cavalcante, A.L.B.: Energy and reliability applied to continuous flight auger pilings - the SCCAP methodology. In: Proceedings of the 18th International Conference on Soil Mechanics and Geotechnical Engineering, Paris, pp. 2807–2810 (2013)

Teale, R.: The concept of specific energy in rock drilling. Int. J. Rock Mech. Min. Sci. **2**, 57–71 (1965)

Warren, T.M.: Penetration rate performance of roller cone bits. Soc. Pet. Eng. (1987). https://doi.org/10.2118/13259-pa

Warren, T.M.: Drilling model for soft-formation bits. Soc. Pet. Eng. (1981). https://doi.org/10.2118/8438-pa

Vibration Wave Displacement Introduced by Driving Steel Open Ended Pipe and H-Piles

Mahmood R. Mahmood(✉) and Shaimaa M. Abdulrahman

Building and Construction Engineering Department,
University of Technology, Baghdad, Iraq
mahmoudal_qaissy@yahoo.com, Shaimaa.m83@gmail.com

Abstract. Driven piles have been used in many civil engineering structures to provide structural loading support. However, the unavoidable vibrations induced by pile driving process may cause varying degrees of damage to adjacent buildings. Dynamic behavior of machinery deep foundation embedded in dry sand was experimentally investigated. In the dynamic analysis, the amplitudes of land vibrations are examined based on the variation of the pile shape. The aim of this work is to find the effects of the vibration wave transmitted from open ended pipe and H-piles during driving on the nearby structures. Moreover, investigate the dynamic response of the soil to record the maximum distance of the damage impact on neighboring buildings in order to reduce the damage.

The results of analysis indicated that the effect of vibration wave speed that transmitted from open ended pipe pile equal to 0.41 mm/s which is less than that for H-pile which was 1.1 mm/s. While the maximum acceleration value of vibration wave was gradually decreased for H-pile more than that of open ended pipe pile. The dynamic response of both models pipe and H-piles of vibration wave displacement gradually increased until the final penetration depth of pile.

Keywords: Open ended pipe pile · H-piles · Vibration amplitudes
Vibration wave displacement · Driving hammer

1 Introduction

Piles are used to support foundations of buildings, bridges, and other structures where the use of shallow foundations would be risky due to excessive settlement, potential scour effects or future environmental conditions (Mallard and Bastow 1979). According to the technical description, the generation of complex wave pattern is due to energy reflection and refraction of the land surface and subsurface interfaces between dissimilar materials (Svinkin 2004). The pile driving introduces vibrations in the ground which create problems particularly in urban areas such as unsound noise, environmental disturbance, and potential hazard for the neighboring structures (Massarsch 2004). Many case studies have shown that land vibrations due to pile driving often cause damage to the adjacent buildings that are vulnerable to the earth shaking (Svinkin 2005; Masoumi et al. 2006). The damage due to pile driving occurs either directly or via settlement of soil beneath foundations in the proximity of pile driving operations (Madheswaran et al. 2005). The potential damage to the adjacent

H. El-Naggar et al. (Eds.): GeoMEast 2018, SUCI, pp. 206–214, 2019.
https://doi.org/10.1007/978-3-030-01902-0_17

structures can be prevented by conducting pre-construction surveys, monitoring and controlling the vibrations on site, and predicting the anticipated vibrations prior to pile driving (Massarschand and Fellenius 2008).

Extensive review of the available works (Madheswaran et al. 2005; Auersch 2010; Khoubani and Ahmadi 2012) and information provided by the manufacturers, indicate that the PPV from impact pile drivers can be estimated by the following equation:

$$PPV_{Impact\ Pile\ Driver} = PPV_{Ref}(25/D)^{n} \times \left(E_{equip}/E_{\mathrm{Ref}}\right)^{0.5}(in/sec) \quad (1)$$

Where:

PPV_{Ref} = 0.65 in/s for a reference pile driver at 25 ft.
D = distance from pile driver to the receiver in ft.
n = 1.1 is a value related to the vibration attenuation rate through ground.
E_{Ref} = 36,000 ft-lb (rated energy of reference pile driver).
E_{equip} = rated energy of impact pile driver in ft-lbs.

The above Equation is based on actual data points at various distances, measured for a wide range of impact pile drivers. The data were measured at the ground surface outside or within various type of buildings. Literatures indicate that the value of "n" in the above equation is generally taken as (1 to 1.5), the suggested value for n is 1.1. The use of values greater than 1.1 would likely result in overestimation of amplitudes at distances closer than 25 ft and would be slightly conservative at distances beyond 25 ft. If vibration impacts, based on the above approach, are expected to exceed the vibration assessment criteria, vibration estimates may be refined further by using values of "n" that are based on soil type classification, ranging from Class I to IV soils as outlined in the National Cooperative Highway Research Program (NCHRP) Synthesis 253 (Woods 1997) and based on data developed by (Woods and Jedele 1985). This step would require detailed information on soil conditions at the site. Table 1 describes soil materials, soil classes, values of "n" measured by Woods and Jedele and suggested values for "n" for the purposes of estimating vibration amplitude.

As indicated by Davis (2010), the use of published attenuation relationships, based primarily on relating hammer energies, scaled distances, and PPVs to predict vibration levels in moderately large commercial buildings or in buried structures would probably result in overly conservative estimates does not report data points for complete evaluation, but rather presents only generalized curves.

The correlation between vibrations and building damage is complicated for many reasons, such as buildings are constructed in so many different ways, with different dimensions, building materials, construction methods foundation types and executions (Mallard and Bestow 1979). Thandavamoorthy (2004) stated that a rule of thumb in pile driving practice is that structures within one pile length from the driven pile can be damaged due to vibrations.

Building damage due to vibration can range from structural damage, such as major failures in the building structure, to architectural damage such as cracking of plaster (Svinkin 2008). Generally damage on buildings and structures can be divided into: Architectural damage, damage to the appearance of surface finishes and fittings

Table 1. Measured and suggested "n" values based on soil class (Woods and Jedele 1985).

Soil class	Description of soil material	Value of "n" measured by Woods and Jedele	Suggested value of "n"
I	Weak or soft soils: loose soils, dry or partially saturated peat and muck, mud, loose beach sand, and dune sand, recently plowed ground, soft spongy forest or jungle floor, organic soils, top soil. (shovel penetrates easily)	Data not available	1.4
II	Competent soils: most sands, sandy clays, silty clays, gravel, silts, weathered rock. (can dig with shovel)	1.5	1.3
III	Hard soils: dense compacted sand, dry consolidated clay, consolidated glacial till, some exposed rock. (cannot dig with shovel, need pick to break up)	1.1	1.1
IV	Hard, competent rock: bedrock freshly exposed hard rock. (difficult to break with hammer)	Data not available	1.0

Serviceability damage, damage to the function of the building Structural damage, damage to structural parts of the building that potentially leads to failure or collapse, and damage to buildings often starts with the development of cracks in the structure.

2 Methodology

2.1 The Container for Modeling Tests

Steel container of (1200 * 1200 * 900) mm dimension was used in this study to conduct all the entire models tests. Each side of the container was made of a steel plate of 3 mm thickness. Cork layer of 20 mm thickness was used to lining the four sides and the base to absorb the vibration waveguide when applied dynamic force.

2.2 Sand Deposit Preparation

The sand deposit was prepared by using a modified compactor to satisfy the selected relative density. The total height of the container of sand was divided into ten equal lifts, of 100 mm. The soil of each layer was compacted to a predetermined depth to a chive the required density. After completing the final layer, the top surface was leveled to get as near as possible a flat surface.

2.3 Investigated Vibration Wave Within the Soil at Different Distances Due to Pile Driving

To measure the maximum impact of the vibration wave for wave propagation in soil, the sensors was installed at different distances from the pile center at a depth of 50 mm below the soil surface (2.5 d, 5 d, 10 d, 17.5 d, 20 d, and 45 d), where; d is the pile width or diameter. Figures 1 and 2, shows the position of sensors below the soil surface to measure the effect of the vibration wave. This method of sensors installation was used with the two types of the pile models.

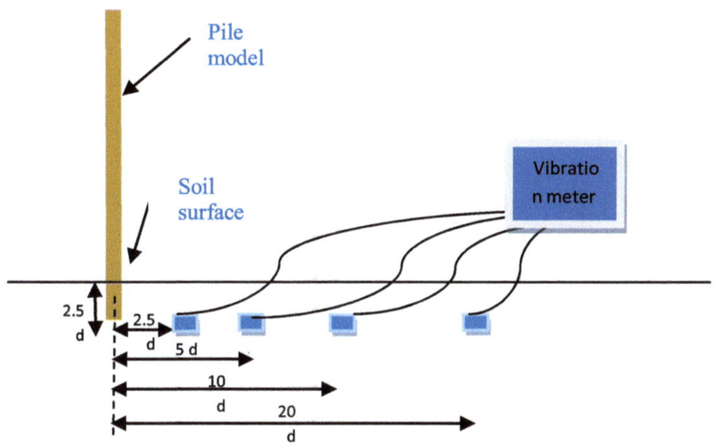

Fig. 1. Sensors location within the soil

The peak particle velocity of vibration wave, the Vibration wave accelerating, and vibration wave displacement during pile model driving was measured.

2.4 Vibration Meter Device

Vibration meter of model BVB-8207SD 4 channels, with SD card data recorder marks a step forward into the field of precision measurement was used to measure vibration wave.

- Frequency range 10 Hz–1 kHz, sensitivity relative meet ISO 2954.
- Professional vibration meter supply with vibration sensor & magnetic base.
- SD card capacity: 1 GB to 16 GB
- Function (acceleration, velocity (RMS, peak, max hold), and displacement (p-p, max hold p-p)).
- Operation temperature: 0 to 50 °C.

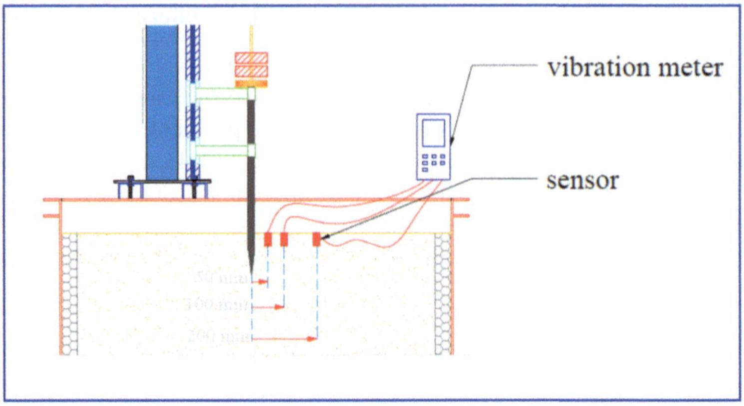

Fig. 2. Details of driving pile model and sensors installation within the soil

3 Results and Discussion

The period of the applied dynamic force and recording of the response of the vibration wave transmit from the hammer to the pile and the dynamic response of each model was recorded for about 2 h. vibration meter was used to measure vibration wave.

Figures 3 and 4, illustrates the dynamic response of pipe and H-pile model respectively. The figures show that the vibration wave velocity gradually increased to reach the maximum vibration wave near the middle of penetration depth of pile. Also, it shows that the peak particle velocity recorded on the surface of soil was not regular with the regular of the energy and increased the penetration depth of pile. The peak particle velocity was increased in some points and decreased in other points because of the dissipation of the stress waves due to the material damping in the ground which caused a significant difference in the magnitudes of the vibrations on the ground surface. The maximum velocity of vibration wave for steel pipe pile was (0.4) mm/s at (20 cm) of pile penetration depth, while the maximum velocity of vibration wave for H-pile (1.1 mm/s) at (25) cm of penetration depth of pile.

Figures 5 and 6 illustrate the dynamic response of vibration wave acceleration for each pile model, which gradually declined until to the end of the pile penetration depth. Also, these figures shows, that the highest value of accelerating of vibration wave at the beginning of penetration of the pipe and H-pile. It seems clearly that when the sensors were located at a different distances, the acceleration value of the vibration wave was gradually decreased for H-pile (40, 25, 20, 15 and 0) mm/s^2 at location of the sensor (2.5 d, 5 d, 10 d, 17.5 d, 20 d). While the acceleration value of the vibration wave for pipe pile was (25, 20, 10, and 0) mm/s^2 at location of the sensors (2.5 d, 5 d, 10 d, and 17.5 d) respectively.

Figures 7 and 8, clearly shows the dynamic response of displacement vibration wave for both models which being gradually increased until the final penetration depth of pile model, where maximum displacement value was 0.065 mm at 25 cm depth for pipe pile and 0.2 mm for H-pile at 35 cm of pile embedded depth.

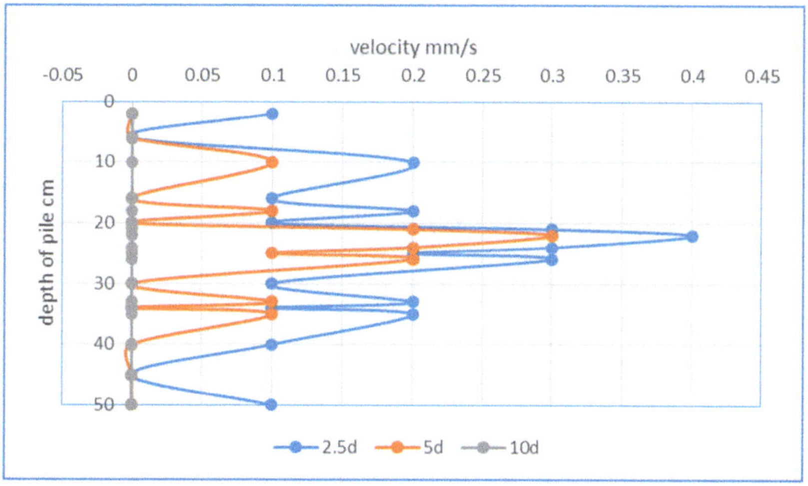

Fig. 3. Behavior of vibration wave velocity with penetration depth of steel hollow pile model during driving

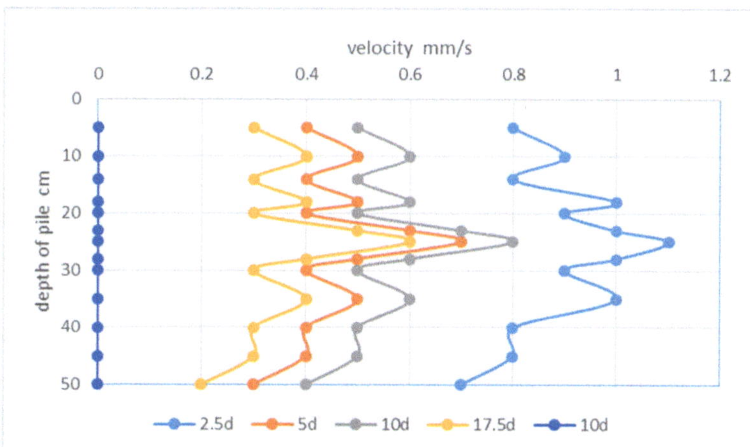

Fig. 4. Behavior of vibration wave velocity with penetration depth of H-pile model during driving

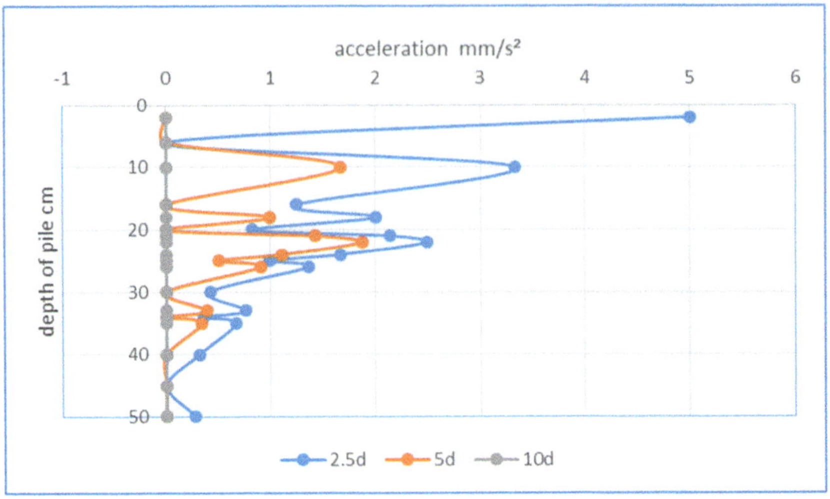

Fig. 5. Behaviour vibration wave acceleration with penetration depth of steel pipe pile model during driving

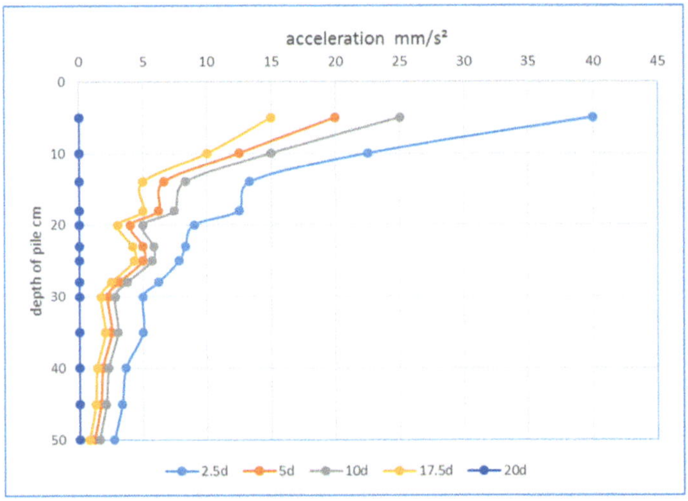

Fig. 6. Behaviour of vibration wave of acceleration with penetration depth of H-pile model during driving

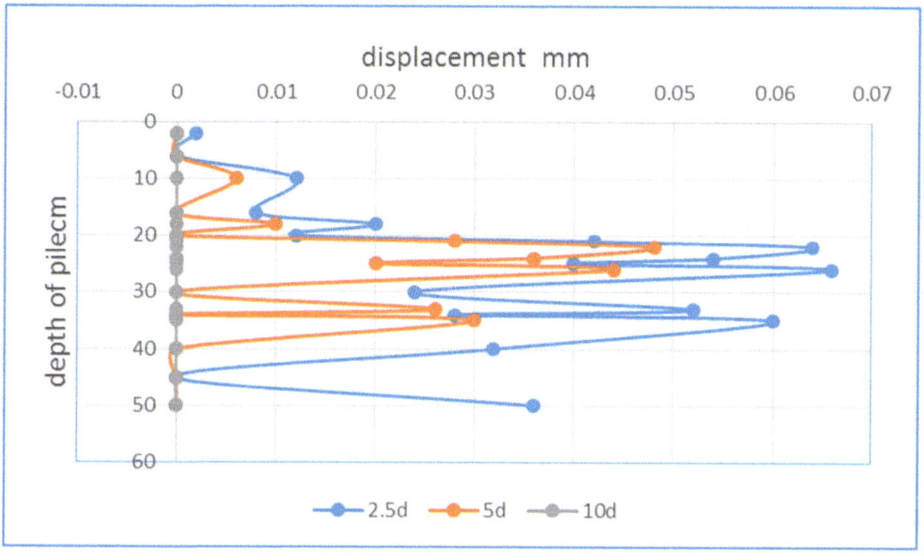

Fig. 7. Behavior of vibration wave displacement with penetration depth of pipe pile during driving

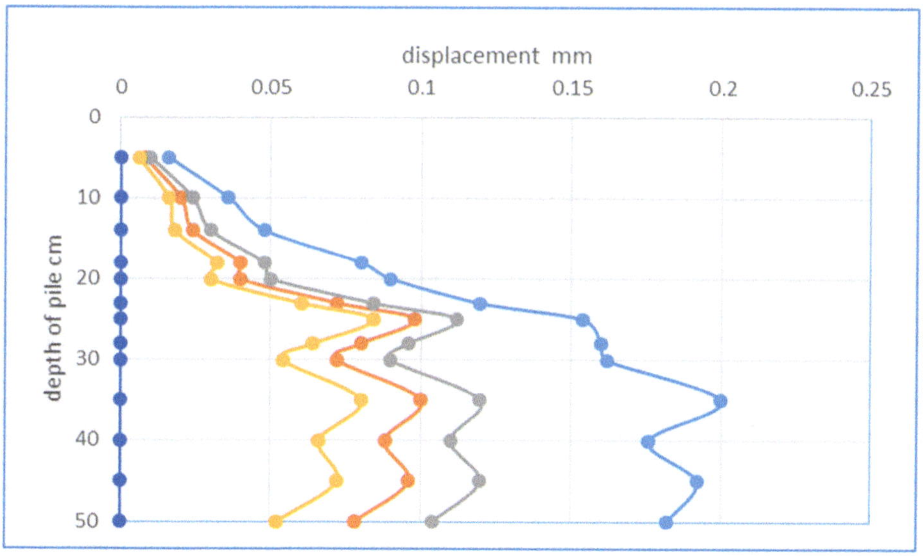

Fig. 8. Behavior of vibration wave displacement with penetration depth of H-pile during driving

4 Conclusions

From the results of the current study, it is possible to conclude the following points:

1. Damage to structural parts of the building may potentially lead to failure or collapse to building, when driving pile approach the middle of the penetration depth with in the soil.
2. It is seems clearly that the acceleration of vibrational waves starts at the highest value while the displacement of the vibration waves was increased with the passage time of various pile models.
3. The maximum of damage distance at the ground surface for H-pile was (20 d) from center of pile. While the maximum of damage distance at the ground surface from steel pipe pile was (17.5 d) from center of pile.
4. The effect of the vibrational wave of steel pipe pile was less than that of H-pile for adjacent buildings.

References

Auersch, L.: Technically induced surface wave fields. Part I: measured attenuation and theoretical amplitude-distance laws. Bull. Seismol. Soc. Am. **100**(4), 1528–1539 (2010)

Davis, D.: A review of prediction methods for ground-borne noise due to construction activities. In: Proceedings of the 20th International Congress on Acoustics, Australia, 23–27 August 2010

Khoubani, A., Ahmadi, M.M.: Numerical study of ground vibration due to impact pile driving. In: Proceedings of the ICE – Geotechnical Engineering, 22 August 2012, p. 12 (2012). https://doi.org/10.1680/geng.11.00094

Madheswaran, C.K., Sundaravadivelu, R., Boominathan, A., Natarajan, K.: Response of ground during pile driving. J. Inst. Eng. (India) Civ. Eng. Div. **86**, 22–27 (2005)

Mallard, D.J., Bastow, P.: Some observations on the vibrations caused by pile driving. In: Proceedings of the ICE Conference on Recent Developments in the Design and Construction of Piles, pp. 261–284 (1979)

Masoumi, H.R., Degrande, G., Lombaert, G.: Free field vibrations due to vibratory pile driving in a layered soil medium. In: Proceedings of TRANSVIB 2006, Paris, 21–22 September 2006

Massarsch, K.R.: Vibrations caused by pile driving. Deep foundations, summer 2004 and fall 2004 (two parts) (2004)

Massarsch, K.R., Fellenius, B.H.: Ground vibrations induced by impact pile driving. In: Proceedings of the 6th International Conference on Case Histories in Geotechnical Engineering, Arlington, 11–16 August 2008

Svinkin, M.R.: Minimizing construction vibration effects. Pract. Period. Struct. Des. Constr. ASCE **9**(2), 108–115 (2004)

Svinkin, M.R.: Environmental vibration problems during construction. In: Proceedings of the International Conference on Soil Mechanics and Geotechnical Engineering, vol. 4, pp. 2453–2456 (2005)

Svinkin, M.R.: Soil and structure vibrations from construction and industrial sources. In: Proceedings of the 6th International Conference on Case Histories in Geotechnical Engineering, Arlington, 11–16 August 2008

Development of Negative Skin Friction and Dragloads on Piles and Pile Groups in Soft Clay

Manal Salem(✉), A'laa Adel, and Mohamed Amer

Faculty of Engineering, Cairo University, Giza, Egypt
manalasalem@eng.cu.edu.eg, alaadel591@gmail.com,
m.amer@cu.edu.eg

Abstract. One of the challenging problems that faces geotechnical engineers in pile design in soft soils is the determination of negative skin friction (NSF) value. Negative skin friction causes additional compressive force on the pile, which decreases the pile carrying capacity. In this paper, a 3D finite element (FE) model is presented to model a field case study for instrumented driven pile in soft Bangkok clay subject to NSF. Assigning short and long term material sets to the soft and medium stiff clay layers was necessary to provide well short and long term settlement predictions. The developed 3D FE model was used to investigate the behavior of pile groups in soft soils for surcharge and embankment loading. Piles were spaced at three times the pile diameter for all investigated cases. For surcharge loading, the maximum group effect occurred for the central pile, while the minimum group effect was observed for the corner pile. Smaller dragloads were observed to develop on piles in pile groups due to the "shielding effect" on the inner piles by the outer piles. The larger the number of piles in a group, the greater the shielding effects on the reduction of dragloads. This reduction suggests using sacrificial piles around pile groups in consolidating soils to obtain more economical designs. The depth to the neutral point was larger for the central piles compared to the corner piles. For embankment loading, the maximum group effect was noted for the central pile. The width of embankment increased as the number of pile rows increased for constant spacing between piles. Increasing the number of pile rows was not associated with further reduction in the dragloads on the central piles, which indicated that the shielding effect was counteracted by the effect of increasing the extent of loading.

Keywords: Piles · Negative skin friction · Dragload · Group effect
Finite element

1 Introduction

Pile foundations in soft soils are often subjected to negative skin friction (NSF), which occurs when the rate of settlement of the surrounding soil is larger than the rate of settlement of the pile. Pile driving in soft soils generally remolds the soft soil around the pile. The subsequent consolidation of the remolded clay may cause NSF. For piles

H. El-Naggar et al. (Eds.): GeoMEast 2018, SUCI, pp. 215–231, 2019.
https://doi.org/10.1007/978-3-030-01902-0_18

subjected to NSF, a neutral plane is defined as the plane at which there is no relative movement between the pile and the surrounding soil. It is worth noting that the development of NSF is time dependent, as it increases with the dissipation of excess pore-water pressure with time.

Several attempts have been reported in the literature to investigate the mechanism of the development of NSF around piles, whether by field monitoring of piles subjected to NSF (Johannessen and Bjerrum 1965; Fellenius 1972, 2006; Blanchet et al. 1980; Bozozuk 1981; Indraratna et al. 1992); or through numerical modeling (Chow et al. 1990; Poorooshab et al. 1996; Lee et al. 2002; Jeong et al. 2004; Lee and Ng 2004; Cheen et al. 2009). Lee et al. (2002) developed 2D and 3D finite element (FE) models using ABAQUS software to investigate the group effects on the distribution of dragloads (defined as the force exerted on the pile due to NSF) in pile foundations. The piles passed through soft clay layer and the pile tips rested an end bearing sand layer. The development of NSF was initiated by applying surcharge loading to the surface of the soft clay layer. The group effects were investigated by comparing the behavior of center, side, and corner piles to the behavior of single pile. The group effect was defined as the reduction in maximum dragload compared to the single pile. Within a group, maximum group effect was observed on the central pile; while, minimum group effect was observed on the corner pile. The group effect increased as the number of piles in the group increased. Lee and Ng (2004) investigated the group effects on the distribution of downdrag (defined as the settlement introduced to the pile due to the effect of NSF) and dragloads in pile foundations. Within a group, maximum reduction in downdrag was observed on the central pile; while, minimum reduction was observed on the corner pile. The authors attributed the reduced downdrag inside the pile group to the pile–soil interaction effect, referred to as "shielding effects" on the inner piles by the outer piles of the pile group. The larger the number of piles in a group, the greater the shielding effects on the predicted dragloads and downdrag. The shielding effects were more pronounced on the dragloads compared to the downdrag.

Fellenius (2006) presented the results of long term measurements in driven piles in different sites. Field measurements showed that the negative skin friction and positive shaft resistance were governed by the effective stress and were fully mobilized with a very small movement of the pile. Reconsolidation after pile installation due to dissipation of pore-water pressures resulted in significant dragload. Cheen et al. (2009) developed a numerical model to describe the behavior of load transfer at pile-soil interface during soil consolidation and loading-unloading conditions. The soil stress history affected the NSF distribution and the position of the neutral plane varied during the process of soil consolidation. El-Mossallamy et al. (2012) simulated the pile-soil interaction problem of NSF for a field case study of instrumented driven precast concrete piles in Bangkok soft clay through an axisymmetric FE model. The ratio of undrained to drained soil stiffness moduli (E_u/E') of 1.03 to 1.36 based on assuming a constant value for the shear modulus in undrained and drained conditions yielded unrealistic results for numerical models involving soft clay. For soft and medium stiff Bangkok clays, a value of 8.75 was recommended for E_u/E'.

In this paper, a 3D FE PLAXIS model is presented to model a field case study for instrumented driven pile in soft Bangkok clay subject to NSF. The developed 3D FE model was used to investigate the group effects on the distribution of dragloads in pile groups in soft soils for surcharge and embankment loading.

2 Case Study by Indraratna et al. (1992)

Indraratna et al. (1992) conducted full-scale field measurements on two instrumented hollow precast concrete driven piles in a site located 10 km east of Bangkok - Thailand. The two piles were loaded with a 2 m-high embankment. The embankment was 24 × 14 m with 2:1 side slopes (Fig. 1) and was constructed in 3 days. The sub-soil consisted mainly of a thick marine clay layer followed by a sand layer. The clay layer was divided into 5 layers based on the variation of water content, bulk unit weight, over-consolidation ratio, and undrained shear strength with depth as follows: Layer (1) of weathered clay crust with a total thickness of 4 m; Layers (2) and (3) were soft clay with a total thickness of 16 m, subdivided into two layers, 6 and 10 m thick, based on differences noted in water content and compressibility; Layer (4) was medium stiff clay with a thickness of 4 m; and Layer (5) was stiff clay with a thickness of 3 m. It is noted that the groundwater table was located at 2 m below the ground surface. The two piles were driven for 25 m below the ground surface. The first pile was uncoated while the second pile was coated with a bitumen layer along the top 20 m of its length. The current study is concerned with studying the behavior of the uncoated pile only. The piles had an outer and inner diameters of 0.4 and 0.25 m, respectively, and were divided into six segments. Load cells were placed at the pile tip and at the connection joints between the segments. A telltale system was implemented to measure the piles compression and movements. The variation of pore-water pressures on the pile surface was measured using hydraulic piezometers installed at each load cell, except at the pile tip. Piezometers were installed 0.5 m and 1 m away from the pile center. The surface settlement at the site was monitored through surface settlement plates (placed on top of embankment) and screw-type deep settlement points. Pile loads and settlements were recorded over a period of 265 days. Indraratna et al. (1992) noted that the developed NSF along the pile-soil interface was primarily generated due to soil consolidation under the effect of embankment load, as well as due to the consolidation of clay surrounding the pile that has been remolded during pile driving. The embankment, piles, and instrumentation arrangement is shown in Fig. 1. The field measurements are presented along with the finite element predictions in the next section (Figs. 3, 4, and 5).

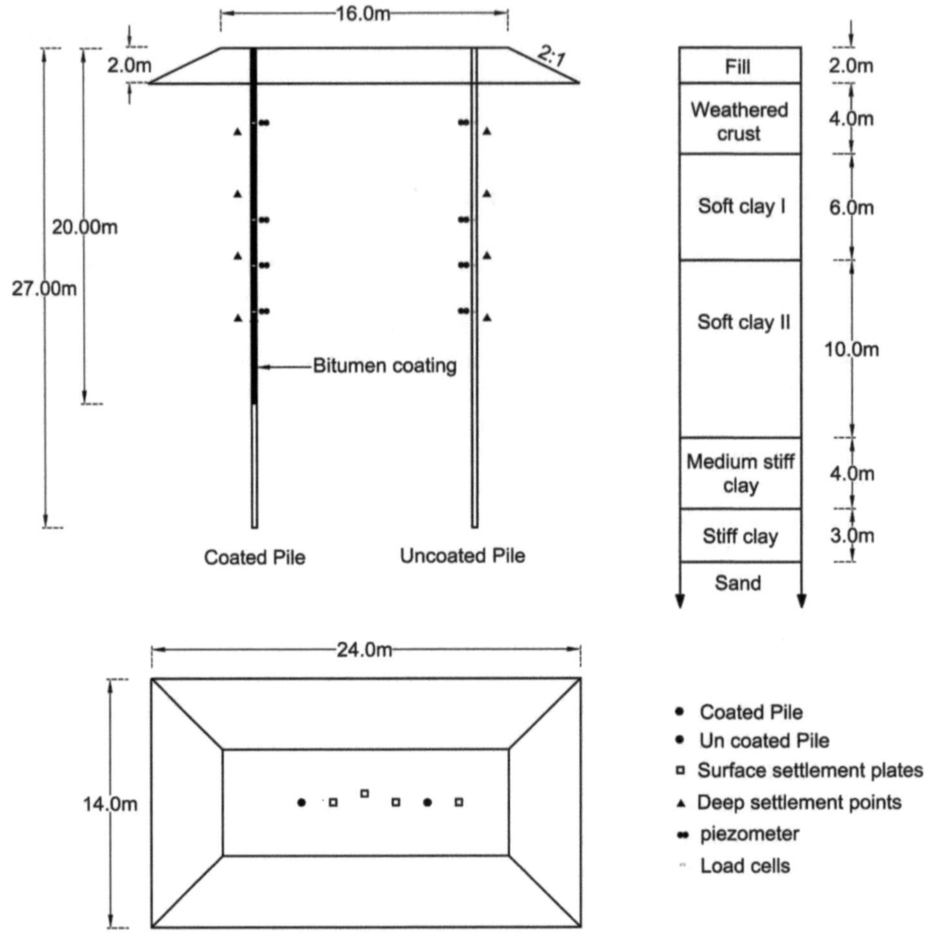

Fig. 1. Embankment, instrumented piles, and instrumentation details (after Indraratna et al. 1992)

3 Finite Element Modeling

A 3D FE model was developed to simulate the behavior of the uncoated pile in the case study presented by Indraratna et al. (1992) in the previous section. The different soil layers were modeled using Hardening soil model, the embankment fill was modeled using Mohr-Coulomb soil model, and the pile was modeled using linear elastic model. Indraratna et al. (1992) reported measured soil properties for the different soil layers such as bulk unit weight (γ), void ratio, and over-consolidation ratio (OCR), which are shown in Table 1. In addition, Indraratna et al. (1992) presented the Modified Cam-clay model parameters used in their FE model for the weathered crust, soft clay, and medium stiff clay. The effective angle of shearing resistance (ϕ') in the Hardening soil model was deduced from the tangent of the critical state line (M) in the Modified Cam-

Clay model, while the stiffness parameters (E_{oed}, E_{50}, and E_{ur}) were deduced from the Modified Cam-Clay compression index (λ^*) and swelling index (κ^*) (Brinkgreve et al. 2013). The stiffness of the stiff clay layer was reported by Indraratna et al. (1992), which was adjusted herein to correspond to a reference pressure (P_{ref}) of 100 kPa defined in the Hardening soil model. The input data for the sand layer was taken equivalent to the data reported by El-Mossallamy et al. (2012) for the same case study. The input Hardening and Mohr-Coulomb soil properties for the soil layers and embankment fill are summarized in Table 1.

El-Mossallamy et al. (2012) noted that for modeling the undrained consolidating soil behavior using effective soil parameters, a constant value for the shear modulus for both undrained and drained conditions is assumed in PLAXIS, which yields unrealistic high elastic deformations in the undrained condition for soft and medium stiff clays. Therefore, El-Mossallamy et al. (2012) suggested assigning two material sets to these soil layers: short term material set assigned to the soil in the immediate settlement calculation phase; and long term material set representing the drained soil properties during the consolidation phase. The difference between the two sets was the stiffness moduli values. The idea of the two sets of materials was implemented in the current study. For Bangkok clay, Parnploy (1985) mentioned that the ratio between the undrained and drained stiffness (E_u/E') for soft and medium stiff Bangkok clay was 3.85 and 1.75, respectively. Bergando et al. (1990) reported (E_u/E') value for the soft Bangkok clay of 10. In the verification FE model developed in the current study, (E_u/E') ratios for the soft and medium stiff clay layers of 4 and 1.75, respectively, yielded immediate settlements that matched the field measurements (after 3 days) as discussed in the next section (Figs. 3 and 4). Lower (E_u/E') ratios for the soft and medium stiff clay layers resulted in overestimated immediate settlements and vice versa. The coefficients of permeability (k) for the different clay layers were back-calculated such that the consolidation settlement of the clay layers matched the field settlement curve after 265 days as presented in the following section (Figs. 3 and 4). Higher k values yielded overestimated consolidation settlements after 265 days and vice versa. Finally, the measured skin friction (interpreted from axial load measurements) reported by Indraratna et al. (1992) was used to back-calculate the interface strength reduction factor (R_{inter}) for the different soil layers. Higher R_{inter} values, resulted in overestimated skin friction and vice versa. An effective soil cohesion (c′) of 0.1 kPa was assumed for all layers. The power for stress level dependency of stiffness (m) was assumed "1" for clay, and "0.5" for sand. The hollow prestressed precast concrete pile was modeled to have a unit weight of 15 kN/m^3 and modulus of elasticity (E) of 3×10^7 kN/m^2 as reported by Indraratna et al. (1992). The model FE mesh is shown in Fig. 2. A relatively fine mesh was used around the piles, and it became coarser further from the piles. The extent of the vertical and bottom boundaries was investigated to assure that that boundary conditions had insignificant effect on the model results.

Table 1. Input Hardening and Mohr-Coulomb soil model parameters

Layer	Depth (m)	γ^a (kN/m^3)	e^a	k^c (m/day)	OCR[a]	ϕ'^b (deg.)	R^c_{inter}	Drained set of parameters[b]			Undrained set of parameters		
								E_{50} (kN/m^2)	E_{oed} (kN/m^2)	E_{ur} (kN/m^2)	E_{50} (kN/m^2)	E_{oed} (kN/m^2)	E_{ur} (kN/m^2)
Fill	−2–0	17.0	1.67	0.006	1.0	30.0	0.60	15000	–	–	15000	–	–
Weathered crust	0–4	17.0	1.67	0.006	3.0	26.5	0.60	1830	1464	4525	1830	1464	4525
Soft clay 1	4–10	15.0	3.05	0.0026	1.1	24.7	0.75	986	789	4346	3942	3154	17384
Soft clay 2	10–20	16.0	2.09	0.0026	1.1	24.9	0.75	1194	955	4404	4776	3820	17616
Medium stiff clay	20–24	19.0	1.20	0.0037	1.8	23.0	0.75	2368	1894	7342	4144	3315	12849
Stiff clay	24–27	19.5	1.20	0.0037	1.8	23.0	0.75	15000	12000	45000	15000	12000	45000
Sand[d]	below 27	20.0	0.5	1	1.0	35.0	–	55000	44000	165000	55000	44000	165000

[a]Data reported by Indraratna et al. (1992) except for Sand layer
[b]Interpreted from Modified Cam-Clay model parameters reported by Indraratna et al. (1992) except for Sand layer
[c]Back-calculated such that predicted values of settlement and skin friction matched field measurements
[d]Data reported by El-Mossallamy et al. (2012) for Sand layer

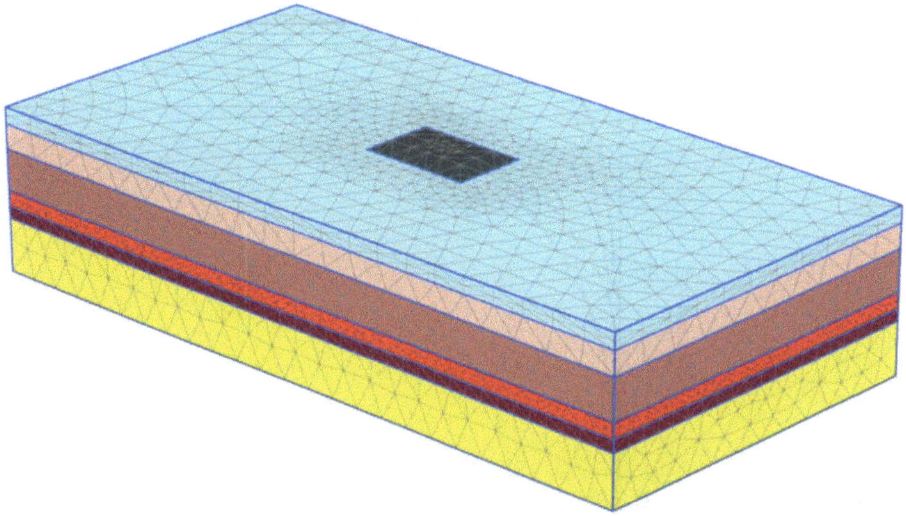

Fig. 2. Finite element model mesh

4 Comparison Between Numerical Predictions and Field Measurements

The variation of predicted ground surface settlement with elapsed time at distance 0.25 m from pile center is compared to field measurements in Fig. 3. The predicted distribution of ground surface settlement with depth at distance 0.25 m from pile center after 3 and 265 days is compared to field measurements in Fig. 4. Figure 3 shows good

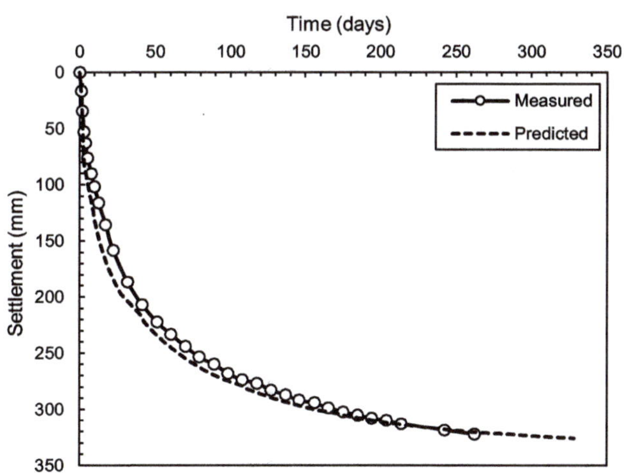

Fig. 3. Variation of predicted and measured ground surface settlement at distance 0.25 m from pile center with time

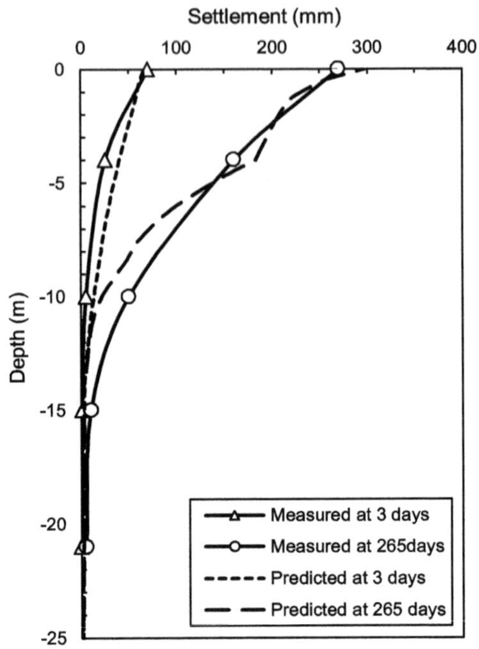

Fig. 4. Predicted and measured distribution of ground surface settlement with depth at distance 0.25 m from pile center after 3 and 265 days

agreement between measured and predicted settlements. Furthermore, relatively good agreement was observed between the measured and predicted settlements at various depths as shown in Fig. 4. The match obtained in Figs. 3 and 4 was achieved by using the (E_u/E') ratios mentioned in the previous section for the soft and medium stiff clay layers. Thus, assigning two material sets to the soft and medium stiff clay layers (short term and long term) was necessary to provide well predictions in the short term (3 days) and long term (265 days), which agrees with the findings of El-Mossallamy et al. (2012).

The predicted distribution of the dragload along the pile length is compared to the measured distribution after 265 days in Fig. 5a. The dragload distribution along the pile length was generally well predicted, with the maximum load slightly higher (by 3%) than the measured value. According to the axial load field measurements, the neutral point occurred at 20 m below the ground surface, which coincided with the boundary between the soft clay and medium stiff clay layers. The location of the neutral point based on predicted dargloads was shifted upwards at approximately 16.5 m below the ground surface. Similar observation was noted by Indraratna et al. (1992) and Lee et al. (2002) in their modeling for the same case study. The variation of predicted and measured skin friction values along the pile length after 265 days are compared in Fig. 5b. Relatively good agreement was observed between the predicted and measured skin friction values along the pile length. Indraratna et al. (1992) mentioned that the pile load distribution curve was used to obtain the skin friction values at the pile-soil

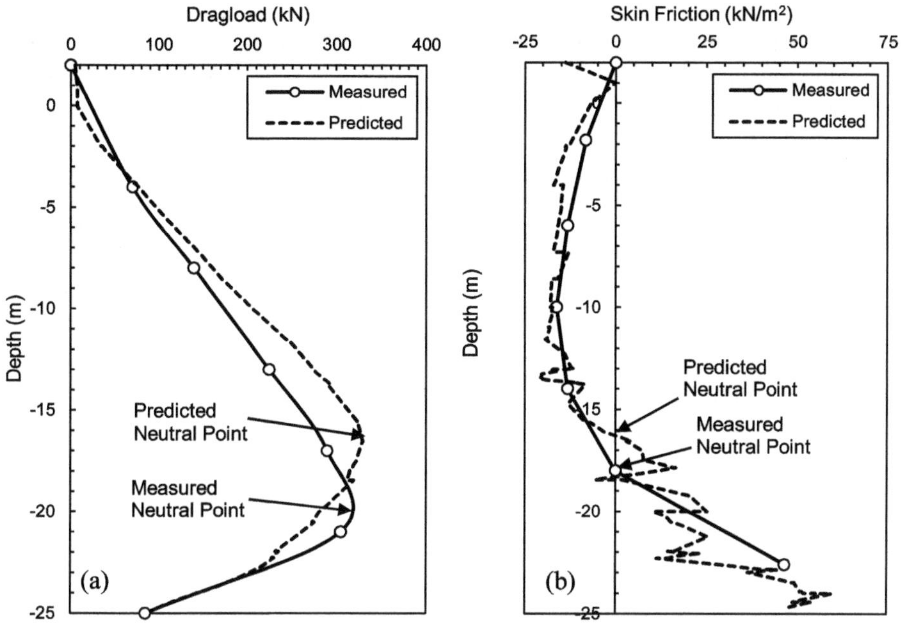

Fig. 5. Predicted and measured distribution of: (a) dragload and (b) skin friction along pile length after 265 days

interface. According to the interpreted field skin friction values, the neutral point occurred at 18.4 m below the ground surface. The reason for the noted difference in the location of the neutral point based on the axial load field measurements and the interpreted field skin friction values reported by Indraratna et al. (1992) was not clear. The location of the neutral point based on predicted skin friction values occurred at approximately 16.5 m below the ground surface, which matched the location of the neutral point based on the predicted dragloads.

5 Behavior of Piles in a Group

The developed 3D FE model was used to investigate the group effects on the distribution of dragloads in pile groups in soft soils under surcharge and embankment loading. Soil settlement, and thus the development of NSF, was initiated by the application of surcharge or embankment loading on top of the ground surface. Surcharge loading is common in coastal areas, which are usually raised by couple of meters to avoid flooding. Unfortunately, raising the ground surface causes settlement in the soft clay layers leading to NSF on piles. Similarly, embankments supported on piles in soft soils are subject to settlement due to consolidation of the soft layers, which generates NSF on piles. In the current study, the surcharge and embankment loads were applied by raising the ground surface by 2 m of fill with unit weight of 17 kN/m^3 (similar to the case study by Indraratna et al. 1992). The embankment had side slopes of 2H:1 V, which started at the boundary of the edge pile as shown in Fig. 6. The piles were modeled as free-headed piles in a group. The piles were not connected to each other with structural elements; thus, each pile responded separately in response to the applied loading. For the surcharge loading, a single pile and 4 pile group configurations were studied: 2 × 2, 4 × 4, 6 × 6, and 8 × 8 pile groups. For the embankment loading, 4 pile group configurations were investigated representing embankments supported on 2, 4, 6, and 8 rows of piles. Piles were spaced at three times the pile diameter in all pile groups considered herein. Typical model FE mesh is shown in Fig. 6 for the 6 × 6 group under surcharge loading and for embankment supported on 6 rows of piles. Position of piles in pile groups are shown in Fig. 7. Due to symmetry, only quarter of the whole mesh was used in the 3D model for surcharge loading; while for embankment loading, half the mesh was used to model a strip of the embankment including the volume served by one row of piles (the thickness of the strip was three times the diameter of the pile).

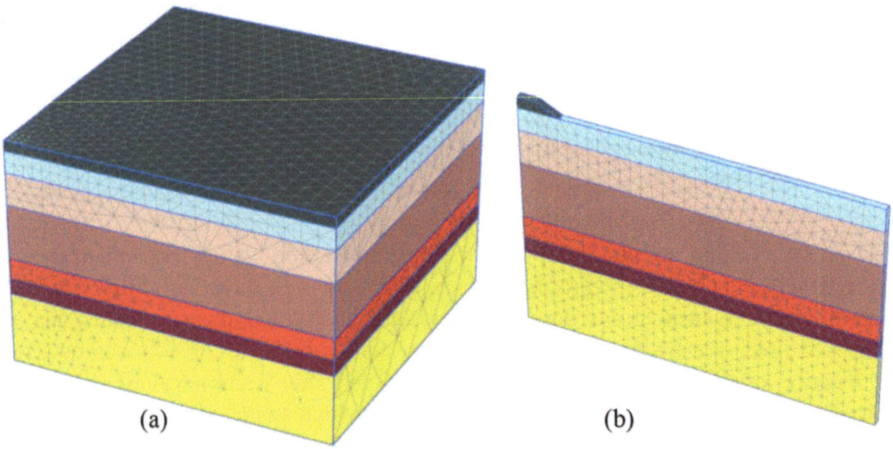

(a) (b)

Fig. 6. Typical FE model mesh for: (a) 6 × 6 pile group under surcharge loading and (b) embankment on 6 pile rows

5.1 Surcharge Loading

The group effect was defined for surcharge loading as the reduction in maximum dragload compared to the single pile. The variation of predicted skin friction and dragload along the pile length in 2 × 2 pile group is compared to that for single pile in Fig. 8. The behavior of piles in 2 × 2 pile group was similar to that for single pile; thus, the group effect was not pronounced in this case. As the number of piles in the group increased, the group effect became noticeable as shown in Figs. 9, 10, and 11 for 4 × 4, 6 × 6, and 8 × 8 pile groups, respectively. Smaller dragloads were observed to develop on piles in pile groups due to pile-soil interaction effect, known as the “shielding effect” on the inner piles by the outer piles, leading to decreased dragload on

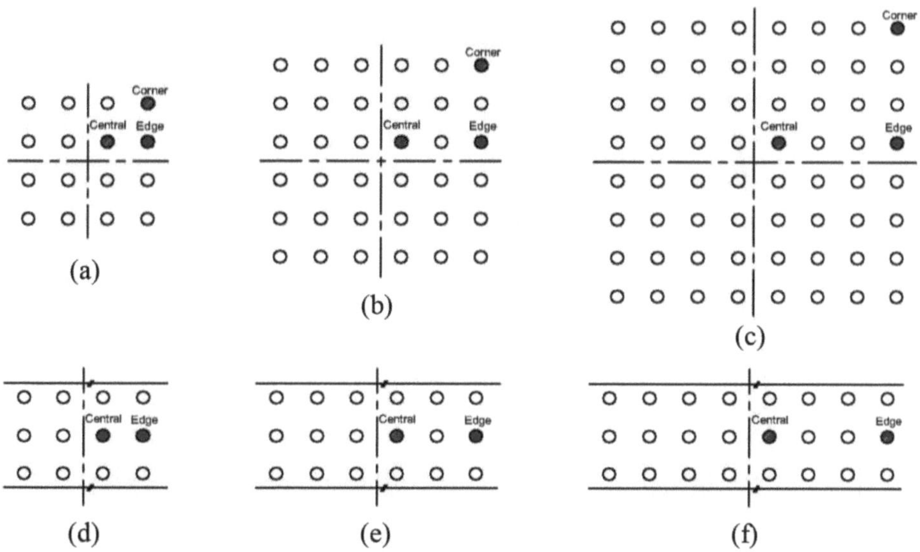

Fig. 7. Position of piles in a group: (a) 4 × 4, (b) 6 × 6, and (c) 8 × 8 under surcharge loading; (d) 4 rows, (e) 6 rows, and 8 rows under embankment loading

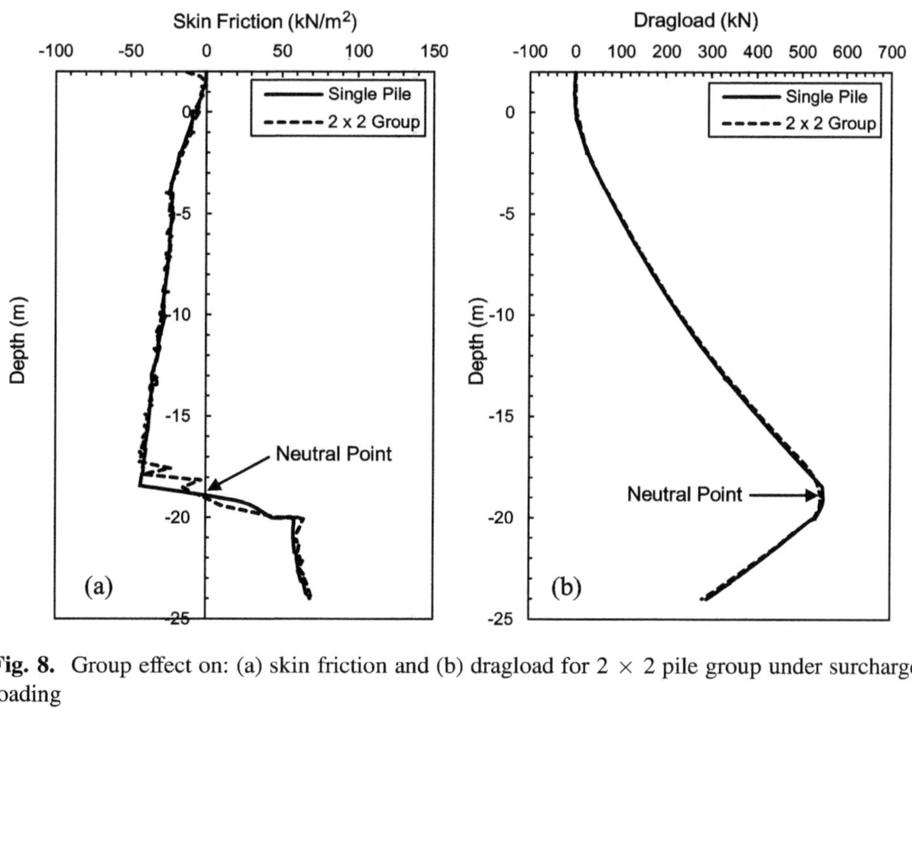

Fig. 8. Group effect on: (a) skin friction and (b) dragload for 2 × 2 pile group under surcharge loading

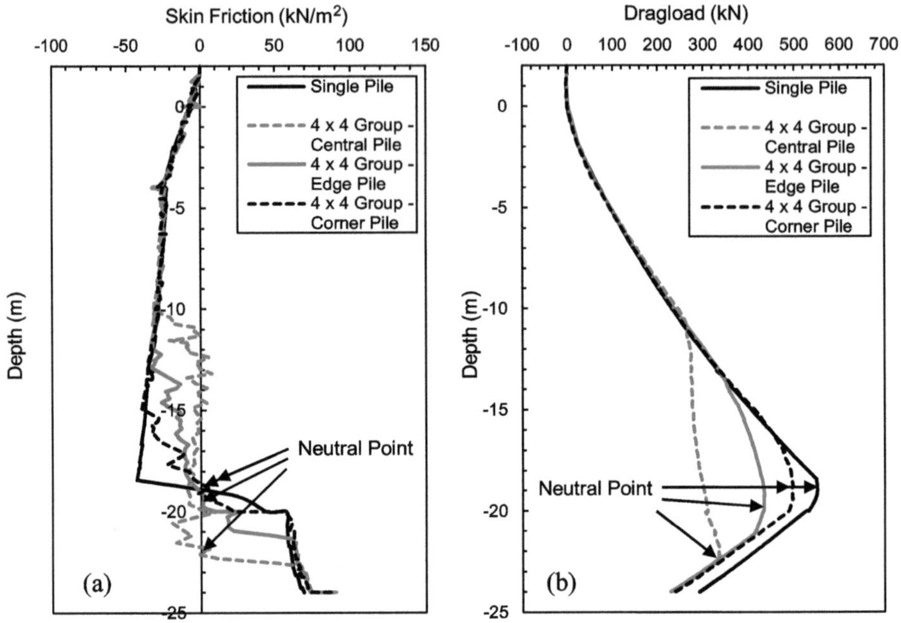

Fig. 9. Group effect on: (a) skin friction and (b) dragload for 4 × 4 pile group under surcharge loading

the inner piles inside the group. The maximum group effect was noted for the central pile, while the minimum group effect was observed for the corner pile. Group effects for the central, edge, and corner piles of 39%, 21%, and 10%, respectively, were observed for the 4 × 4 pile group. As the number of piles increased in the 6 × 6 pile group, the group effects increased to 60%, 36%, and 13% for the central, edge, and corner piles, respectively. Similarly, as the number of piles increased in the 8 × 8 group, the group effects increased to 70%, 41%, and 22% for the central, edge, and corner piles, respectively. The variation of group effects with number of piles within the group is shown in Fig. 12. As the number of piles increased within the pile group, the group effects increased for the central, edge, and corner piles. The larger the number of piles in a group, the greater the shielding effects on the reduction of dragloads. This reduction suggests using sacrificial piles around pile groups in consolidating soils, which would result in more economical designs. Similar observations were reported by Chow et al. (1990), Lee et al. (2002), and Lee and Ng (2004). The depth to reach the neutral point was compared for the different pile positions within the pile group. It was observed that the depth to the neutral point was similar for the single pile, 2 × 2 pile group, and edge and corner piles in the 4 × 4, 6 × 6, and 8 × 8 pile groups and occurred at an average depth of 76% of the pile length as shown in Figs. 8, 9, 10, and 11. The neutral point was translated downwards for the central piles in the 4 × 4, 6 × 6, and 8 × 8 pile groups to occur at an average depth of 89% of the pile length as shown in Figs. 9, 10, and 11. The minimum NSF and dragloads were noted for the central pile, while the maximum NSF and dragloads were observed for the corner pile.

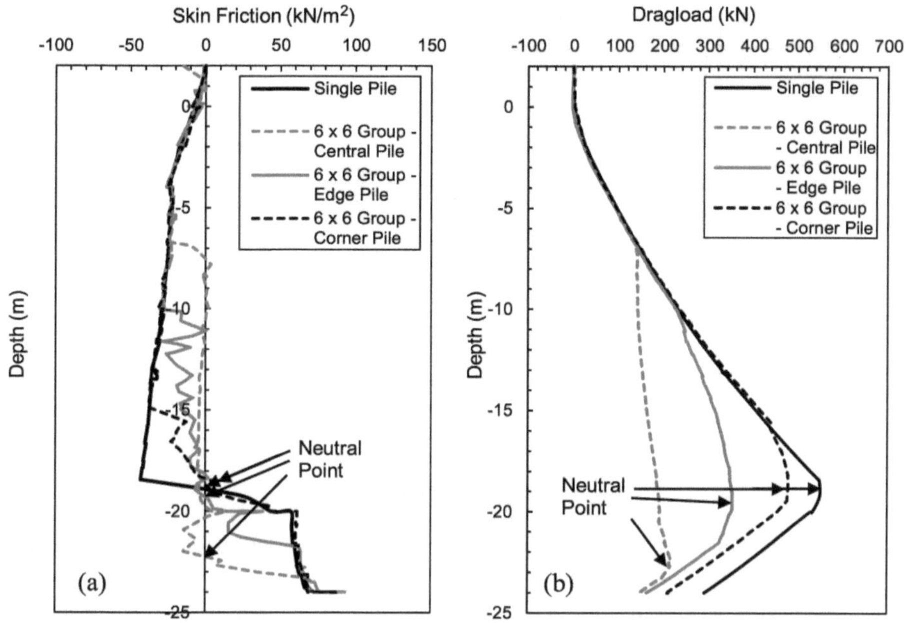

Fig. 10. Group effect on: (a) skin friction and (b) dragload for 6 × 6 pile group under surcharge loading

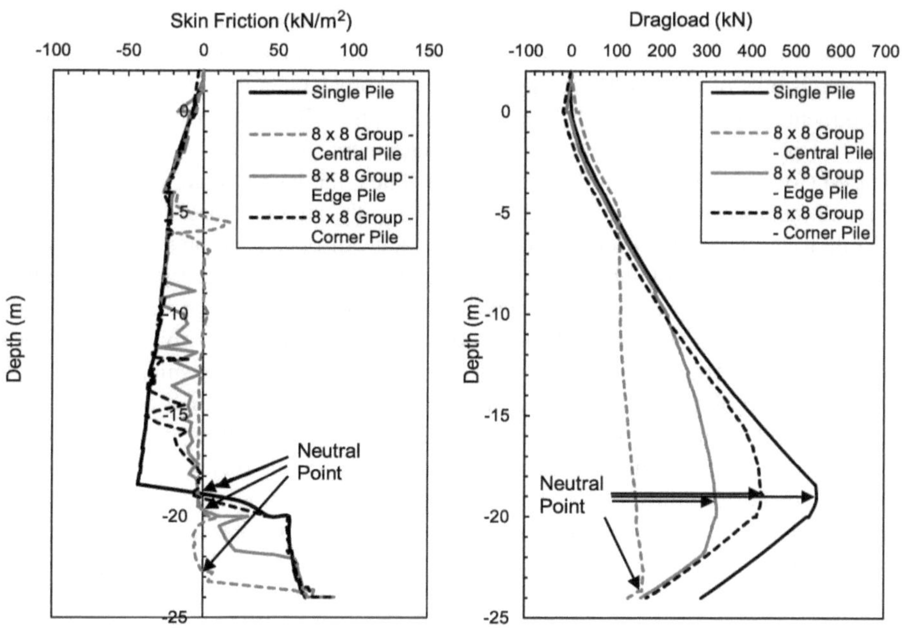

Fig. 11. Group effect on: (a) skin friction and (b) dragload for 8 × 8 pile group under surcharge loading

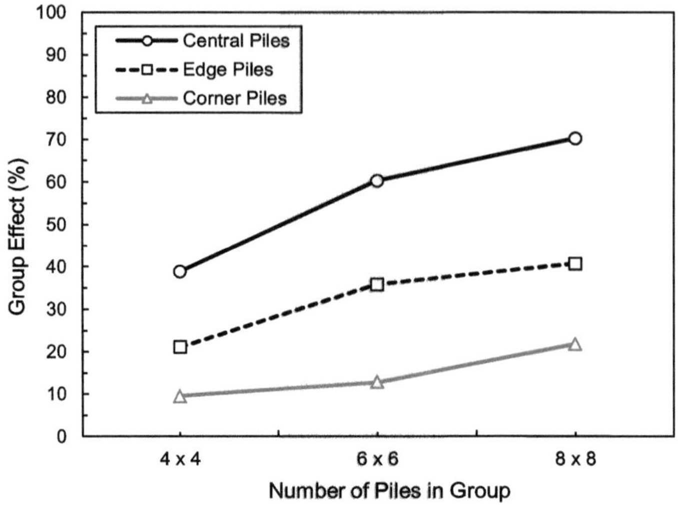

Fig. 12. Variation of group effect with number of piles in pile groups under surcharge loading

5.2 Embankment Loading

The effect of extent of loading was investigated by comparing the behavior of the 2 rows of piles under embankment loading to the 2 × 2 pile group under surcharge loading as shown in Fig. 13. Increasing the loading extent in the case of surcharge loading increased the load influence depth; and thus, lead to more soil consolidation, increased NSF, and downward shifting of the neutral plane. The embankment supported on 2 rows of piles was used as a reference to calculate the group effects for the embankments supported on 4, 6, and 8 rows of piles. It is worth noting that as the number of pile rows increased under the embankment, the width of the embankment increased (for constant spacing between piles). Thus, the largest loading extent in case of embankment loading was for the 8 pile rows and the smallest loading extent was for the 2 pile rows. The behavior of the 4, 6, and 8 rows of piles is compared to that of the 2 rows of piles in Fig. 14. The behavior of the central piles was observed to be similar in the 4, 6, and 8 rows of piles, where the dragloads decreased compared to that for the 2 rows of piles with a group effect of about 35% for the three investigated cases. Increasing the number of pile rows was expected to increase the shielding effect as noted in the case of surcharge loading; however, this behavior was not observed in the case of embankment loading. This may be attributed to the effect of increasing the loading extent of the embankment as the number of pile rows increased, which counteracted the shielding effect. For the edge piles, the dragloads increased as the number of pile rows increased, which was attributed to the increased extent of loading. The variation of group effects with number of pile rows is shown in Fig. 15. The depth to reach the neutral point was compared for the central and edge piles. Practically, the depth to the neutral point was similar for all the piles within the 2, 4, 6, and 8 pile rows, which occurred at a depth of about 45% of the pile length as shown in Fig. 14.

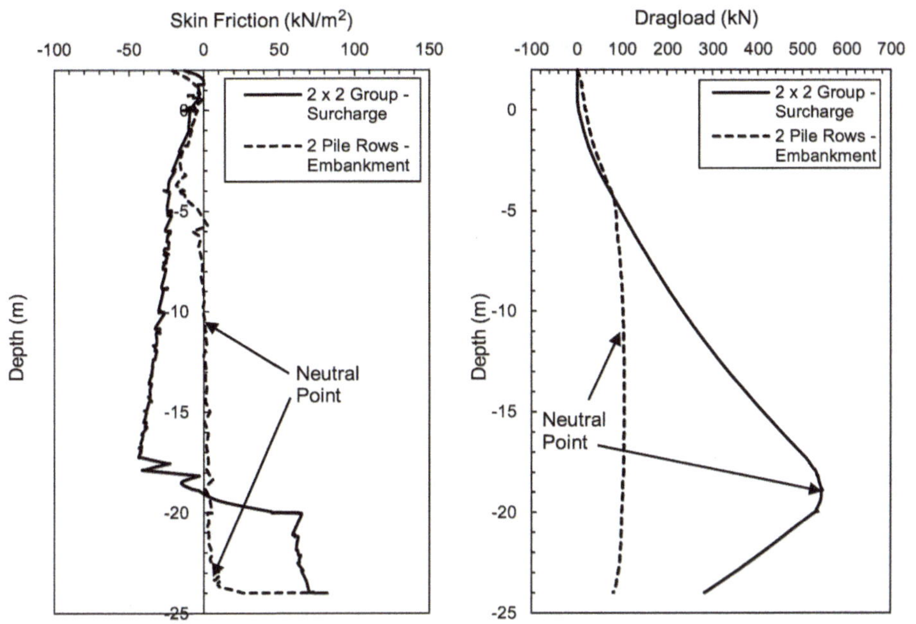

Fig. 13. Effect of load extent on: (a) skin friction and (b) dragload for 2 × 2 pile groups under surcharge loading and 2 pile rows under embankment loading

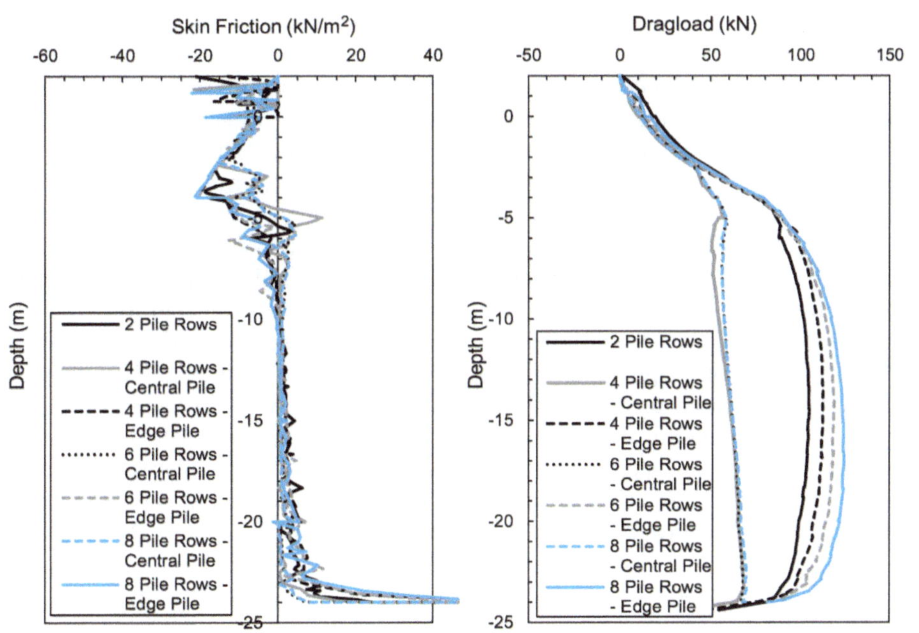

Fig. 14. Group effect on: (a) skin friction and (b) dragload for 2, 4, 6, and 8 pile rows under embankment loading

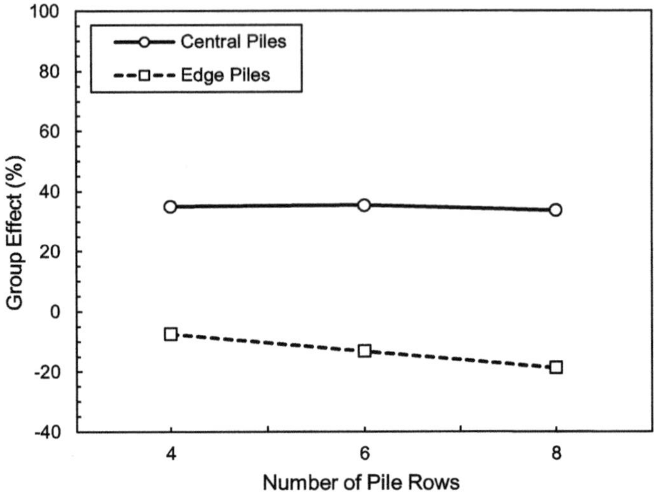

Fig. 15. Variation of group effect with number of pile rows in a group under embankment loading

6 Conclusions

A 3D FE model was developed to model a field case study for instrumented driven pile in soft Bangkok clay subjected to NSF. Assigning short and long term material sets to the soft and medium stiff clay layers was necessary to well predict the short and long term settlements. The developed 3D FE model was used to investigate the behavior of pile groups in soft clays under surcharge and embankment loading. Piles were spaced at three times the pile diameter for all investigated cases. Smaller dragloads were observed to develop on inner piles in pile groups due to the "shielding effect" by the outer piles. For surcharge loading, the maximum group effect was noted for the central pile, while the minimum group effect was observed for the corner pile. The larger the number of piles in a group, the greater the shielding effects on the reduction of dragloads. This reduction suggests using sacrificial piles around pile groups in consolidating soils, which may result in more economical designs. The depth to the neutral point was larger for the central piles compared to the corner piles. For embankment loading, the maximum group effect was noted for the central pile. The width of embankment increased as the number of pile rows increased for constant spacing between piles. Increasing the number of pile rows was not associated with further reduction in the dragloads on the central piles, which indicated that the shielding effect was counteracted by the effect of increasing the extent of loading. The depth to the neutral point was practically similar for all piles under the embankment. By comparing the surcharge loading to embankment loading, increasing the loading extent increased the load influence depth, and thus lead to more soil consolidation and increased NSF and dragloads.

Acknowledgments. Sincere appreciation is expressed to Dr. Rami El-Sherbiny, Professor at Cairo University, for his valuable contribution during conducting the numerical analysis.

References

Bergando, D.T., Ahmed, S., Sampco, C.L., Balasurbamaniam, A.S.: Settlements of Banga-Bangpakong highway on Bangkok clay. J. Geotech. Eng. ASCE **116**(1), 136–155 (1990)

Blanchet, R., Tavenas, F., Garneau, R.: Behavior of friction piles in soft sensitive clays. Can. Geotech. J. **17**, 203–224 (1980)

Brinkgreve, R.B.J., Engin, E., Swolfs, W.M.: PLAXIS 3D—Materials Model Manual. PLAXIS bv, Delft (2013)

Cheen, R.P., Zhou, W.H., Chen, Y.M.: Influences of soil consolidation and pile load on the development of negative skin friction of a pile. Comput. Geotech. **36**, 1265–1271 (2009)

Chow, Y.K., Chin, J.T., Lee, S.L.: Negative skin friction on pile groups. Int. J. Numer. Anal. Meth. Geomech. **14**(2), 75–91 (1990)

Fellenius, B.H.: Downdrag on piles due to negative skin friction. Can. J. **9**(4), 323–337 (1972)

Fellenius, B.H.: Results from long-term measurement in piles of dragload and downdrag. Can. Geotech. J. **43**(4), 409–430 (2006)

Indraratna, B., Balasubramaniam, A.S., Phamvan, P., Wong, Y.K.: Development of negative skin friction on driven piles in soft Bangkok clay. Can. Geotech. J. **29**(3), 393–404 (1992)

Jeong, S., Lee, J., Lee, C.J.: Slip effect at the pile-soil interface on dragload. Comput. Geotech. **31**, 115–126 (2004)

Johannessen, I.J., Bjerrum, L.: Measurement of the compression of steel pile to rock due to surrounding clay. In: Proceeding of 6th International Soil Mechanics and Foundation Engineering, Montreal, Canada, vol. 2, pp. 261–264 (1965)

Lee, C.J., Ng, W.W.: Development of downdrag on piles and pile groups in consolidating soil. J. Geotech. Geoenviron. Eng. **130**(9), 905–914 (2004)

Lee, C.J., Bolton, M.D., Al Tabbaa, A.: Numerical modelling of group effects on the distribution of dragloads in pile foundations. Geotéchnique **52**(5), 325–335 (2002)

Parnploy, U.: Deformation Analysis and Settlement Reduction of Bangna-Bangpakong Highway (Section I), 8840th edn. Asian Institute of Technology, Bangkok (1985)

Poorooshab, A.B., Alagmir, M., Miura, N.: Negative skin friction on rigid and deformable piles. Comput. Geotech. **18**(2), 109–126 (1996)

El-Mossallamy, Y.M., Hefny, A.M., Demerdash, M.A.: Numerical analysis of negative skin friction on piles in soft clay. Hous. Build. Natl. Res. Center **9**, 68–76 (2012)

Shaft Resistance of Displacement Piles in Overconsolidated Cohesionless Soils

Yasir M. Alharthi[1,2(✉)] and Adel M. Hanna[1]

[1] Department of Building, Civil and Environmental Engineering, Concordia University, 1455 de Maisonneuve Blvd. W., Montreal, QC H3G 1M8, Canada
harthiym@hotmail.com

[2] Department of Civil Engineering, Taif University, P.O. Box 888, Taif 2197, Saudi Arabia

Abstract. The conventional theories for predicting the capacity of a single pile in sand produce wide range of discrepancies. This is mainly due to the complexity of modeling cohesionless material, collecting field data, and ignoring the effect of overconsolidation in cohesionless soils.

This paper presents an experimental investigation on the capacity of closed-ended displacement piles (CEDP) in overconsolidated cohesionless soils with an emphasis on the shaft resistance. A large testing tank was instrumented to measure the overconsolidation ratio in the laboratory sand bed of homogeneous loose, medium, and dense sands. Tests were conducted on a long model pile at various embedment depths.

The resulted shaft resistance was found to be much higher than that calculated using conventional methods for homogeneous sand. By incorporating the overconsolidation ratio in the design, unique values were obtained for the shaft resistance of these piles.

The results of this experimental investigation are presented in the form of the measured shear stress along the pile's shaft and its unique distribution along its shaft. The critical depth for piles in cohesionless soils will be also addressed.

Keywords: Displacement piles · Shaft resistance · Stress distribution
Critical depth · Overconsolidation ratio (OCR) · Sand
Experimental investigation

1 Introduction

Piles are structural members that transfer the applied load of superstructures to deep supportive layers of soil or bedrock to provide sufficient support both axially and laterally. Conventional theories for predicting the shaft resistance of a single pile in sand have generated a range of discrepancies (Hanna and Nguyen 2002; Poulos and Davis 1980). This is due to the complexity of modeling cohesionless materials and collecting field data (Randolph et al. 1994).

In addition, the role of overconsolidation in such soils is often neglected. Whether occurring naturally or artificially, overconsolidation in cohesionless soils directly affects the lateral earth pressure that acts upon the pile shafts and thus upon pile

H. El-Naggar et al. (Eds.): GeoMEast 2018, SUCI, pp. 232–246, 2019.
https://doi.org/10.1007/978-3-030-01902-0_19

capacity. Overconsolidation can occur when the ground surface is subjected to erosion, excavation, or unloading, often due to glacial melting, the demolition of structures, raised water tables (Coduto 2001), compaction, or vibration (Hanna and Soliman-Saad 2001; El-Emam 2011).

Conventionally, the unit shaft resistance at failure (τ_f) is calculated a function of the lateral earth pressure applied on the pile and the interface friction coefficient as given in the following equation:

$$\tau_f = \sigma'_{rf} \tan \delta_f = K_s \sigma'_v \tan \delta_f = \beta \sigma'_v = K_0 (K_s/K_0) \sigma'_v \tan \delta_f \quad (1)$$

Where (σ'_{rf}) is radial effective stresses at failure, (K_s) is lateral earth pressure coefficient at failure, (σ'_v) is average theoretical effective overburden pressure, and (δ_f) is pile-soil interface friction angle at failure. (β) is a coefficient that combines the mobilized lateral earth pressure coefficient and the pile–soil friction coefficient ($K_s \tan \delta_f$). (K_0) is at-rest lateral earth pressure coefficient which can be calculated as follows:

$$K_0 = 1 - \sin \Phi' \quad \text{(Jaky 1944)} \quad (2)$$

$$K_0 = (1 - \sin \Phi') OCR^{\sin \Phi'} \quad \text{(Mayne and Kulhawy 1982)} \quad (3)$$

Where (Φ') is the effective angle of shearing resistance.

It can be noted from Eq. 1 that the lateral earth pressure coefficient is presented in different forms; (K_s), (β), and (K_s/K_0). The value of these coefficients are mostly found empirically. The (K_s) value proposed to estimate the shaft resistance of CEDP varies widely from one researcher to another (Broms 1966; Meyerhof 1976; Coyle and Castello 1981; API 2000), and varies according to the pile material, relative density, angle of shearing resistance, and pile embedment depth. The (β) value suggested to estimate the shaft resistance of CEDP also shows some variance from one researcher to another (Poulos and Davis 1980; Meyerhof 1976; Toolan et al. 1990). Similarly, different values proposed for the (K_s/K_0) value to estimate the shaft resistance of CEDP (Kulhawy 1991; Tomlinson and Woodward 2014; Das 2011).

Despite the variation observed in the proposed lateral earth pressure coefficients (K_s, β, and K_s/K_0), these values are much lower than that needed to estimate the shaft resistance of CEDP in overconsolidated cohesionless soils. In the field, Beringen et al. (1979) performed compressive and tensile field tests on opened- and closed-ended pile load at a site with overconsolidated sand and observed that shaft resistance exceeded recommended design limits by 100–200%. More recently, Foray et al. (1998) conducted laboratory tests on piles driven in both normally consolidated and overconsolidated sand and observed that shaft resistance values of piles driven into normally consolidated sand were almost half those driven into overconsolidated sand.

The distribution of shaft resistance was long believed to increase linearly at a constant rate down to a certain embedment depth, also known as *critical depth* (Vesic 1964; Meyerhof 1976). However, recent field pile tests have dispelled that theory (Fellenius 2002; Kulhawy 1995), and researchers now attribute critical depth to the effect of the angle of shearing resistance, the fact that it decreases with depth, residual forces, and soil arching development due to driving, and the tendency of the OCR to decrease with depth as well.

Because the overconsolidation of cohesionless soil and its variation with depth influences shaft resistance, this paper investigates experimentally the influence of such overconsolidation, its effects upon shaft resistance, the distribution of shaft resistance along the pile shaft, and the possibility of critical depth.

2 Experimental Investigation

A prototype experimental setup was developed in the laboratory, which consists of a tank made of thick steel (1 m long, 1 m wide, and 1.25 m deep), loading system of 25 kN capacity equipped with an AKD servo driver which allowed for strain-controlled testing, and a pile model 0.8 m long and 55 mm in diameter, to perform pile load tests in sand. The set-up can measure the total and the tip resistance of the pile, and the stresses on the pile's shaft and accordingly the total shaft resistance. Furthermore, the OCR and the stresses in the sand mass in the tank can also be measured. A schematic sketch of the set-up is presented in Fig. 1a.

Two load cells were installed at the top and the bottom of the pile model to measure the total applied load and the tip resistance, respectively. Two other load cells were placed inside the pile to measure the total load at their levels. The load cells are located at 120 mm and 240 mm from the pile base. The load cells used for this pile had a maximum capacity of 22.25 kN with an accuracy of ±0.25%. Several direct shear tests were performed to establish the degree of roughness of a sandpaper to be glued to the pile shaft to maintain a ratio of interface angle to angle of shearing resistance (δ/Φ') equal to 1. As a result of those tests, Grade 150 sandpaper satisfied the unit ratio at different relative densities. Figure 1b and c show a photograph of the pile model, and a cross section for the pile model, respectively.

In the present investigation, the ratio of the distance from the pile center to the tank sides to the pile diameter (B/D) was taken as 9.1. This ratio was sufficient to avoid any boundary effect from the sides of the testing tank (Chari and Meyerhof 1983; Shalabi and Bader 2014). Pile load tests are also affected by the mean particle size (d_{50}) of the soil. Fioravante (2002) has suggested that the scale effect can be avoided when the (D/d_{50}) ratio exceeds 50. In the study reported here, the (D/d_{50}) ratio was 211.

The soil used in this experiment is commercially known as "Silica Sand 4010". The series of tests performed on this sand revealed a mean effective particle size d_{50} of 0.26 mm, a uniformity coefficient C_u of 1.88, a specific gravity G_s of 2.62, and a minimum and a maximum void ratios e_{min} and e_{max} of 0.50 and 0.84, respectively. The sand can be categorized as sub-rounded material with a peak angle of shearing resistance of 32.96°, 34.93°, and 36.80° at 30%, 45%, and 60% relative density, respectively.

The pile load test performed at relative densities of 30%, 45%, and 60%. To fill up the tank up to 1.2 m height, eight soil layers of equal height were placed in the tank. To achieve the desired relative density, each soil layer were compacted. To achieve the desired relative density uniformly across the tank, preliminary tests were performed using trial and error approach where less energy (drops) was applied to the lower layers

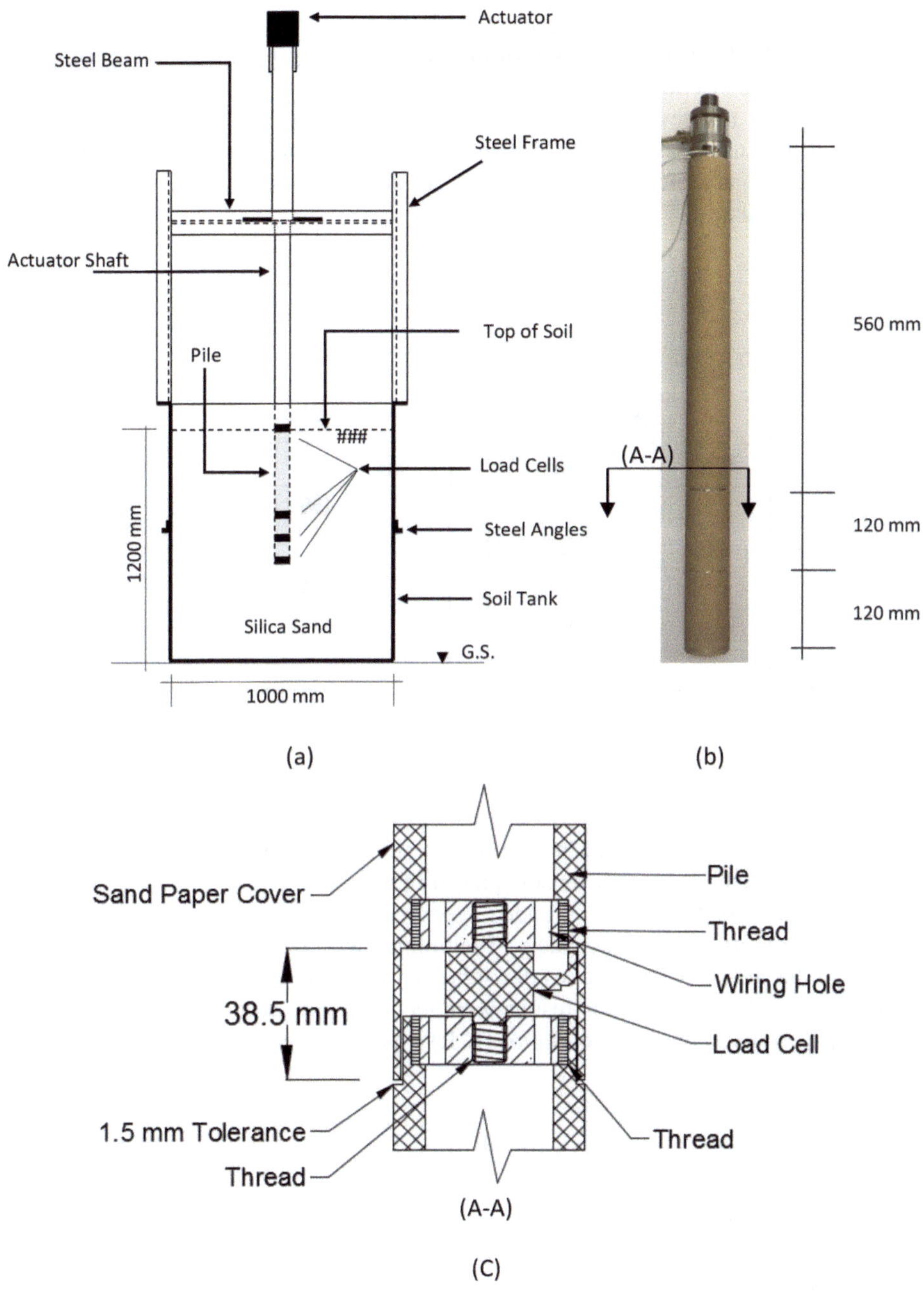

Fig. 1. (a) Schematic sketch of the experimental setup, (b) photograph of the pile model, and (c) section of the pile model

and more energy applied to the upper layers (Hanna and Al Khoury 2005; Hanna and Soliman-Saad 2001). The unit weight of the sand in the testing tank was measured by placing small cans of known volumes in strategic locations in the sand. Two cans were embedded in the top seven layers. Each one has a diameter of 110 mm, height of 50 mm and thickness of 3 mm. A schematic sketch for the cans arrangement and a photo of the can are presented in Fig. 2a. At the end of each test, the cans were retrieved to calculate the unit weight.

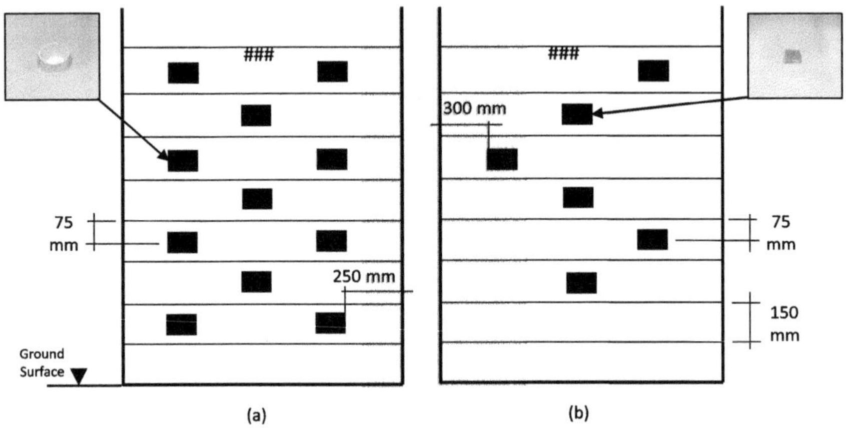

Fig. 2. (a) Unit weight can, and (b) sensors arrangement in the testing tank

The vertical stresses induced in the sand due to the compaction were measured for the first six soil layers using six sensors placed in the middle of each layer as shown in Fig. 2b. The sensors, Single Tact branded, were capacitive force sensors which differ in size according to its maximum capacity. Preliminary tests were performed to verify the readings of the sensors where the results of the sensors agreed with that obtained from a pressure load cell. Therefore, the results were recorded after each compaction test, and accordingly, the overconsolidation ratio (OCR) was calculated for each layer as follows (Hanna and Al Khoury 2005):

$$\mathrm{OCR} = \frac{\sigma_c}{\sigma'_v} \tag{4}$$

Where (σ_c) is vertical pressure at a given depth as measured by the corresponding sensor, and (σ'_v) is theoretical effective overburden pressure at that depth. Figure 3a and b present the produced relative density of the sand in the testing tank and the deduced OCR values for each relative density, respectively.

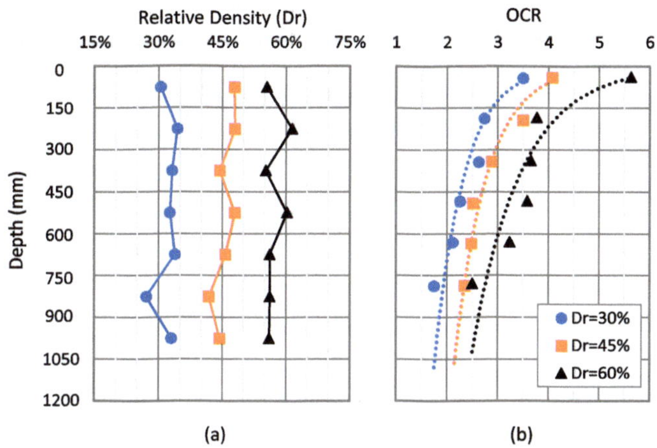

Fig. 3. Compaction results for (a) relative density; and (b) overconsolidation ratio versus depth

To perform the pile load test, the pile was driven vertically at the center of the tank, at a rate of 25 mm/min, to the desired depth. At this stage, all sensors were set to zero, and the pile load test started when the actuator drove the pile at a rate of 5 mm/min (Le Kouby et al. 2013). The pile load test ended when the pile was displaced by 20 mm. During the test, readings of the load cells were collected continuously by the DAS. Table 1 summarizes the experimental testing program followed. A photograph taken during the pile load testing is presented in Fig. 4.

Table 1. Experimental testing program

Testing program			
Test No.	Relative density	Nominal depth	Depth
–	Dr	L/D	L
–	(%)	–	(mm)
1	30	5	275
2		8	440
3		10	550
4		13	715
5	45	5	275
6		8	440
7		10	550
8		13	715
9	60	5	275
10		8	440
11		10	550
12		13	715

Fig. 4. Photo of pile during testing

3 Results and Analysis

Figure 5 presents typical test results (test 3). The total shaft resistance was found as the difference between the pile and the tip resistance. It can be noted that load-settlement curve for the total shaft resistance reached a peak value then dropped slightly as the pile displacement increased and reached a constant value at the end of the test. This behavior is in line with the results of Foray et al. (1998) for model pile tested in overconsolidated sand.

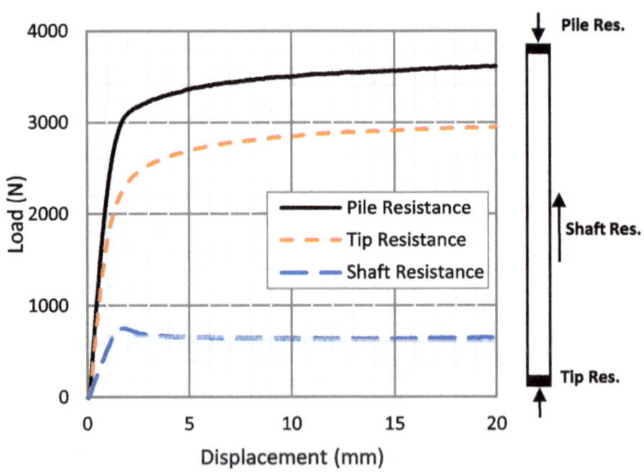

Fig. 5. Typical results (test 3): total, tip, and shaft resistance versus head displacement

Based on the load-settlement curve of the total pile resistance, the ultimate capacity was found using the tangential method. Accordingly, the ultimate shaft resistance was found. The (K_s) was back-calculated using Eq. 1 for all the tests performed and depicted in Fig. 6. It is evident that (K_s) values for CEDP in OC cohesionless soils have higher values than those traditionally recommended. It is noted that the (K_s) values increased linearly with depth, and remained almost constant after nominal depth (L/D = 10). This supports Vesic's (1967) and Meyerhof's (1976) findings that the critical depth exists when the total shaft resistance analyzed as an average even when the soil is overconsolidated. When compared to Vesic's (1967) results, as shown in Fig. 7, the critical depth appeared in overconsolidated cohesionless soils and was significantly influenced at 30% and 45% relative densities.

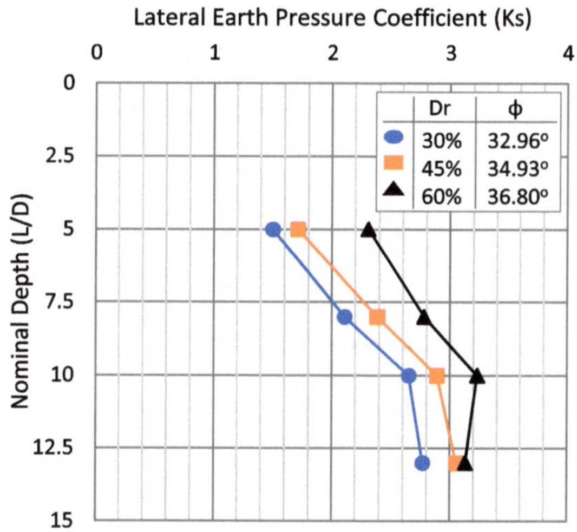

Fig. 6. *Ks* values versus ratio of depth to diameter (L/D)

The shaft resistance for each section of the pile was found as the difference between the two load cells at each end. Figure 8 presents typical test results (test 3) for the shaft resistance of each pile section. The load-settlement response was almost the same at the initial stages of the loading for the three sections. On the other hand, the peak shaft resistance for each section were different. The highest shaft resistance was for Sect. 1, whereas the lowest shaft resistance was for Sect. 2. This indicates a nonlinear distribution of the shear stress along the pile's shaft.

The local shear stress distribution (measured load divided by the corresponding area) on the total length of the pile model is presented in Fig. 9. The local shear stress is presented as an average value at the middle of each section. It is evident that distribution of the shear stress along the pile shaft is nonlinear. It is noted that the local shear stress for tests performed at the same nominal depth (L/D) behave similarly where the stress value increased as the relative density increased. It is also noted that the local shear stress at a certain depth decreased as the pile depth increased which is, generally,

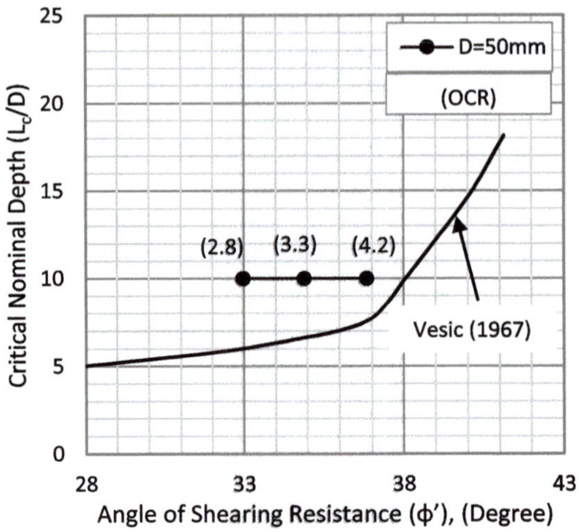

Fig. 7. Experimental critical depth results

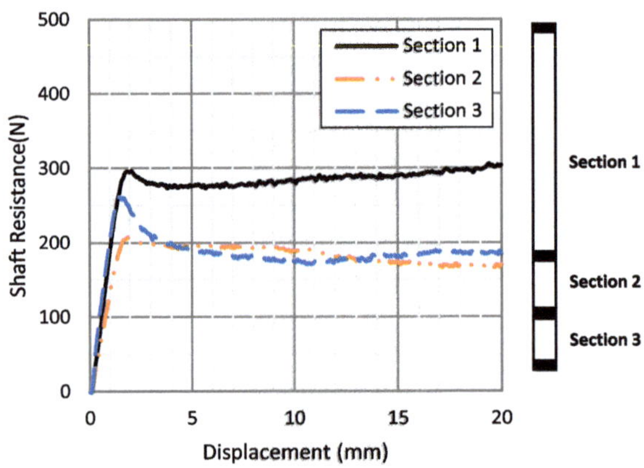

Fig. 8. Typical results (test 3): sections shaft resistance versus head displacement

in accordance with the results of Vesic (1970), Lehane et al. (1993), and Flynn and McCabe (2015). Because of this reduction, the average total shaft friction remains relatively constant as the embedment depth increases, and consequently, the critical depth appears. It can be concluded that the critical depth (*Lc*) appears only when the total shaft resistance is analyzed as an average. Evidently, the critical depth appeared in overconsolidated cohesionless soils and was significantly influenced at 30% and 45% relative density.

Figure 10 presents a comparison between the test results and that found using the lateral earth pressure coefficients (K_s, β, and K_s/K_0) proposed by different authors without limiting the shear stress value. It can be noted that the shaft resistance is highly under estimated.

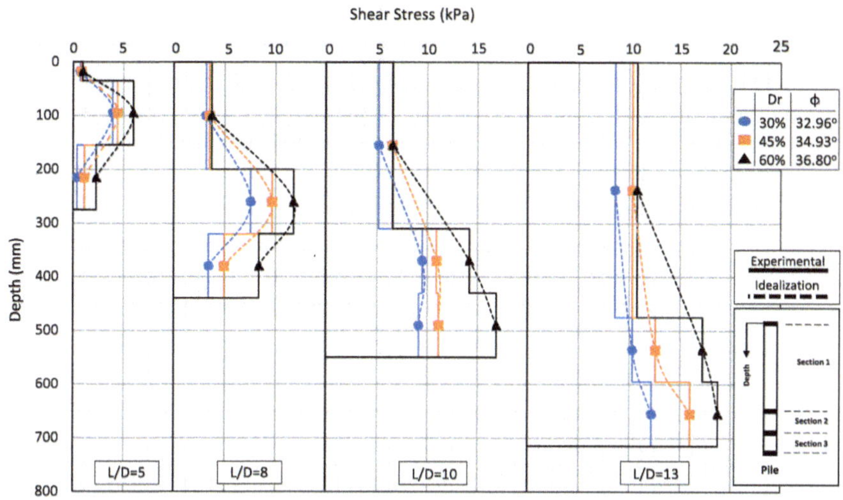

Fig. 9. Test results: shear stress on pile sections versus depth

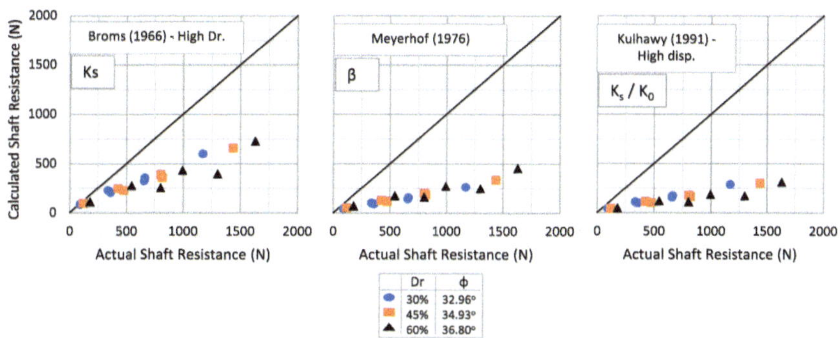

Fig. 10. Comparison between the actual and the calculated shaft resistance using values proposed by different authors

For practicality, the experimental results were correlated to the cone penetration test (CPT) end resistance (q_c). Because the CPT was not performed, the end resistance (q_c) was calculated according to Mayne's (1991) equation as follow:

$$K_0 = \frac{(P_a/\sigma'_v)(q_c/P_a)^{1.6}}{145\, exp\left\{\left[\frac{(q_c/P_a)/(\sigma'_v/P_a)^{0.5}}{12.2\, OCR^{0.18}}\right]^{0.5}\right\}} \tag{5}$$

Where (K_0) is at rest lateral earth pressure coefficient which is calculated according to Eq. 3, (q_c) is cone tip resistance, (P_a) is a reference stress equal to one atmosphere (1 bar = 100 kPa), and (σ'_v) is vertical effective stress. The (OCR) was found according to the experimental results presented in Fig. 3. Knowing the soil properties and the OCR values, Eq. 5 was solved to find the qc value at different levels across the testing tank at 30%, 45%, and 60% soil relative densities.

Knowing the value of shear stress and the value of the interface angle, the (σ'_{rf}) value was calculated according to Eq. 1 for every section of the pile model and normalized by the average end cone resistance (q_c) between the corresponding load cell levels to present it in unitless values. At the same nominal depth, the (σ'_{rf}/q_c) values for each section at 30% relative density was close to that at the other relative densities. Accordingly, the (σ'_{rf}/q_c) were averaged. Figure 11a presents the (σ'_{rf}/q_c) values against the normalized distance from the pile base (*h*/*D*). It can be noted that, for (*h*/*D*) higher than 3.3, the (σ'_{rf}/q_c) values decreased with increasing (*h*/*D*) for piles tested at nominal depth (*L*/*D* = 5). As the nominal depth increased (*L*/*D* = 8 and 10), the (σ'_{rf}/q_c) decreased at a slower rate. For nominal depth (*L*/*D* = 13), the (σ'_{rf}/q_c) increased. For

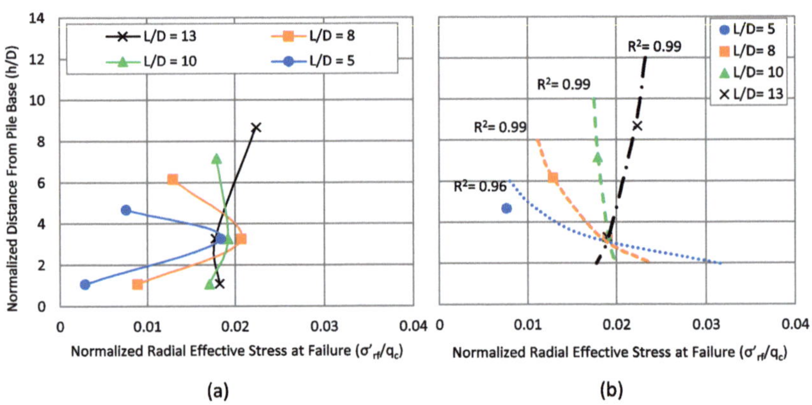

Fig. 11. (a) Results of (σ'_{rf}/q_c) versus Ratio of (h/D), and (b) Best fitting lines for the ratio of (σ'_{rf}/q_c)

(*h*/*D*) lower than 3.3, the (σ'_{rf}/q_c) values decreased with decreasing (*h*/*D*) up to a certain (*L*/*D*) where (σ'_{rf}/q_c) seemed to reach a constant value. Therefore, it can be concluded that the (σ'_{rf}/q_c) values depend on the nominal depth (*L*/*D*) ratio.

For (*h*/*D*) higher than 3.3, the (σ'_{rf}/q_c) values seem to be matched by a power function as expressed in Eq. 6. According to this equation, a regression analysis was performed, using the dataset for the twelve pile load tests, to determine the values of factor a and b that adequately described the distribution of (σ'_{rf}/q_c) values along the pile shaft at different nominal depths (*L/D*).

$$\left(\sigma'_{rf}/q_c\right) = a(h/D)^b \tag{6}$$

Where (h/D) is normalized distances from the pile base. The (σ'_{rf}/q_c) values at (h/D = 3.3) are close to each other, and accordingly, they were averaged to simplify the analysis. The values of factor a and b were found, as presented in Fig. 11b, based on the best fitting lines. Figure 12a and b present the value of "a" and "b", respectively, versus the nominal depth (*L/D*). Accordingly, the following empirical relationships were found:

$$a = 1.0843(L/D)^{-1.667} \tag{7}$$

$$b = 0.48e^{-7.372E-5(L/D)} - 4.83e^{-0.2(L/D)} \tag{8}$$

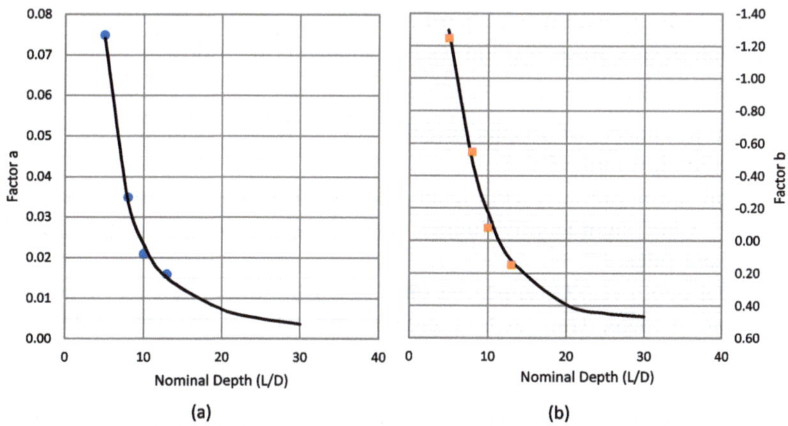

Fig. 12. (a) Factor a, and (b) Factor b versus the ratio of depth to diameter (L/D)

For (*h/D*) lower than 3.3, the (σ'_{rf}/q_c) value was assumed to be half of that found at *h/D* = 3.3 for nominal depth (*L/D*) equal to or less than 8, and same (σ'_{rf}/q_c) value as that found at *h/D* = 3.3 for higher nominal depth (*L/D*) ratios.

The experimental results were compared to those predicted by the proposed equations. It can be noted from the comparison presented in Fig. 13 that the predicted shear stresses agree with the actual values.

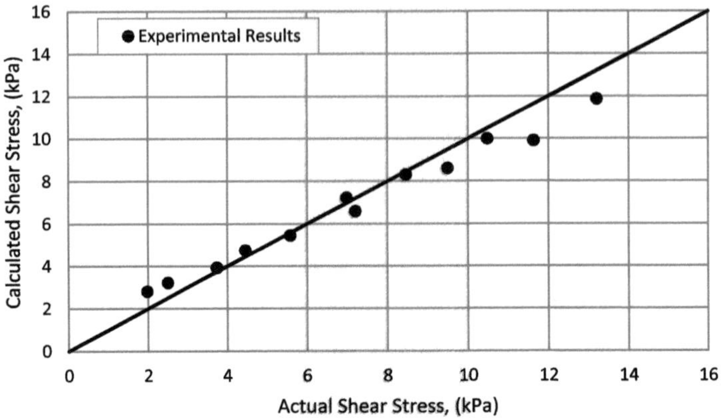

Fig. 13. Comparison between the actual and the calculated shaft resistance using the proposed equations

4 Conclusion

An experimental investigation was performed on displacement piles in overconsolidated cohesionless soil. Based on the results obtained, the following conclusions were drawn:

1- The shaft resistance of displacement piles increases with the increase of the overconsolidation ratio (OCR).
2- The lateral earth pressure coefficient (K_s) back calculated from the total shaft resistance, were significantly higher than those traditionally recommended for displacement piles.
3- The critical depth (L_c/D) appeared only when the total shaft resistance was analyzed as an average. The (OCR) increased the value of the critical depth at loose and medium dense cohesionless soils which in turn increased the shaft resistance significantly.
4- The shear stress distribution showed some dependancy on the nominal depth (*L*/*D*).
5- Empirical equations were proposed for displacement piles in overconsolidated cohesionless soils to estimate the shear stress and its distribution along the pile's shaft.

Acknowledgements. The first author gratefully acknowledges the financial support he received from the Saudi Arabian Cultural Bureau in Canada and the scholarship sponsored by Taif University. The financial support from the Natural Sciences and Engineering Research Council of Canada (NSERC) and Concordia University are acknowledged as well.

References

API: Recommended practice for planning, designing and constructing fixed offshore platforms-working stress design. 2A-WSD, 21 (2000)

Beringen, F., Windle, D., Van Hooydonk, W.: Results of loading tests on driven piles in sand. In: International Conference of the Institution of Civil Engineers on Recent Developments in the Design and Construction of Piles, London, pp. 213–225 (1979)

Broms, B.B.: Methods of calculating the ultimate bearing capacity of piles - a summary. Sols-Soils 18–32 (1966)

Chari, T.R., Meyerhof, G.G.: Ultimate capacity of rigid single piles under inclined loads in sand. Can. Geotech. J. **20**(4), 849–854 (1983)

Coduto, D.P.: Foundation Design: Principles and Practices. Prentice Hall, Upper Saddle River (2001)

Coyle, H.M., Castello, R.R.: New design correlations for piles in sand. J. Geotech. Geoenviron. Eng. **107**(7), 965–986 (1981)

Das, B.M.: Prinsiple of Foundation Engineering. Global Engineering: Christopher M. Shortt, Cengage Learning, Boston (2011)

El-Emam, M.: Experimental and numerical study of at-rest lateral earth pressure of overconsolidated sand. Adv. Civ. Eng. (2011)

Fellenius, H.B.: Determining the resistance distribution in piles. Part 1: notes on shift of no-load reading and residual load. Geotech. News Mag. **20**(2), 35–38 (2002)

Flynn, K.N., McCabe, B.A.: Shaft resistance of driven cast-in-situ piles in sand. Can. Geotech. J. **53**(1), 49–59 (2015)

Foray, P., Balachowski, L., Colliat, J.L.: Bearing capacity of model piles driven into dense overconsolidated sands. Can. Geotech. J. **35**(2), 374–385 (1998)

Hanna, A.M., Nguyen, T.Q.: An axisymmetric model for of ultimate capacity of a single pile in sand. Jpn. Geotech. Soc. **42**, 47–58 (2002)

Hanna, A.M., Soliman-Saad, N.: Effect of compaction duration on the induced stress levels in a laboratory prepared sand bed. Geotech. Test. J. ASTM **24**(4), 430–438 (2001)

Hanna, A., Al Khoury, I.: Passive earth pressure of overconsolidated cohesionless backfill. J. Geotech. Geoenviron. Eng. **131**(8), 978–986 (2005)

Jaky, J.: The coefficient of earth pressure at rest. J. Soc. Hung. Archit. Eng. 335–358 (1944)

Kulhawy, F.H.: Drilled shaft foundations. In: Fang, H.Y. (ed.) Foundation Engineering Handbook, pp. 537–552. Springer, New York (1991)

Kulhawy, H.: Discussion of the critical depth – how it came into being and why it does not exist. Civ. Eng. Geotech. Eng. J. 244–245 (1995)

Lehane, B., Jardine, R., Bond, A., Frank, R.: Mechanisms of shaft friction in sand from instrumented pile tests. J. Geotech. Eng. 19–35 (1993)

Mayne, P.W.: Tentative method for estimating $\sigma v'$ from qc data in sands. In: Proceedings on 1st International Symposium on Calibration Chamber Testing, Potsdam, NY, pp. 249–256 (1991)

Mayne, P.W., Kulhawy, F.H.: Ko-OCR relationships in soil. J. Geotech. Eng. Div. **108**(6), 851–872 (1982)

Meyerhof, G.: Bearing capacity and settlement of pile foudations. J. Geotech. Eng. Div. ASCE **102**(GT3), 195–228 (1976)

Poulos, H.G., Davis, E.H.: Pile Foundation Analysis and Design. Wiley, New York (1980)

Randolph, M.F., Dolwin, J., Beck, R.: Design of driven piles in sand. Geotechnique **44**(3), 427–448 (1994)

Shalabi, F., Bader, T.A.: Effect of sand densification due to pile-driving on pile resistance. Int. J. Civ. Eng. (IJCE) **3**(1), 17–30 (2014)

Tomlinson, M., Woodward, J.: Pile Design and Construction Practice. CRC Press, Boca Raton (2014)
Toolan, F.E., Lings, M.L., Mirza, U.A.: An appraisal of API RP2A recommendations for determining skin friction of piles in sand. In: Offshore Technology Conference. Offshore Technology Conference (1990)
Vesic, A.S.: Investigations of bearing capacity of piles in sand. In: Proceedings of National American Conference on Deep Foundation, Mexico City (1964)
Vesic, A.S.: Tests on instrumented piles, Ogeechee River site. J. Soil Mech. Found. Div. (1970)

Influence of Exposure Temperature on Shaft Capacity of Steel Piles in Ice-Poor and Ice-Rich Frozen Soils

Abdulghader A. Aldaeef(✉) and Mohammad T. Rayhani

Geoengineering Research Group, Carleton University, Ottawa, Canada
{abdulghader.abdulrahman, mohammad.rayhani}@carleton.ca

Abstract. An experimental investigation was carried out to study the performance of steel piles in ice-rich and ice-poor frozen soils. The behavior of pile-soil interface was characterized using interface element test using a modified direct shear test apparatus. The interface testing was conducted in an environmental chamber to enable testing at various temperatures below the freezing point. A series of stress-strain curves were established at different temperatures and under various normal stresses. The results showed a significant reduction in ultimate shaft capacity of the pile-soil interface as the exposure surface temperature increased. The pile loses approximately 75% of its ultimate shaft capacity when the exposure temperature increased from −10 °C to −1 °C in both granular and cohesive frozen soils. The shear stress–strain curves showed a brittle behavior followed by significant loss of pile-soil interface strength. Steel piles installed in frozen ice-rich soils would exhibit higher load carrying capacities compared to the piles in frozen ice-poor soils at the examined normal stresses.

1 Introduction

Performance of pile foundations in frozen ground has been evaluated in both field (e.g., Vialov 1959; Crory 1966; Aldaeef and Rayhani 2015, 2018) and laboratory (e.g., Parameswaran 1978; Weaver and Morgenstern 1981; Nixon and McRoberts 1976; Morgenstern et al. 1980; Ladanyi and Theriault 1990; Aldaeef and Rayhani 2017). However, most design approaches were often developed based on data collection and interpolations, and there has been almost no comprehensive study on performance of pile foundations in different frozen soil types when exposed to different levels of temperature and stress conditions.

Parameswaran (1978) reported that ultimate bearing capacity of piles depends mainly on the shaft resistance along the permanently frozen depth. The strength gained from end bearing could be negligible for piles installed in ice-rich soils, and may only be significant if the pile is situated on ice-poor frozen soils such as bedrock or granular soils (Crory 1966). The US Army/Air Force (1967) suggested that the end bearing

H. El-Naggar et al. (Eds.): GeoMEast 2018, SUCI, pp. 247–257, 2019.
https://doi.org/10.1007/978-3-030-01902-0_20

capacity could be neglected from design considerations for piles with base diameter smaller than 150 mm regardless of the soil type, attributing that to the greater movement that is required for mobilizing the end bearing compared to the much smaller movement needed to overcome the adfreeze bond.

The principle of the design criteria for pile foundations in frozen soils was initially based on the allowable adfreeze strength obtained by dividing the ultimate adfreeze bond at the pile-soil interface by a factor of safety (e.g., Vialov 1959; Crory and Reed 1965; Penner 1970, 1974; Penner and Irwin 1969; Penner and Gold 1971; Dalmatov et al. 1973). Johnston and Ladanyi (1972), in contrast, determined the allowable load capacity based on adfreeze strength that would cause an allowable displacement (or creep) rate over the life time of structures to accommodate for the time-dependent behavior of frozen soils. Other researchers including Nixon and McRoberts (1976), Morgenstern et al. (1980) and Weaver and Morgenstern (1981) suggested to check both adfreeze strengths and design based on the smaller value amongst them.

An empirical equation proposed by Vialov (1959) was one of the earliest theoretical approach for estimating long-term adfreeze strength of piles in frozen grounaa function of ground temperature. The equation was derived based on experimental data collected from full scale loading tests on hand-driven timber piles into pre-steamed holes in silty sandy loam and argillaceous loam as follow,

$$\tau_{lt} = \sqrt{1.65\ \theta(^{\circ}\mathrm{C})} - 0.3 \tag{1}$$

where τ_{lt} is the long-term adfreeze strength (kg/cm^2) and θ is the positive value of the temperature below the freezing point (°C). For piles installed in dry-augered holes with diameters 10 to 40% smaller than the pile diameter, Vialov (1959) proposed the values obtained from this equation to be increased by 30 to 50%.

In Alaska, Crory (1966) conducted a series of full scale loading tests on 200 mm diameter steel pipe piles installed in predrilled frozen silty sand and shaft-grouted with silt-water slurry. He was successful in providing graphical correlation between the ultimate adfreeze strength and the average ground temperature in the range of −4 °C to 0 °C. These results are in a very good agreement with the result reported by Aldaeef and Rayhani (2017) for adfreeze strength of steel piles in ice-poor soils using interface element test. The results from both studies may be expressed in the following form:

$$\tau_{ul} = 0.098T(^{\circ}\mathrm{C}) + 0.0581 \tag{2}$$

where, τ_{ul} is the ultimate adfreeze strength (MPa) and T is the positive value of the freezing temperature (°C).

Parameswaran (1978) tested model piles made from different materials in frozen sand at −6 °C and reported that the peak adfreeze strength was at maximum for wood piles, followed by concrete piles and then steel and coated piles and generally increased with increasing loading rate following this empirical equation:

$$\tau_1 \propto (\ell)^m \tag{3}$$

or

$$(\ell) \propto \tau_a^n \tag{4}$$

where, m is pile surface roughness factor and $n = \frac{1}{m}$, and it ranges from 4.48 to 10.23 for different piles in frozen sand.

Weaver and Morgenstern (1981) reported that adfreeze strength of piles can be directly correlated to the strength of the surrounding soil by using surficial roughness factor "m" and can be expressed in the following form:

$$\tau_{a,lt} = m\,C_{lt} \tag{5}$$

where, $\tau_{a,lt}$ is the long-term adfreeze strength and C_{lt} is the long-term cohesion of the frozen soil. The roughness factor "m" was extrapolated from different field and laboratory test data and proposed to be 0.7 for timber piles based on the ratio between adfreeze strength of timber piles in ice to the long-term cohesion of ice obtained from field and experimental data of Voitkovskii (1960) and Vialov (1959). A roughness factor of 0.6 was inferred for steel and concrete piles based on data from field experiments conducted by Johnston and Ladanyi (1972) and Crory (1966). Weaver and Morgenstern (1981) stated that the long-term adfreeze strengths of steel and concrete piles in the used data were not well defined, however, 0.6 represented the lower boundary and could still be acceptable for conservative pile design.

Ladanyi and Theriault (1990) evaluated the soil-metal adfreeze bond at −2 °C and reported that the long-term shaft resistance of piles in frozen ground does not solely depends on the long-term adfreeze, but also on the long-term friction angle at the interface, and respectively on the total lateral ground stress. Therefore, they improved the Weaver and Morgenstern's (1981) equation by adding the contribution of the long-term friction angle of frozen soil and proposed the following formula:

$$\tau_{a,lt} = m\,C_{lt} + \sigma_{n_{total}} \tan \emptyset_{lt} \tag{6}$$

Adfreeze strength of pile foundations in frozen ground has been intensely studied, however, each study tackled limited range of temperatures, soil types, or/and loading rates. The current paper evaluates the load carrying capacity of steel piles in two different ground materials when exposed to different temperature levels, and under different normal stresses.

2 Physical Properties of the Test Soil and Steel-Soil Interface

A poorly graded non-plastic sand (SP, in accordance to the USCS classification system, ASTM D2487, 2005) was used to represent ice-poor frozen soil. The maximum dry density (ρ_d max) and optimum moisture content (Wopt) were determined using Standard Proctor test procedure outlined in ASTM D698 (2005) and found to be 1.850 Mg/m^3 and 10% respectively. The particle size distribution for the sand was determined in accordance with ASTM D422, (2005). The sand contained approximately 5% fines passing through 0.075 mm sieve.

A marine soil known as Leda clay was also used in this experiment to represent ice-rich frozen materials. The clay soil was collected intact from the Navan landfill in Ottawa. Using Atterberg limits test in accordance with ASTM D4318-10, Leda clay showed a liquid limit of 51%, and plasticity index of 23%. The maximum dry density and the corresponding optimum moisture content for the clay was determined using Standard Procter test (ASTM D 698-12) and found to be 1.41 Mg/m^3 and 30% respectively. The particle size distribution for Leda clay was determined in accordance with ASTM D422, (2005) where the clay content was found to be 71%.

For pile-soil interface testing, steel plates were used to simulate the shaft surface of a typical steel pile. Steel piles are commonly used in permafrost region with different geometries including pipe piles, H-section piles, and helical piers. The interface test specimens were 90 mm by 90 mm square steel plates with a thickness of 25.4 mm, machined to couple with the upper half of the direct shear box apparatus and provide a steel-soil interface area of 60 mm by 60 mm (Fig. 1). The total and the average surface

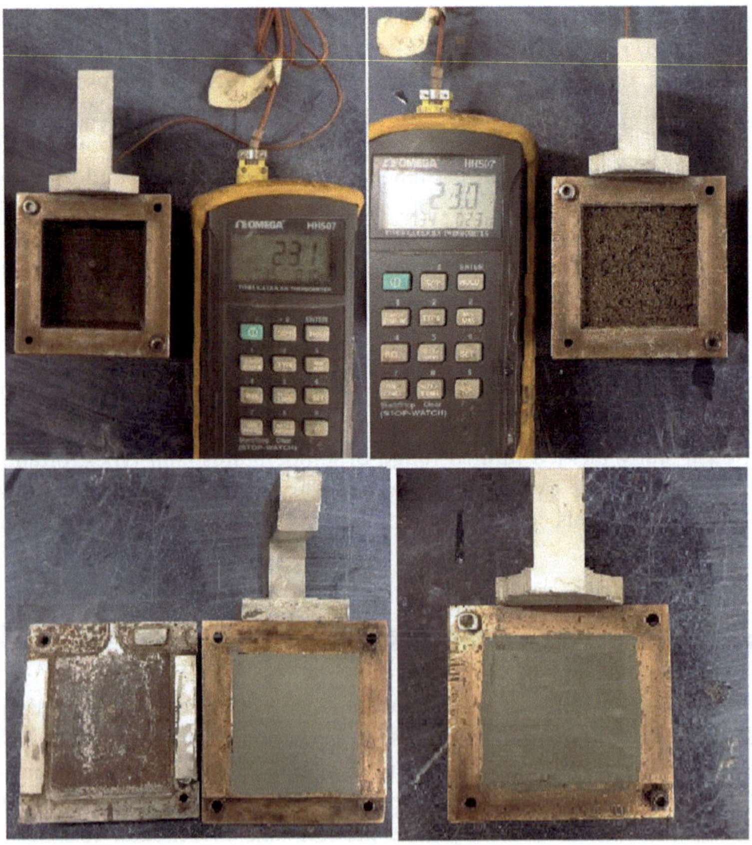

Fig. 1. Modified shear box assembly for interface testing.

roughness values for this particular type of steel were reported by Giraldo and Rayhani (2013) to be 9.7 μm and 11.3 μm respectively. The steel plates were equipped with thermocouples inserted in tiny holes underneath the upper surface of the plates in order to track the temperature change at the steel-soil interface.

3 Experimental Program

3.1 Test Equipment

All shear and interface tests were conducted using a direct shear test apparatus in accordance to ASTM D3080/D3080M-11 and ASTM D5321-12. The apparatus was placed inside a fully controlled environmental chamber to enable testing at various freezing temperatures. The direct shear test apparatus consists of an electrical motor that enables applying a constant shear velocity to the lower part of the shear box, while the upper part is restrained by a digital load cell connected horizontally to it. The electrical motor is connected to a gearbox to enable adjustment of the shear velocity to the desired rate. Horizontal and vertical displacements are measured through a Linear Variable Differential Transducers (LVDT) connected to a digital logging station using LabView software. The apparatus frame facilitates applying normal stresses to the top of the test specimen by incorporating a steel bearing arm.

For the interface testing, a modified shear box was utilized to simulate pile-soil interface characteristics. The modified shear box incorporates a 90 mm by 90 mm by 25.4-thick steel plate which was machined to couple with the upper half of the regular direct shear box. This facilitated a steel-soil interface area of 60 mm by 60 mm (Fig. 1).

3.2 Sample Preparation

The sand soil was prepared at a bulk density corresponding to its field capacity moisture content. This bulk density was determined by pouring sand-water slurry into a known volume container and permitting water drainage and soil settling under gravity. When water flow stops, the known volume soil sample was then weighed, and its density was determined to be 2.06 Mg/m^3 at a field capacity of 13.5%. To prepare the test samples, dry sand was hydrated to the predefined field capacity and compacted to its corresponding density into the interface shear box.

Leda clay soil, on the other hand, was sampled using a Shelby tube sampler with an inner diameter of 101.6 mm. The soil was then extracted from Shelby tube using a mechanical extruder and cut into cylindrical specimens of 30 mm height. For steel-clay interface testing, the 30 mm-height cylindrical specimens were trimmed to a square shape with dimensions of 62 mm × 62 mm and placed on a plastic plate. The upper part of the shear box was then gently pushed through the clay specimen and placed on the steel plate interface. The final shape of the interface test samples for sand and clay was square with dimensions of 60 mm × 60 mm × 20 mm height.

The shear box was then mounted on the direct shear test apparatus inside the environmental chamber. The desired normal stress was respectively applied simultaneously with setting the temperature of the chamber to the desired level in order to permit soil consolidation and freezing.

3.3 Test Procedure

After keeping the test setup for 24 h in the cold room, the shear load was applied at a constant shear velocity of 0.011 mm/min (≅16.3 mm/day). The shear tests were conducted at different temperature levels including −10 °C, −7 °C, −4 °C, and −1 °C, with an accuracy of ±0.5 °C. At each temperature level, the shear strength parameters were determined under different normal stresses including 25 kPa, 50 kPa, 100 kPa, and 200 kPa.

4 Test Results

The interface testing program was aimed at studying the strength behaviors of pile foundations in frozen grounds. Stress-strain curves at different temperatures and normal stresses for steel-sand interface and steel-clay interface are presented in Fig. 2(a) and (b) respectively. Most of the interface samples failed in brittle mode at relatively small displacements where significant strength reductions observed right after the peak. Subsequently, the interface samples demonstrated very small residual strengths that represented a small fraction of the peak strength. However, the interface tests conducted at −1 °C showed less pronounced brittle failures compared to the other tests conducted at colder temperatures. The brittle behavior could be associated to the breakage of ice bonding at the pile-soil interface. Since the ice content decreases as temperature increases, the ice bonding would respectively decrease resulting in plastic-brittle transition and less fragile failure mode.

At any given temperature, steel-clay interface samples showed higher peak strengths compared to steel-sand interface samples. Their shear strain at peak ranged between 0.8% and 3.5% compared to 1% to 3% shear strains at peaks for steel-sand interface samples. Higher shear strain at peak was usually recorded for interface tests conducted at lower freezing temperatures and/or under higher normal stresses. This usually would refer to a higher ice bonding potential and larger frictional resistance which leads to greater energy to be overcome.

5 Discussion

5.1 Shear Strength Parameters at Pile-Soil Interface in Ice-Rich and Ice-Poor Soils

Performance of pile foundations is directly influenced by the shear strength parameters of the surrounding soil. In saturated unfrozen ground, a soil would resist the applied shear load by its internal friction and/or by its cohesion strength. Many factors, including soil type and loading rate, determine which strength parameters would be exerted by the soil toward resisting a certain applied load. Generally, strength behaviors

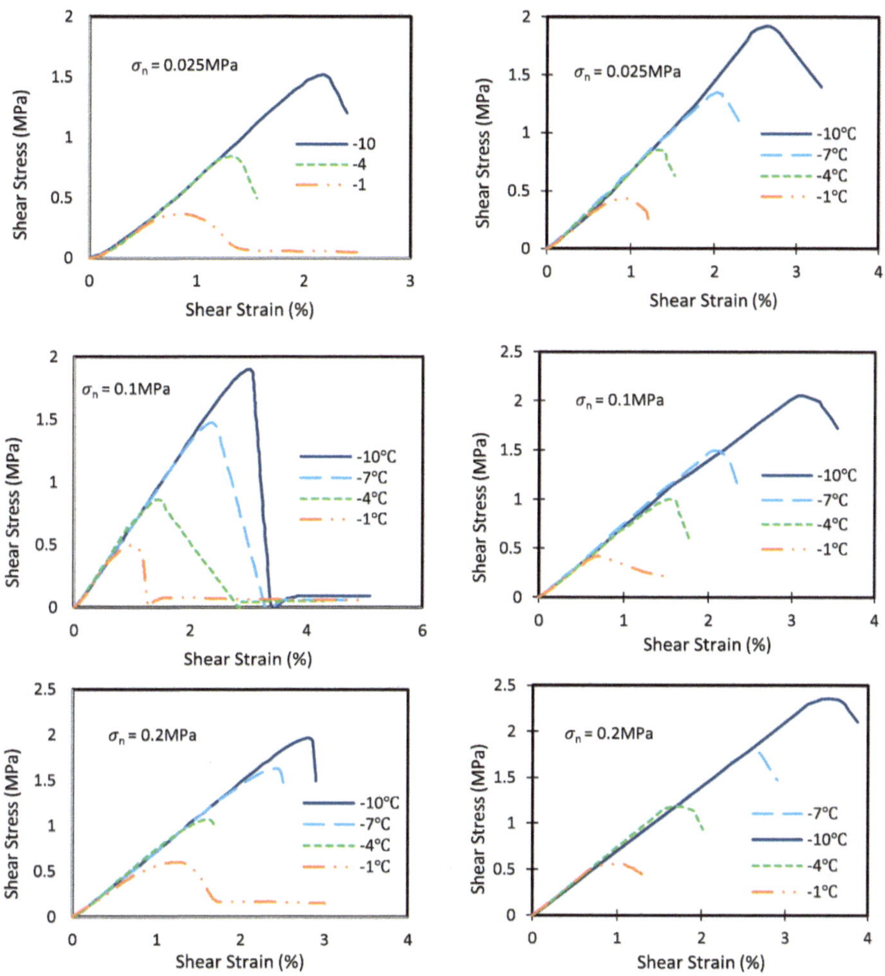

a. Steel-sand interface b. Steel-clay Interface

Fig. 2. Stress-strain curves for interface samples at different temperatures and normal stresses.

of coarse-grained soils, such as sand, is dominated by frictional resistance with zero or minimal cohesion strength. Fine-grained soils, in contrast, are distinguished with their predominant cohesion strength with little to zero internal friction. In frozen grounds, however, the existence of ice content changes the working mechanism of soil strength. The transformation of liquid water to solid ice within soil voids during the process of freezing significantly enhances the strength behavior of frozen grounds by incorporating the strength of ice cementation (ice bonding). The formation of crystalline ice results in increasing the internal friction and cohesion strength in both coarse- and fine-grained soils. Pile foundations, as a result, are expected to exhibit greater bearing capacity when installed in frozen soils compared to unfrozen soils. However, the degree of increase depends on the quantity of crystalline ice and unfrozen water content at the pile-soil interface, thus on freezing temperature. This section evaluates the strength parameters that contribute to pile-soil interface resistance in frozen sand and clay.

Using Mohr-Coulomb criterion, peak interface shear stresses versus normal stresses at different freezing temperatures were plotted for both steel-sand and steel-clay interfaces (Fig. 3). The peak strengths of steel-soil interfaces were resultant of adfreeze bond and interface frictional resistance. The test temperatures, the corresponding adfreeze strength, and interface friction angles are all presented in Table 1.

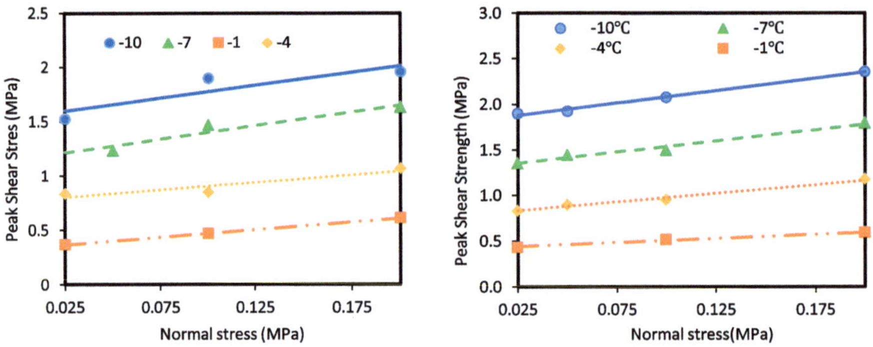

a. Steel-sand interface b. Steel-clay interface

Fig. 3. Variation of adfreeze strength and frictional resistance of pile-soil interface with ground temperature.

Table 1. Interface strength parameters of steel pile in frozen clay and frozen sand.

T (°C)	Steel-sand interface		Steel-clay interface	
	C_a (MPa)	δ (°)	C_a (MPa)	δ (°)
−10.0	1.54	67	1.81	70
−7.0	1.15	68	1.30	67.8
−4.0	0.77	54.5	0.78	62.6
−1.0	0.33	54.5	0.42	42.3

At any given temperature, adfreeze strength of the steel-clay interface was always higher than that for the steel-sand interface. Since sand is a cohesion-less soil, its adfreeze strength would be probably gained from ice formation and the resulted ice cementation. On the other hand, the higher adfreeze strength of steel-clay interface could be attributed to a higher ice content compared to steel-sand interface. Unfrozen clay soils usually contain larger amount of water compared to unfrozen sand soils. Upon freezing, therefore, larger volume of liquid water transforms to ice within clay soils resulting in a greater ice content compared to frozen sand. The solid crystalline ice, with respect to its quantity, would increase the cohesion strength in both soils and subsequently the adfreeze strength at the pile-soil interface. The ice in its solid state, furthermore, would also enhance the interlock force with soil particles resulting in a higher internal friction and similarly pile-soil interface frictional resistance. This phenomenon explains the reduction in both adfreeze strength and the interface frictional resistance as the test temperature increases toward freezing point due to the reduction in ice content and respectively the ice bond.

Frictional resistance is predominantly due to the soil particles interlock force which increases with increasing normal stress. Fractured ice crystals, moreover, contributes to the overall frictional resistance. At lower temperatures, therefore, higher friction angles were recorded for both steel-sand and steel-clay interface samples due to a higher ice content. The frictional resistance of the pile-soil interface refers to the positive contribution of the normal stress to the overall shear strength. Increasing the normal stress perpendicular to the shear plane could densify the soil texture and enhance the water flow towards the shear plane during consolidation. This, therefore, increases the quantity of the ice fraction and the particles interlock force at the pile-soil interface as the normal stress increases. Although, limited information is available about in-situ normal stress in frozen ground due to the absence of a systematic measurement technique, installation of pile foundations in frozen grounds could induce horizontal stress increase at any depth along the pile shaft. This happens following to slurry freezing around piles installed in oversized holes, or by soil freeze-back occurrence after driving a pile directly in frozen ground or in a slightly undersized hole. Ladanyi and Theriault (1990) suggested that lateral stresses around a pile in frozen ice-rich soil are improbable to become less than the total lateral stress in the ground. Subsequently they proposed that the ratio between total lateral stress acting on the pile shaft to the total vertical stress could vary from k_0 for dense frozen soils to 1 for ice-rich soils and crystalline ice.

5.2 Impact of Ground Worming on Ultimate Shaft Capacity of Steel Piles in Frozen Sand and Clay

Freezing temperature is directly proportional to the quantity of ice content and ice bonding concentration, thus to peak strengths too. This was evidence of the higher interface peak strengths witnessed at lower freezing temperature (Fig. 4). At a normal stress of 25 kPa and a test temperature of −10 °C, the interface peak strength was around 1.9 MPa and 1.52 MPa for the steel-clay and steel-sand respectively. However, peak strengths under a similar normal stress but at a temperature of −1 °C was significantly smaller recording 0.45 MPa and 0.35 MPa respectively. This thermal change led to reducing shear strength of the steel-soil interface by almost 75%. This, indeed, highlights the significant effect of ground warming on the ultimate bearing capacity of steel piles in frozen grounds. Although, such dramatic temperature increase may not be encountered in short-term especially when considering macroscale level (i.e., global warming effects), it may happen in the microscale level due to different causes including improper thermal insulation measures and/or induced warming due to

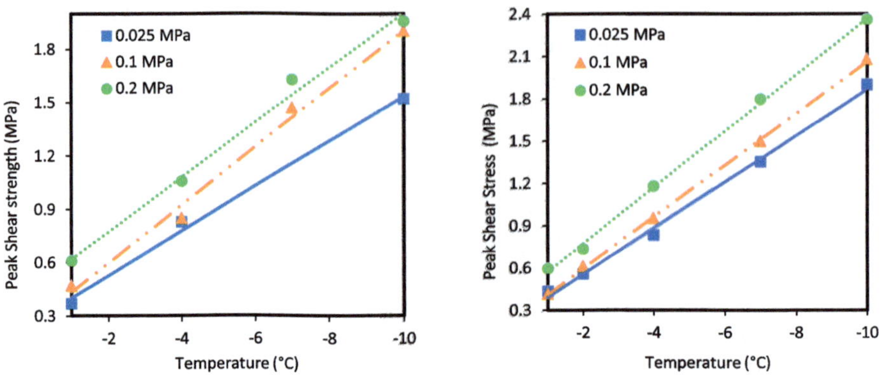

a. Steel-sand interface b. Steel-clay interface

Fig. 4. Effects of temperature and normal stress on ultimate strength of pile-soil interface

improper pile installation techniques. In discontinues permafrost zone, change from frozen to unfrozen condition is more likely to happen as the ground exists at relatively warm temperature.

6 Conclusions

This study evaluates the thermal effects on performance of steel piles in ice-rich and ice-poor frozen soils. The adfreeze strength of steel piles in frozen soils is a resultant of adfreeze strength and frictional resistance at the pile-soil interface. Steel piles in frozen clay showed larger load carrying capacity compared to the one in frozen sand. However, the ultimate bearing capacity decreases linearly as the freezing temperature increases toward freezing point. Piles in frozen ground fail in brittle mode when the peak strength is exceeded and subsequently lose significant capacity after failure. Failure can happen also because of permafrost degradation especially those in warm conditions at relatively high temperatures ranging from −3 °C to 0 °C.

Acknowledgments. This study was financially supported by Carleton University and the Natural Sciences and Engineering Research Council of Canada (NSERC). The authors would like to thank both agencies for this support.

References

Aldaeef, A.A., Rayhani, M.T.: Load transfer of pile foundations in warming frozen ground. In: The 7th Canadian Permafrost Conference, Quebec City, Canada (2015)

Aldaeef, A.A., Rayhani, M.T.: Adfreeze strength and creep behavior of pile foundations in warming permafrost. In: International Congress and Exhibition on Sustainable Civil Infrastructures: Innovative Infrastructure Geotechnology. Springer, Cham (2017)

Aldaeef, A.A., Rayhani, M.T.: Load transfer of pile foundations in frozen and unfrozen soft clay. Cold Reg. Sci. Technol. (2018, Submitted)

ASTM D3080/D3080M-11: Standard test method for direct shear test of soils under 494 consolidated drained conditions, In: Annual Book of ASTM Standards. ASTM International, West Conshohocken (2012)

ASTM D422-63: Standard test method for particle-size analysis of soils. In: Annual Book of 484 ASTM Standards. ASTM International, West Conshohocken (2007)

ASTM D698: Standard Test Methods for Laboratory Compaction Characteristics of Soil Using Standard Effort (12,400 ft-lbf/ft3 (600 kN-m/m^3). ASTM International, West Conshohocken, PA (2012). https://doi.org/10.1520/d0698-12

Crory, F.E.: Pile foundations in permafrost. In: Proceedings of 1st Permafrost International Conference, Lafayette, Indiana, NAS-NRC Publ. No. 1287, pp. 467–472 (1966)

Crory, F.E., Reed, R.E.: Measurement of frost heaving forces on piles. Cold Regions Research and Engineering Laboratory (CRREL), Technical Report, 145 (1965)

Dalmatov, B.I., Karlov, V.D., Turenko, I.I., Ulitskiy, V.M., Kharlab, V.D.: Interaction of freezing heaving with foundations. In: Proceedings of 2nd International Conference on Permafrost, Yakutsk, USSR Contribution, pp. 572–576. National Academy Press, Washington, DC (1973)

Giraldo, J., Rayhani, M.T.: Influence of fiber-reinforced polymers on pile–soil interface strength in clays. Adv. Civ. Eng. Mater. **2**(1), 1–17 (2013). https://doi.org/10.1520/ACEM20120043. ISSN 2165-3984

Johnston, G.H., Ladanyi, B.: Field tests of grouted rod anchors in permafrost. Can. Geotech. J. **9**(2), 176–194 (1972)

Ladanyi, B., Theriault, A.: A study of some factors affecting the adfreeze bond of piles in permafrost. In: Proceedings of Geotechnical Engineering Congress GSP, vol. 27, pp. 213–224 (1990)

Morgenstern, N.R., Roggensack, W.D., Weaver, J.S.: The behavior of friction piles in ice and ice-rich soils. Can. Geotech. J. **17**(3), 405–415 (1980)

Nixon, J.F., McRoberts, E.C.: A design approach for pile foundations in permafrost. Can. Geotech. J. **13**(1), 40–57 (1976)

Parameswaran, V.R.: Adfreeze strength of frozen sand to model piles. Can. Geotech. J. **15**(4), 494–500 (1978)

Penner, E.: Frost heaving forces in Leda clay. Can. Geotech. J. **7**(1), 8–16 (1970)

Penner, E.: Uplift forces on foundations in frost heaving soils. Can. Geotech. J. **11**, 323 (1974)

Penner, E., Gold, L.W.: Transfer of heaving forces by adfreezing to columns and foundation walls in frost susceptible soils. Can. Geotech. J. **8**, 514 (1971)

Penner, E., Irwin, W.W.: Adfreezing of Leda clay to anchored footing columns. Can. Geotech. J. **6**, 327 (1969)

U.S. Army/Air Force: Arctic and Subarctic Constructiotz: Structure Fozmdntions. Technical Manual TM5-852-4/ AFM 88-19, chap. 4 (1967)

Vialov, S.S.: Rheological properties and bearing capacity of frozen soils, Transl. 74, US. Army CRREL, Hanover, NH 1965 (1959)

Voytkovskiy, K.F.: Mekjanicheskiye svoystva lda. Izvestiya Akademii Nauk, Moscow. [The mechanical properties of ice. U.S. Air Force Cambridge Research Laboratory, Bedford, MA, translation No. AFCRL-62–838.] (1960)

Weaver, J.S., Morgenstern, N.R.: Pile design in permafrost. Can. Geotech. J. **18**(3), 357–370 (1981)

Effect of Scour Depth on Ultimate Capacity and Settlement of Piled Foundation

Wael Nashaat Abd El Samee(✉) and Mohamed Kassem Elsamny

Faculty of Engineering, Beni Suef University, Beni Suef 62512, Egypt
Waelnashat@eng.bsu.edu.eg

Abstract. Scour may cause significant effect on the bearing capacity, settlement as well as load distribution (skin friction and end bearing) of the pile foundation. In the present study, finite element analysis has been selected to develop numerical models to study the effect of scour depth on piled foundation. The analysis program consists of five groups of piles with different pile diameter from 0.50 m to 0.90 m. Each group is analyzed with different scour depths of pile dimeter from 0.00 D to 4.00 D while keeping the pile length constant. It is concluded that, the settlement of piled foundation increases with increasing scour depths. However, a decrease in the embedded depth of pile of foundation due to scour decreases skin friction. In addition, conical local scour hole causes the negative skin friction (additional vertical stress). However, the values of ultimate load capacities of piles foundation decrease with increasing scour depths. In addition, at scour depth = 4.0 pile diameter the of ultimate load capacities decrease by about 39% to 44% from ultimate load capacities of zero scour depth. However, fair agreement has been obtained between theoretical and finite element analysis.

Keywords: Scour · Skin friction · Depth · Conical · Ultimate capacity

1 Introduction

The flow of water in rivers and streams causes scour around piled foundations. Rivers have aggressive flow regime. During major flood events, the volume and velocity of flood waters can cause considerable scouring. Scour may cause significant effect on the bearing capacity, settlement as well as load distribution (skin friction and end bearing) of the pile foundation.

1. Sham et al. (2010) described how the foundation design of the Padma Main Bridge poses many challenges since, in addition to the large live loads it has to support, it must also be able to withstand significant forces and conditions imposed by the environment, such as wind, earthquake and very deep river bed scour [1].
2. Liang (2011) investigated the effect of scour on the bearing capacity of the piled foundation by using finite element analysis. The effects of scour depths on the behavior of piled foundations were analyzed. The numerical results showed that scour had some negative effects on the bearing capacity of the piles [2].

H. El-Naggar et al. (Eds.): GeoMEast 2018, SUCI, pp. 258–284, 2019.
https://doi.org/10.1007/978-3-030-01902-0_21

3. Erion and Yavuz (2012) investigated the effect of the scour on pile bent load carring capacity. Load carring capacities by piles were calculated. Site inspections on Mat bridge show that 19 from 32 pile bents have serious scour problems. Analysis results indicate that pile bents due to scouring height have loosed a load carry capacity varying from 17.64 to 32.11% of that designed [3].
4. Jae et al. (2012) evaluated scour bridges founded on pile foundations registered in the National Highway Bridge Inventory of the capital region. The case studies consist of site investigation including boring test. Three of 12 pile foundation bridges showed the potential future vulnerability to scour with considerable decrease in the embedded depth of foundations due to scour and the remaining 9 bridges were expected to maintain their stability to resist the effects of scour [4].
5. Ibrahim (2013) investigated the variation of scour depths that may occur at bridge piers. The maximum depth of scour has been experimentally investigated. It was found that the maximum depth of scour increases with time [5].
6. Cheng (2017) investigated vertical soil stress at the pile under different scour-hole dimension conditions. An analytical solution based on Boussinesq's theory was developed for the stress distribution. The effects of scour-hole dimensions, including scour depth, width, and slope angles on the vertical soil stress distribution were discussed. As the scour depth or width increases, the vertical soil stress decreases [6].
7. Elsamny et al. (2018) investigated experimentally the effect of the load distribution of friction along the pile shaft on the ultimate capacity of piles [7].

Thus, it can be reported that no research works were done to study the effect of scour depth on ultimate capacity, settlement as well as load distribution (skin friction and end bearing) of the piled foundation.

2 Effect of Scour on Pile Shaft Resistance

The flow of water in rivers and streams causes scour around piled foundations. Scour damage at piled foundation is called local scour as shown in Fig. 1. A conical hole is then formed at the upper soil surface, which is usually called conical local scour as shown in Fig. 2 [8].

Scour can result in soil loss around pile's foundation, thereby forming a conical local scour hole with a depth of Sd as shown in Fig. 3. Scour reduces the embedded pile length of the pile foundation. Scour could influence the soil properties within 2–4 times the pile diameter D. However, the conical local scour hole causes negative skin friction (additional vertical stress).

Fig. 1. Local scour damage at piled foundation.

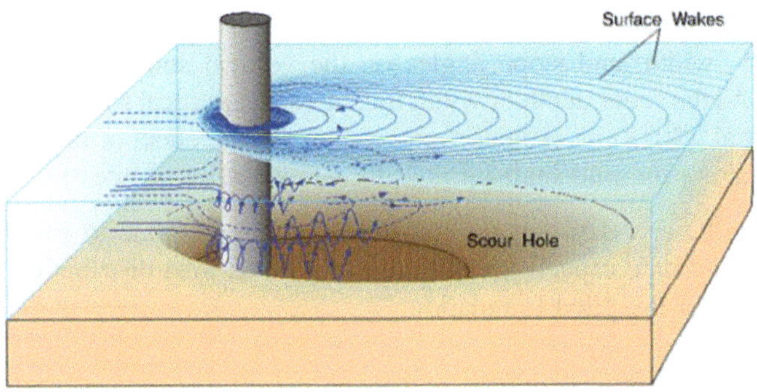

Fig. 2. Diagram shows the formation of a local scour hole around a pile.

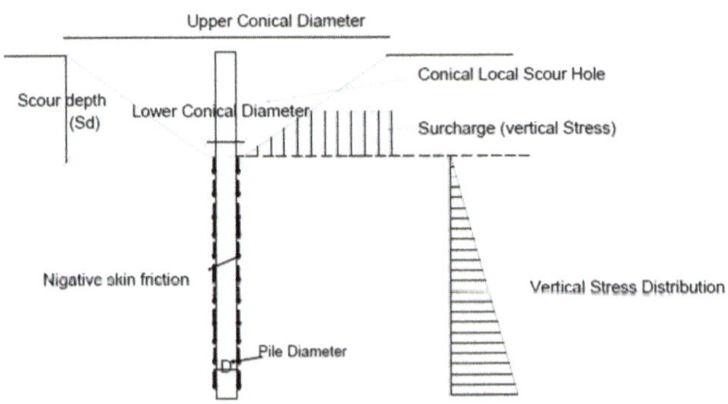

Fig. 3. Effective stress distribution.

3 Finite Element Analysis

The finite element method was selected in the present study to develop numerical models to study the effect of scour depth on ultimate capacity, settlement as well as load distribution (skin friction and end bearing) of the pile foundation. The finite element analysis has been performed using software PLAXIS 3D 2014. The soil is simulated by isotropic homogeneous semi-infinite elastic material.

3.1 Numerical Program

A numerical analysis was conducted to study the effect of scour depth on ultimate capacity, settlement as well as load distribution (skin friction and end bearing) of the pile foundation. A borehole for the investigation is presented in Fig. 4. The analysis program consists of five groups; first group is a pile cap with one piles pile diameter = 0.50 m. Second group is a pile cap with pile diameter = 0.60 m. Third group is a pile cap with pile diameter = 0.70 m. Fourth group is a pile cap with pile diameter = 0.80 m and finally fifth group is a pile cap with pile diameter = 0.90 m. Each group is analyzed with different scour depths of 0.00 D 0.50 D, 1.00 D, 2.00 D, 3.00 D and 4.00 D where D = pile diameter. The analysis program is shown in Table 1.

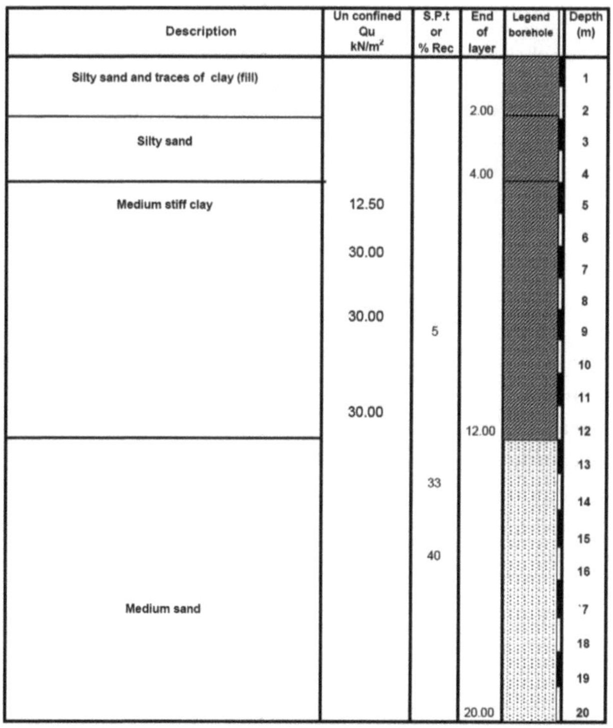

Fig. 4. Borehole log for soil.

Table 1. Numerical program.

Group no.	Pile diameter (D)	Scour depth	Pile length
1	D = 0.50 m	0.00 D, 0.50 D, 1.00 D, 2.00 D, 3.00 D and 4.00 D	L = 15 m
2	D = 0.60 m		
3	D = 0.70 m		
4	D = 0.80 m		
5	D = 0.90 m		

3.2 Material Properties Used in the Analysis

The soil is simulated by a semi-infinite element isotropic homogeneous elastic material simulates the soil and the material model used is Mohr-Coulomb. The used material properties are listed in Tables 2 and 3.

Table 2. Soil properties.

Parameters	Thickness (m)	Description	Unit weight, γ (kN/m^3)	Young's modulus, Es (kN/m^2)	Poisson's ratio, υ	Cohesion, Cu (kN/m^2)	Friction angle, Ø (deg)	Dilatancy angle, ¥ (deg)
Layer 1	2.0	Silly sand and trace of clay (fill)	15.0	2000	0.3	0	0	0
Layer 2	2.0	Silty sand	16.6	4000	0.4	12.5	0	0
Layer 3	8.0	Medium to stiff clay	17.0	3000	0.3	25	0	0
Layer 4	8.0	Medium sand	18.0	15000	0.25	0	33	0

Table 3. Pile and raft material concrete properties.

Parameters	Pile	Raft
Material model	Elastic	Elastic
Type of material	Concrete	Concrete
Thickness/diameter (m)	0.50, 0.60, 0.70, 0.80 and 0.90	0.60
Unit weight, γ (kN/m^3)	25	25
Young's modulus. E_s (kN/m^2)	24×10^6	24×10^6
Poisson's ratio, υ	0.2	0.2

3.3 Theoretical Prediction of Pile Load Capacity

The pile ultimate capacity used in this analysis has been estimated through theoretical method by using Egyptian code (2001) [9]. The theoretical pile capacities have been calculated for pile diameters = 50, 60. 70, 80 and 90 cm and constant pile length = 15 m. Figures 5, 6, 7, 8 and 9 show the theoretical relationship between load and settlement. Table 4 shows the theoretical ultimate bearing capacities.

Table 4. The theoretical ultimate bearing capacities

Pile diameter (D)	Theoretical ultimate bearing capacities as well as pile tip and around pile shaft friction					
	Around pile (friction) (kN)	At pile tip (kN)	Q_{ult} (kN)	F.S	Q_{all} (kN)	Taken Q_{ult} (kN)
D = 0.50 m	215.98	926.80	1142.78	1.50	761.83	750.00
D = 0.60 m	311.02	1112.10	1423.14		948.76	900.00
D = 0.70 m	423.33	1297.5	1720.81		1147.21	1100.00
D = 0.80 m	525.92	1482.80	2035.75		1357.17	1300.00
D = 0.90 m	699.79	1668.20	2367.98		1578.65	1500.00

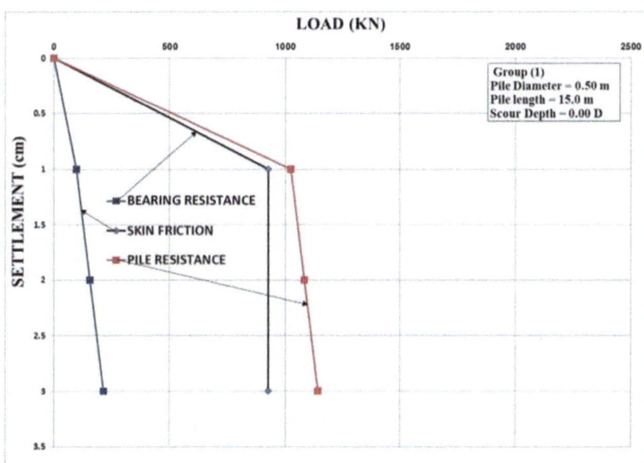

Fig. 5. The theoretical relationship between load and settlement of 50 cm pile diameters.

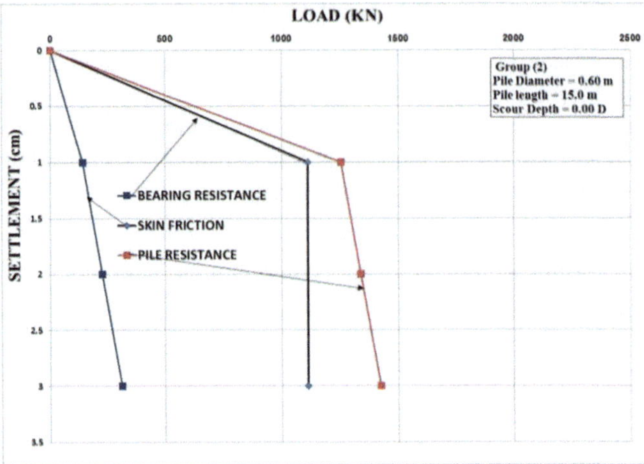

Fig. 6. The theoretical relationship between load and settlement of 60 cm pile diameters.

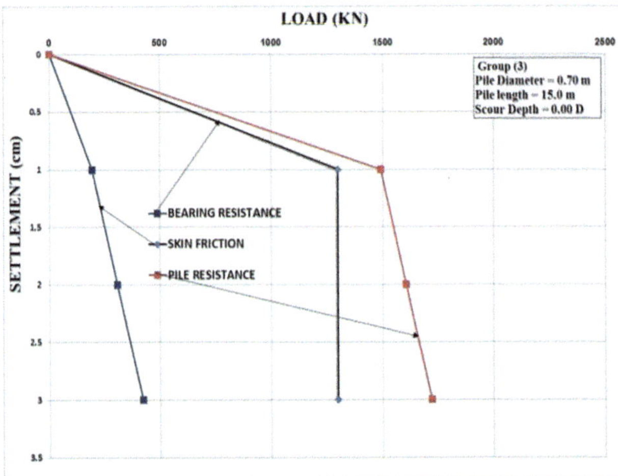

Fig. 7. The theoretical relationship between load and settlement of 70 cm pile diameters.

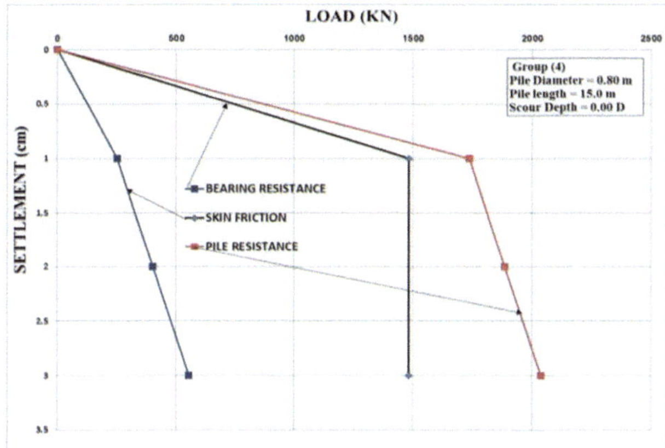

Fig. 8. The theoretical relationship between load and settlement of 80 cm pile diameters.

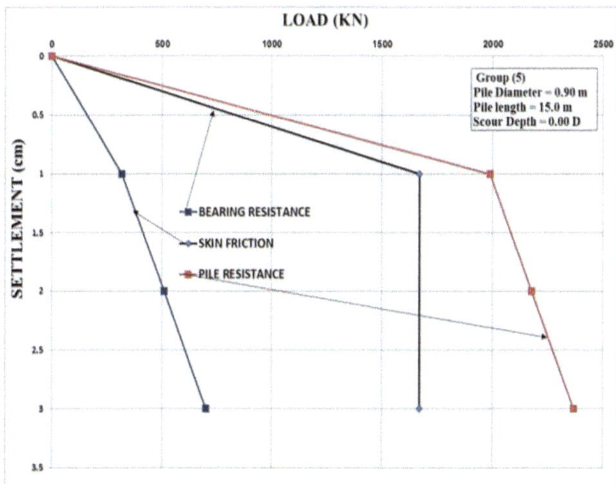

Fig. 9. The theoretical relationship between load and settlement of 90 cm pile diameters.

3.4 Finite Element Models

A finite element model was developed to predict the effect of scour depth on ultimate capacity, settlement as well as load distribution (skin friction and end bearing) of the pile foundation for the groups of piles with different scour depths while keeping the pile length constant. Figures 10, 11, 12, 13, 14 and 15 show some examples of deformed mesh.

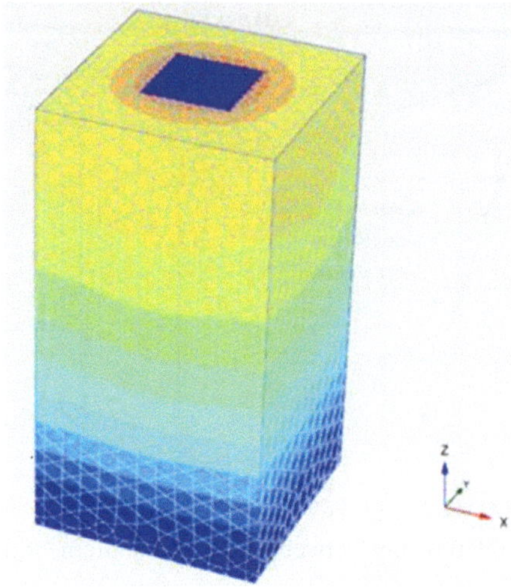

Fig. 10. Deformed mesh – for piles diameter = 0.60 m at scour depth = 0.00 D (control).

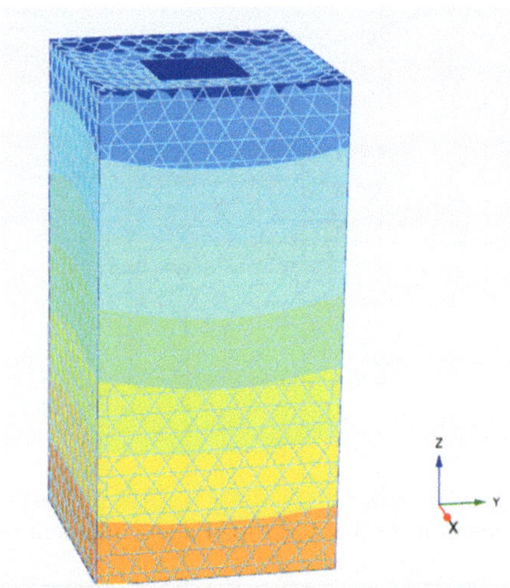

Fig. 11. Deformed mesh – for piles diameter = 0.60 m at scour depth = 0.50 D.

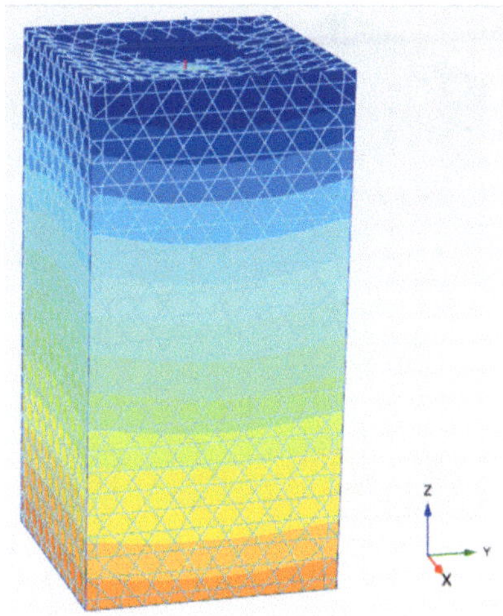

Fig. 12. Deformed mesh – for piles diameter = 0.60 m at scour depth = 1.00 D.

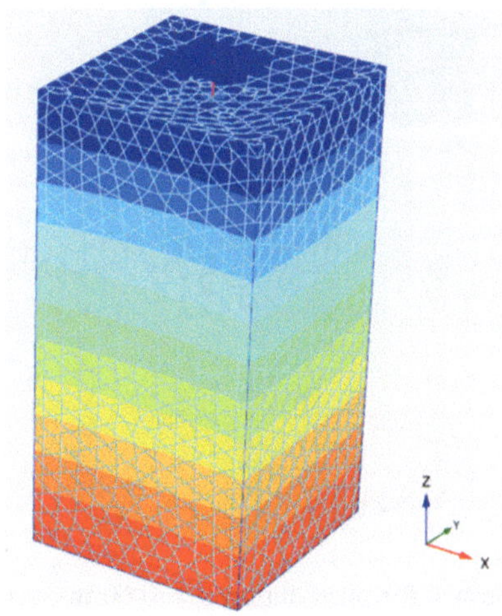

Fig. 13. Deformed mesh – for piles diameter = 0.60 m at scour depth = 2.00 D.

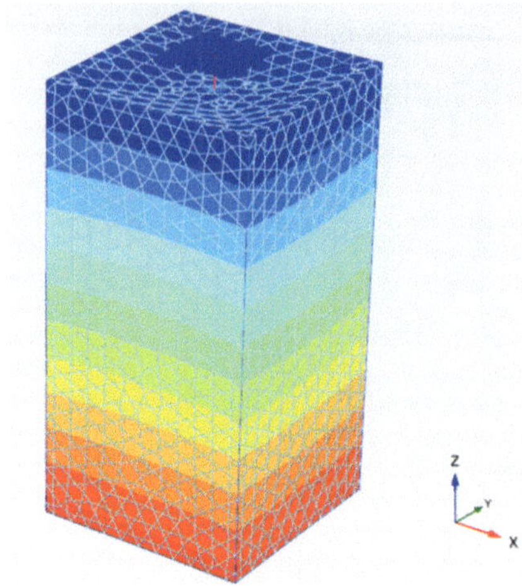

Fig. 14. Deformed mesh – for piles diameter = 0.60 m at scour depth = 3.00 D.

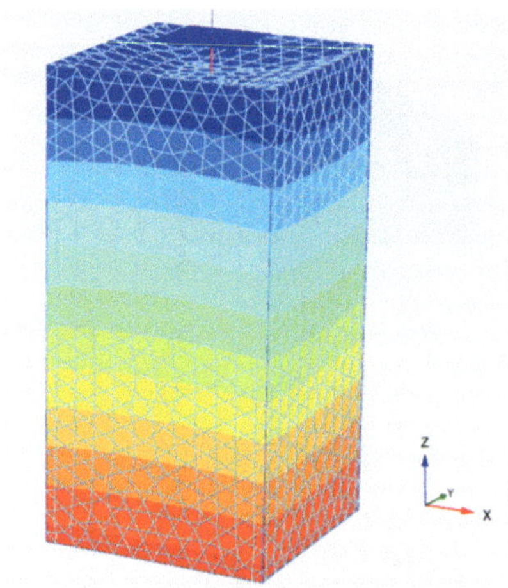

Fig. 15. Deformed mesh – for piles diameter = 0.60 m at scour depth = 4.00 D.

4 Analysis of Results

The obtained results from the numerical analysis are presented as follows:

4.1 Effect of Scour Depth on Settlement

Ultimate capacity and pile tip resistance and shaft friction as well as settlement at different scour depths are listed in Tables 5, 6, 7 and 8. Figure 13 shows the relationship between the values of settlement at ultimate load capacities and scour depth at different pile diameters.

Table 5. Ultimate capacities at different scour depths and different pile diameters.

Pile diameter (D)	Ultimate capacities (kN) at scour depth S					
	S = 0.00 D	S = 0.50 D	S = 1.00 D	S = 2.00 D	S = 3.00 D	S = 4.00 D
D = 0.50 m	1245.48	1083.57	986.05	877.58	789.83	679.25
D = 0 60 m	1551.08	1380.46	1242.42	1130.60	1034.50	951.74
D = 0 70 m	1870.52	1608.65	1463.87	1244.29	1144.75	1018.82
D = 0.80 m	2214.90	1938.03	1763.61	1516.71	1365.03	1208.06
D = 0.90 m	2571.63	2237.31	1991.21	1752.27	1524.47	1372.02

Table 6. Pile friction resistance at different scour depths.

Pile diameter (D)	Pile friction resistance (kN) at scour depth S					
	S = 0.00 D	S = 0.50 D	S = 1.00 D	S = 2.00 D	S = 3.00 D	S = 4.00 D
	Pile friction resistance	Pile friction resistance	Pile friction resistance	Pile friction resistance	Pile friction resistance	Pile friction resistance
D = 0.50 m	429.59	373.75	340.11	302.70	272.43	234.29
D = 0.60 m	535.00	462.28	393.87	333.86	273.32	227.71
D = 0.70 m	645.18	554.86	504.92	429.18	394.85	351.41
D = 0.80 m	763.96	668.47	608.31	523.14	470.83	416.68
D = 0.90 m	887.01	771.70	686.81	604.39	525.82	473.24

Table 7. Pile tip resistance at different scour depths.

Pile diameter (D)	Pile tip resistance (kN) at scour depth S					
	S = 0.00 D	S = 0.50 D	S = 1.00 D	S = 2.00 D	5 = 3.00 D	S = 4.00 D
	Pile rip resistance	Pile rip resistance	Pile tip resistance	Pile rip resistance	Pile rip resistance	Pile tip resistance
D = 0.50 m	815.89	709.82	645.94	574.89	517.40	444.96
D = 0.60 m	1016.08	918.18	848.55	796.74	761.18	724.03
D = 0.70 m	1225.34	1053.79	958.95	815.11	749.90	667.41
D = 0.80 m	1450.93	1269.57	1155.30	993.56	894.21	791.37
D = 0.90 m	1684.62	1465.62	1304.40	1147.87	998.65	898.78

Table 8. Settlement at ultimate capacity at different scour depths.

Pile diameter (D)	Settlement at ultimate capacity (mm) at different scour depths S					
	S = 0.00 D	S = 0.50 D	S = 1.00 D	S = 2.00 D	S = 3.00 D	S = 4.00 D
D = 0.50 m	6.88	7.35	7.87	8.47	9.15	9.92
D = 0.60 m	7.56	8.08	8.65	9.30	10.05	10.90
D = 0.70 m	8.42	9.00	9.63	10.36	11.19	12.14
D = 0.80 m	9.34	9.98	10.68	11.50	12.41	13.47
D = 0.90 m	10.47	11.19	11.98	12.89	13.92	15.10

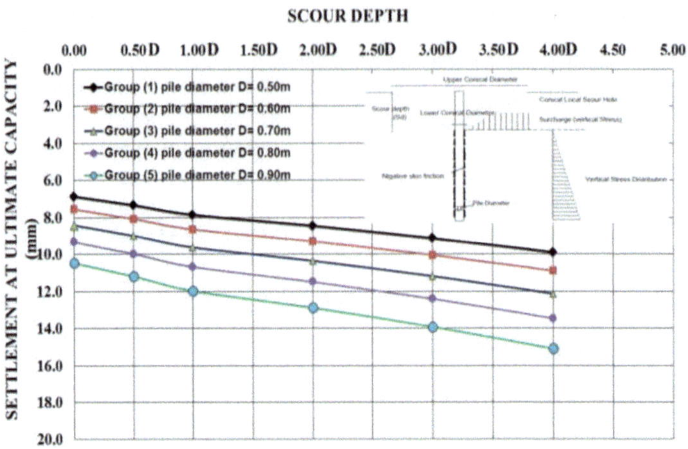

Fig. 16. The relationship between scour depth and settlement at ultimate load capacities for different pile diameters.

From Fig. 16 it can be noticed that, the obtained values of settlement increases with increasing scour depths. However, settlement of pile diameter = 90 cm is higher than that of piles diameter = 80, 70, 60, and 50 cm.

4.2 Comparison of Pile Load Between Egyptian Code and Finite Element Analysis

In the present study The comparison between the values of total pile load resistance, skin friction as well as bearing resistance and effective pile length at zero scour depth obtained from Egyptian Code (2001) and finite element analysis are shown in Figs. 17 and 18.

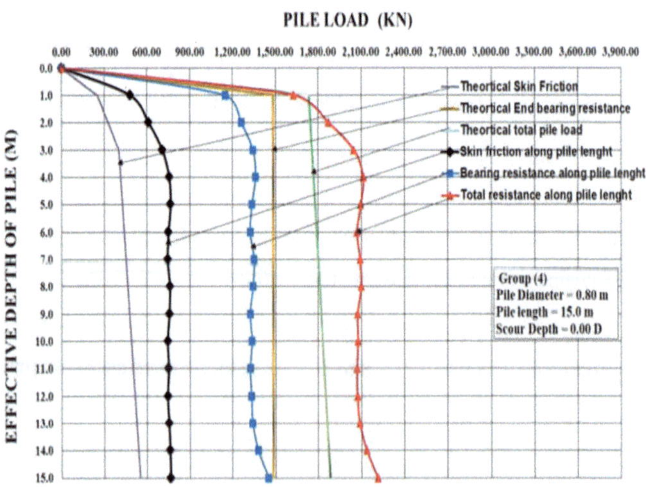

Fig. 17. Relationships between the values of total pile load resistance, skin friction as well as bearing resistance and effective pile length at zero scour depth for pile diameter = 0.80 m.

From Figs. 17 and 18 fair agreement has been obtained between theoretical and finite element analysis.

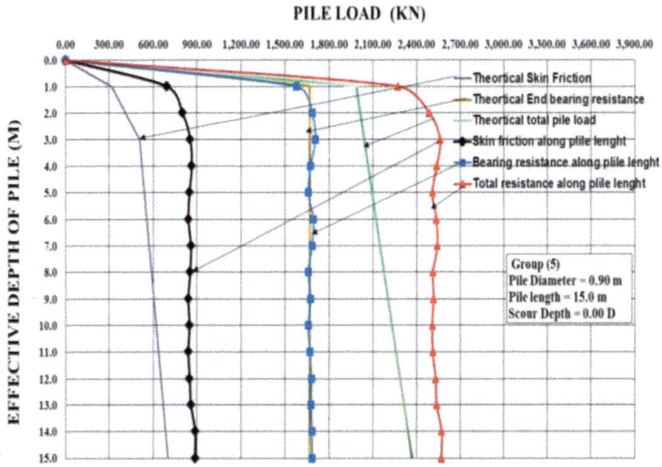

Fig. 18. Relationships between the values of total pile load resistance, skin friction as well as bearing resistance and effective pile length at zero scour depth for pile diameter = 0.90 m.

4.3 Effect of Scour Depth on Friction Around Shaft of Pile

A finite element model was developed to predict the effect of scour depth on friction around shaft of piles (negative skin friction and positive skin friction) of the pile foundation. Figures 19, 20, 21, 22, 23 and 24 show some examples of the relationships between friction around shaft of piles and effective pile depth. Figure 25 shows the relationship between net skin friction around shaft of piles and effective pile depth at different pile diameters.

From these fingers, it can be noticed that, a decrease in the embedded depth of pile of foundation due to scour decreases skin friction. In addition, conical local scour hole causes negative skin friction (additional vertical stress) along pile shaft.

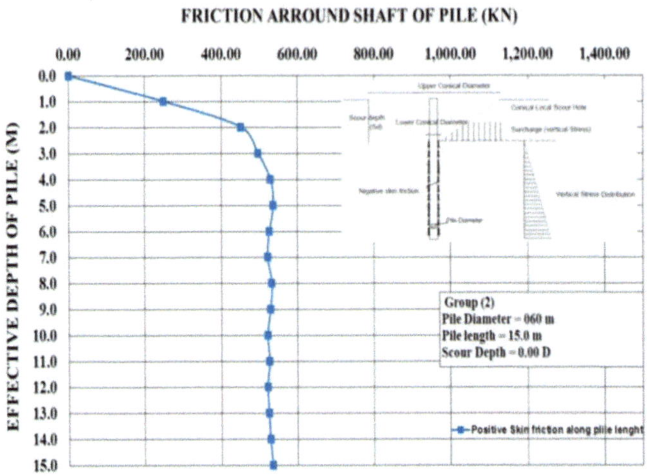

Fig. 19. Relationship between friction around shaft of piles and effective pile depth for group (1) pile diameter = 0.60 m at scour depth = 0.0 D.

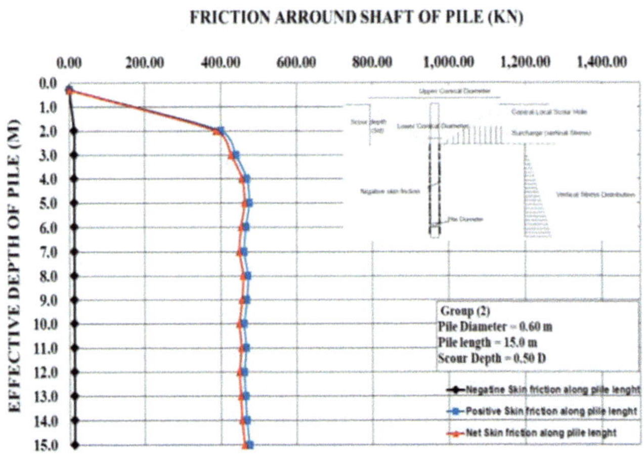

Fig. 20. Relationship between friction around shaft of piles and effective pile depth for group (1) pile diameter = 0.60 m at scour depth = 0.5 D.

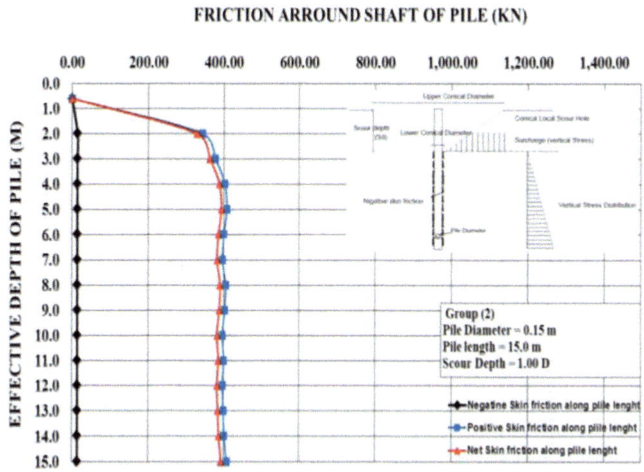

Fig. 21. Relationship between friction around shaft of piles and effective pile depth for group (1) pile diameter = 0.60 m at scour depth = 1.0 D.

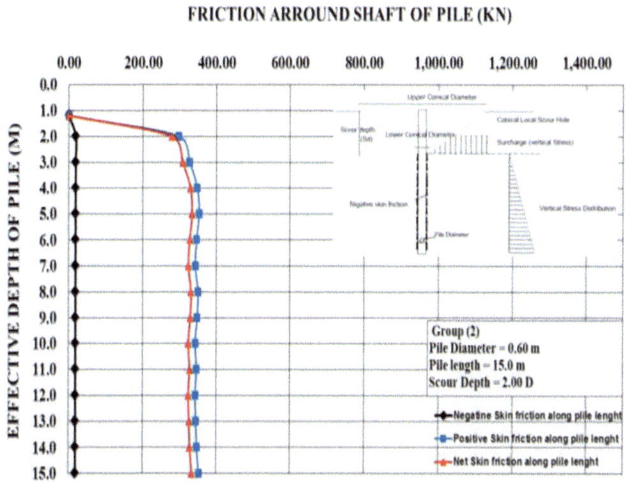

Fig. 22. Relationship between friction around shaft of piles and effective pile depth for group (1) pile diameter = 0.60 m at scour depth = 2.0 D.

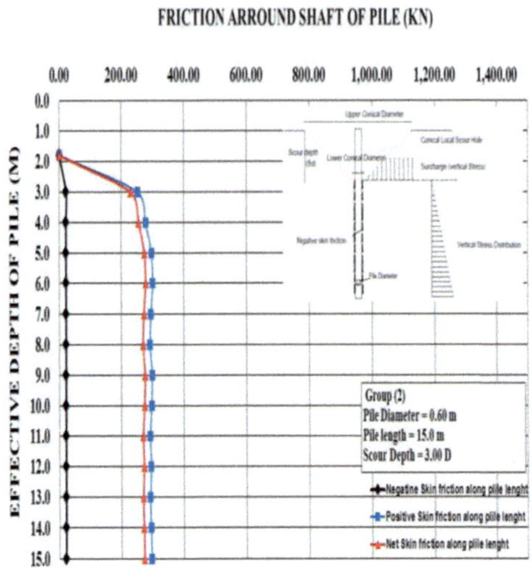

Fig. 23. Relationship between friction around shaft of piles and effective pile depth for group (1) pile diameter = 0.60 m at scour depth = 3.0 D.

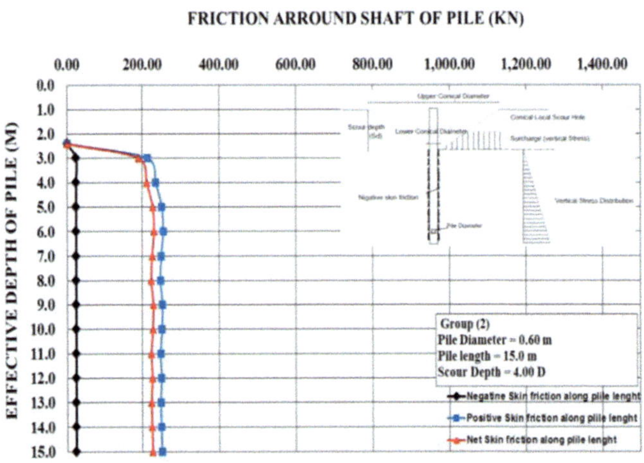

Fig. 24. Relationship between friction around shaft of piles and effective pile depth for group (1) pile diameter = 0.60 m at scour depth = 4.0 D.

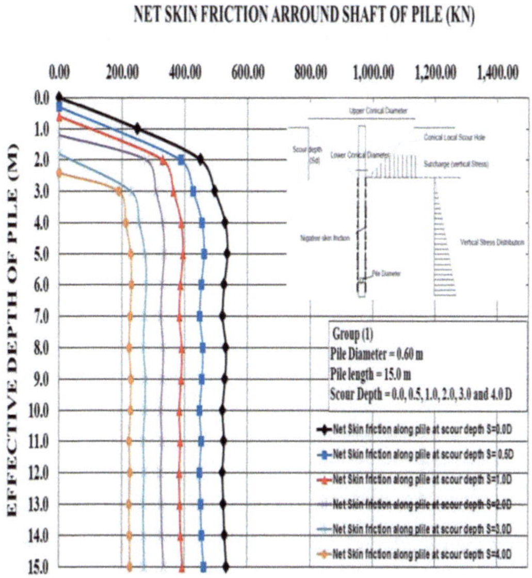

Fig. 25. The relationship between the net skin friction around shaft of piles and effective pile depth for different pile diameters.

4.4 Effect of Scour Depth on End Bearing Resistance of Pile

A finite element model was developed to predict the effect of scour depth on total load and end bearing (at pile tip) of the pile foundation. Figures 26, 27, 28, 29, 30 and 31 show some examples of relationship between total load, end bearing resistance of the pile as well as net friction and effective pile depth. Figure 32 shows the relationship between end bearing resistance and effective pile depth at different scour depths.

From these figures, it can be concluded that the total pile load and end bearing (at pile tip) of the pile foundation values decrease with increasing scour depths.

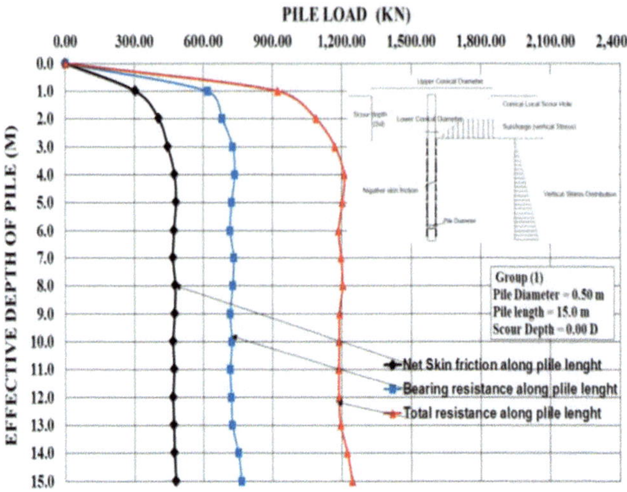

Fig. 26. Relationship between load, end bearing resistance of the pile as well as net friction and effective pile depth for group (1) pile diameter = 0.50 m at scour depth = 0.0 D.

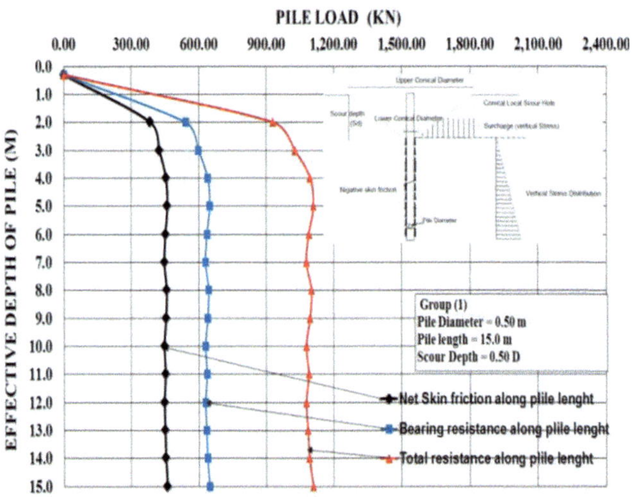

Fig. 27. Relationship between load, end bearing resistance of the pile as well as net friction and effective pile depth for group (1) pile diameter = 0.50 m at scour depth = 0.5 D.

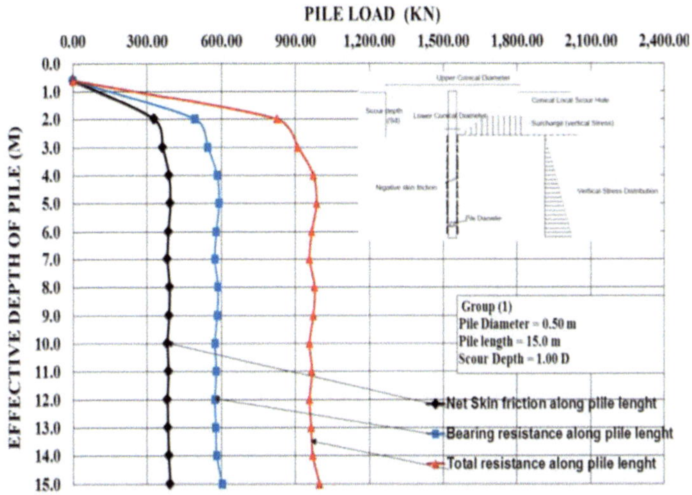

Fig. 28. Relationship between load, end bearing resistance of the pile as well as net friction and effective pile depth for group (1) pile diameter = 0.50 m at scour depth = 1.0 D.

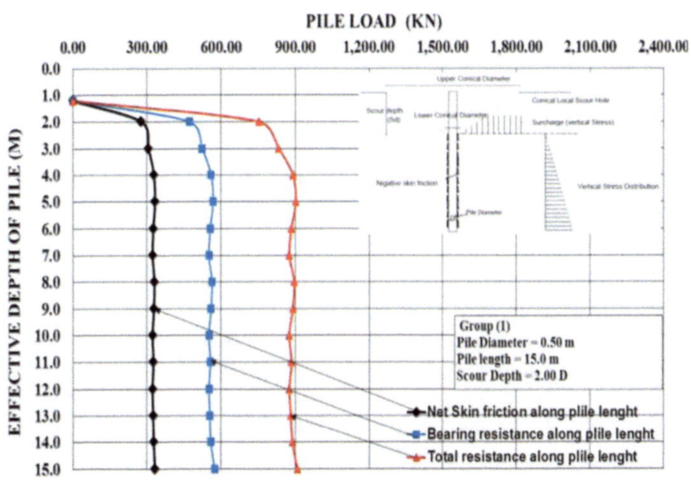

Fig. 29. Relationship between load, end bearing resistance of the pile as well as net friction and effective pile depth for group (1) pile diameter = 0.50 m at scour depth = 2.0 D.

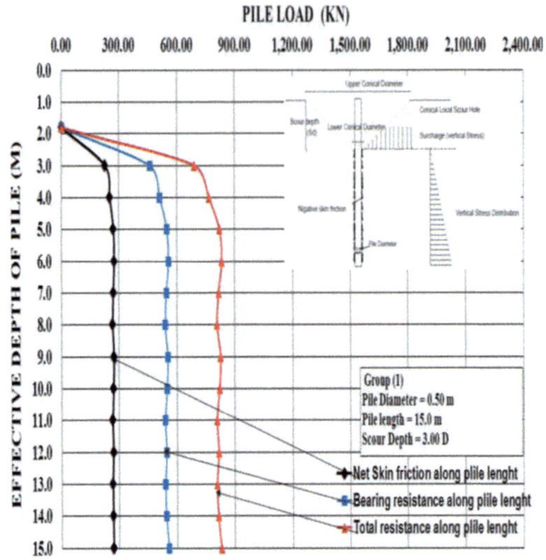

Fig. 30. Relationship between load, end bearing resistance of the pile as well as net friction and effective pile depth for group (1) pile diameter = 0.50 m at scour depth = 3.0 D.

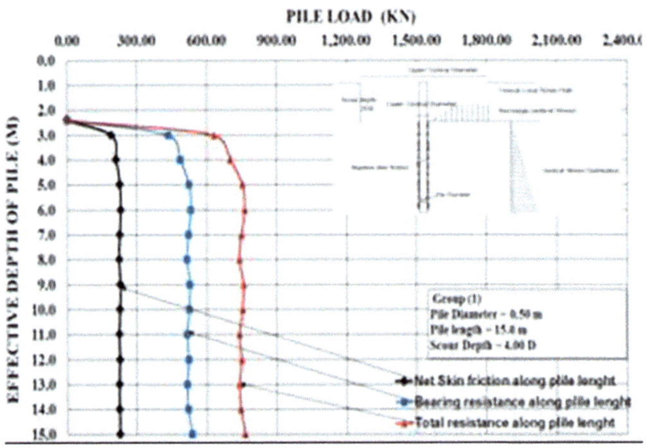

Fig. 31. Relationship between load, end bearing resistance of the pile as well as net friction and effective pile depth for group (1) pile diameter = 0.50 m at scour depth = 4.0 D.

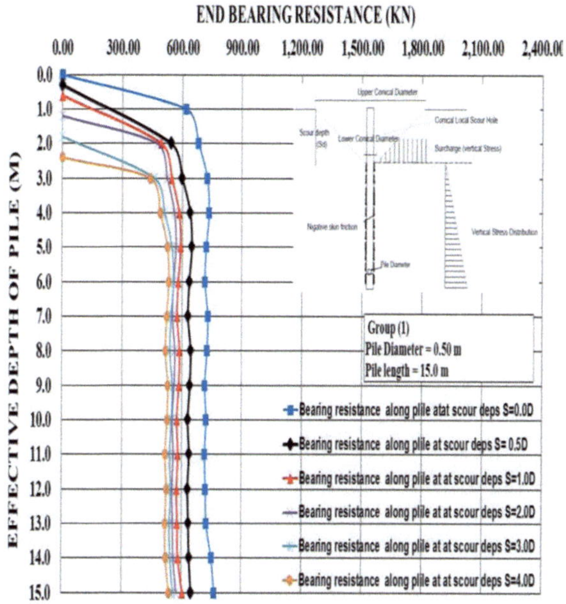

Fig. 32. Relationship between end bearing resistance and effective pile depth at different scour depths.

4.5 Effect of Scour Depth on the Ultimate Pile Load Capacity

The ultimate load capacities obtained from the numerical analysis at different scour depths and at different pile diameters are listed in Table 5. Figure 33 shows the relationship between the values of ultimate load capacities and scour depths at different pile diameters. Figures 34, 35, 36, 37 and 38 show a comparison between the values of ultimate load capacities as percentage of ultimate load capacities of zero scour depth at different pile diameters.

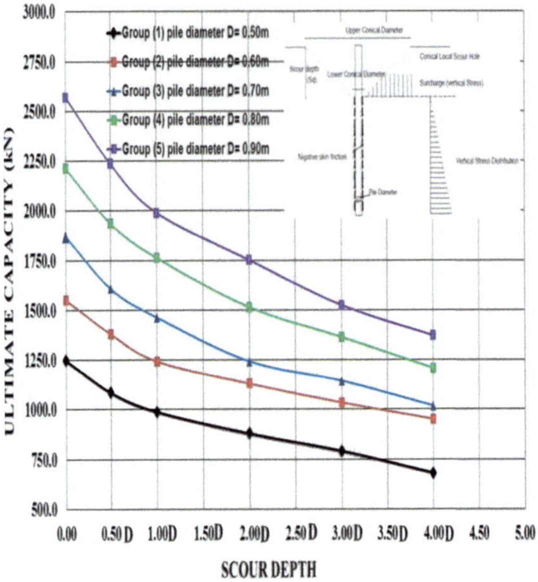

Fig. 33. The relationship between the values of ultimate load capacities and scour depths at different pile diameters.

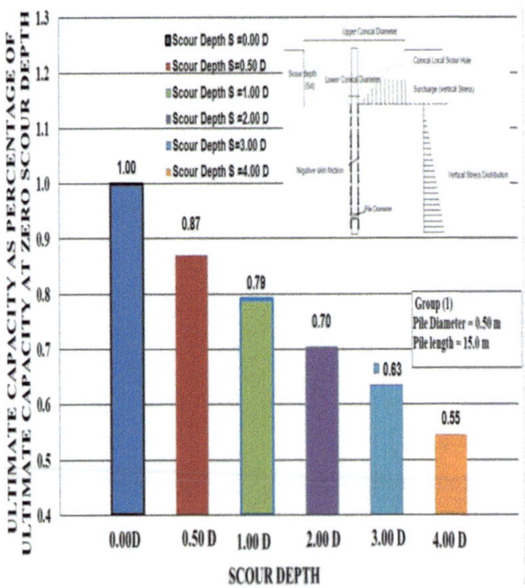

Fig. 34. Comparison between the values of ultimate load capacities as percentage of ultimate load capacities of zero scour depth for pile diameter D = 0.50 m.

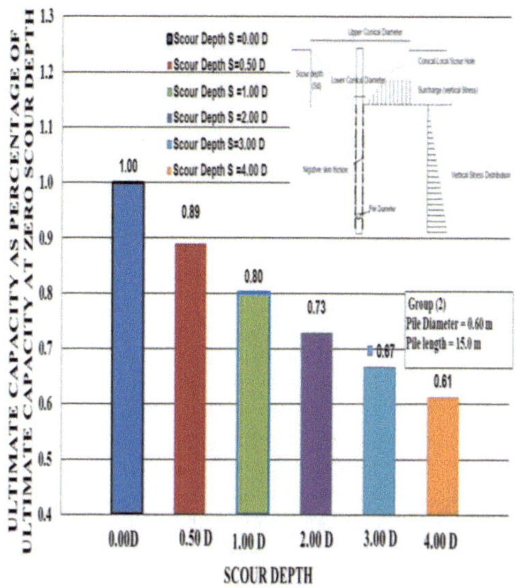

Fig. 35. Comparison between the values of ultimate load capacities as percentage of ultimate load capacities of zero scour depth for pile diameter D = 0.60 m.

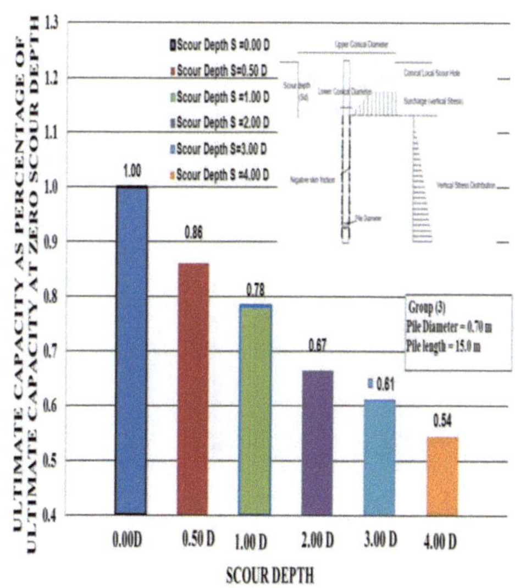

Fig. 36. Comparison between the values of ultimate load capacities as percentage of ultimate load capacities of zero scour depth for pile diameter D = 0.80 m.

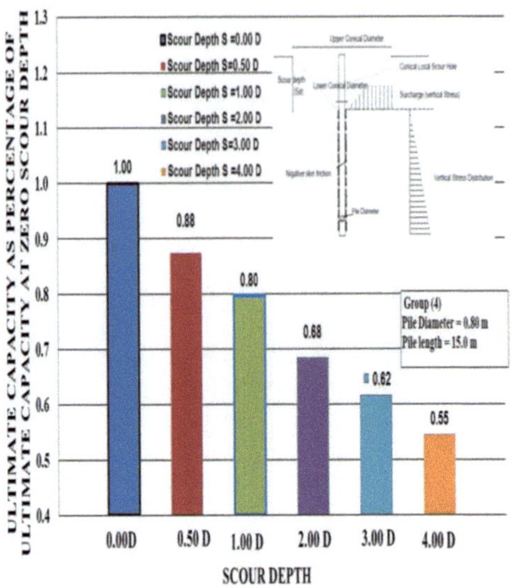

Fig. 37. Comparison between the values of ultimate load capacities as percentage of ultimate load capacities of zero scour depth for pile diameter D = 0.70 m.

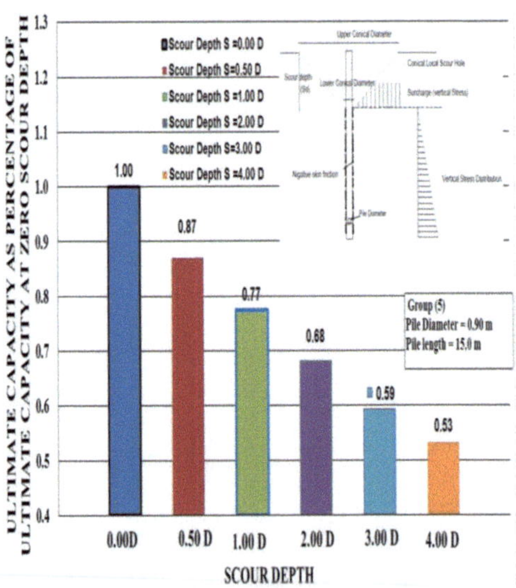

Fig. 38. Comparison between the values of ultimate load capacities as percentage of ultimate load capacities of zero scour depth for pile diameter D = 0.90 m.

5 Conclusions

From the present study, the following conclusions are obtained:

i. The settlement of piled foundation increases with increasing scour depths.
ii. A decrease in the embedded depth of pile of foundation due to scour decreases skin friction.
iii. Conical local scour hole causes the negative skin friction (additional vertical stress).
iv. The values of transferred loads at pile tip (end bearing) decrease with increasing scour depths.
v. The values of ultimate load capacities of piles foundation decrease with increasing scour depths.
vi. At scour depth = 4.0 pile diameter the of ultimate load capacities decrease by about 39% to 44% from ultimate load capacities of zero scour depth.
vii. Fair agreement has been obtained between theoretical and finite element analysis.
viii. For non rested pile cap on soil the pile cap dimensions have no significant effect.

References

1. Sham, S.H.R., Yu, G.X., De Silva, S.: Foundation design methodology for the Padma Main Bridge. In: Amin, Okui, Bhuiyan (eds.) IABSE-JSCE Joint Conference on Advances in Bridge Engineering-II, Dhaka, Bangladesh, 8–10 August 2010. www.iabse-bd.org, ISBN 978-984-33-1893-0
2. Liang, F.-Y., Wang, Y.-Q., Han, J.: Numerical analysis of scouring effects on the behavior of pile foundations with the Mohr-Coulomb model. Geotechnical Special Publication No. 214 © ASCE 2011 Downloaded 09 Jun 2011 to 111.187.79.161. Redistribution subject to ASCE license or copyright (2011). http://www.ascelibrary.org
3. Periku, E., Yardim, Y.: Effect of scour on load carry capacity of piles on mat bridge. In: International Students' Conference of Civil Engineering, ISCCE 2012, Epoka University, Tirana, Albania, 10–11 May 2012. (Department of Civil Engineering, EPOKA University, Albania)
4. Park, J.H., Kwak, K., Lee, J.H. Chung, M.: Scour vulnerability evaluation of pile foundations during floods for national highway bridges. In: ICSE6 Paris, 27–31 August 2012
5. Ibrahim H.E. An experimental study of local scour around circular bridge pier in sand soil. Int. J. Civ. Environ. Eng. IJCEE-IJENS **13**(01), 23–28 (2013). (134901-2525-IJCEE-IJENS © February 2013) IJENS
6. Lin, C.: The loss of pile axial capacities due to scour: vertical stress distribution. In: International Conference on Transportation Infrastructure and Materials (ICTIM 2017). ISBN 978-1-60595-442-4

7. Kassem, E.M., Abd El Samee, W.N., Essa, T.A.: Analysis of pile-raft foundations non-rested and directly rested on soil. ARPN J. Eng. Appl. Sci. **13**(6) (2018). ISSN 1819-6608©2006-2018, Asian Research Publishing Network (ARPN). All rights reserved www.arpnjournals.com
8. Zaaijer, M.B.: Design methods for offshore wind turbines at exposed sites (OWTES) sensitivity analysis for foundations of offshore wind turbines. Delft University of Technology, Delft (2002)
9. Egyptian Code.: Soil mechanics and design and construction of foundation, Part 4, “Deep Foundation” (2001)

Author Index

A
Abd El Samee, Wael Nashaat, 258
Abdulrahman, Shaimaa M., 206
Adel, A'laa, 215
Aldaeef, Abdulghader A., 247
Alharthi, Yasir M., 232
Alzo'ubi, A. K., 49
Amer, Mohamed, 215
Awwad, Talal, 36, 69, 77

B
Badr, Atef, 125
Balg, Christophe, 61

C
Campos, Darym J. F., 191
Cavalcante, Andre L. B., 191
Crispin, Jamie, 162, 181

D
de Carvalho, José Camapum, 191
Dong, Yunxiu, 141

E
Elsamny, Mohamed Kassem, 258

F
Feng, Zhongju, 141

G
Ganiyu, Abideen Adekunle, 125
Gilder, Charlotte, 162

H
Hamidou, Hamadoum, 85
Hangartner, Thomas, 61
Hanna, Adel M., 232
Hanna, Adel, 104
Hao, Yumeng, 141

I
Ibrahim, Farid, 49
Isobe, Koichi, 85

K
Kodsi, Salma Al, 36
Koteswara Rao, L., 26

L
Lin, Lan, 104
Lukpanov, R. E., 69, 77

M
Madhav, M. R., 26
Mahmood, Mahmood R., 206
Meguid, Mohamed A., 1
Miyamoto, Yuji, 14
Mussabayev, T., 116
Mussina, S., 116
Mylonakis, George, 181

N
Nakano, Takaharu, 14
Nowak, Paul, 162

O
O'Riordan, Nick, 162
Osman, Mohd Hanim, 125
Ozelim, Luan C. S. M., 191

H. El-Naggar et al. (Eds.): GeoMEast 2018, SUCI, pp. 285–286, 2019.
https://doi.org/10.1007/978-3-030-01902-0

P
Padmavathi, M., 26
Padmavathi, V., 26
Patel, Dinesh, 162

R
Rashid, Ahmad Safuan A., 125
Rayhani, Mohammad T., 247
Round, Katherine, 181

S
Salem, Manal, 215
Sanjeeva, P., 26
Shakhmov, Zh. A., 69
Shashkin, Alexey, 36
Sinha, Anup, 104
Su, Huize, 181

T
Tsigulyov, D. V., 77
Tulebekova, A. S., 116

V
Vardanega, Paul J., 162
Voyagaki, Elia, 162

W
Wan, Qi, 141

Y
Yao, Hongsheng, 141
Yenkebayev, S. B., 69, 77

Z
Zhussupbekov, A. Zh., 116

MIX
Papier aus verantwortungsvollen Quellen
Paper from responsible sources
FSC® C105338

If you have any concerns about our products,
you can contact us on
ProductSafety@springernature.com

In case Publisher is established outside the EU,
the EU authorized representative is:
Springer Nature Customer Service Center GmbH
Europaplatz 3, 69115 Heidelberg, Germany

Printed by Libri Plureos GmbH
in Hamburg, Germany